LABORATORY EXERCISES IN MICROBIOLOGY SIXTH EDITION

E. C. S. CHAN
Professor of Microbiology, Faculty of Medicine, Faculty of Dentistry
McGill University

Michael J. PELCZAR, Jr.
Emeritus Vice President, Graduate Studies and Research
Emeritus Professor of Microbiology
University of Maryland

Noel R. KRIEG
Professor of Microbiology and Immunology
Virginia Polytechnic Institute and State University

McGRAW-HILL, Inc.
New York St. Louis San Francisco Auckland Bogotá Caracas Lisbon London Madrid Mexico
Milan Montreal New Delhi Paris San Juan Singapore Sydney Tokyo Toronto

LABORATORY EXERCISES IN MICROBIOLOGY

Copyright © 1993, 1977, 1972, 1965, 1958 by McGraw-Hill, Inc.
All rights reserved. Copyright reserved 1986 by Michael J. Pelczar, Jr.
Printed in the United States of America. Except as permitted under
the United States Copyright Act of 1976, no part of this publication
may be reproduced or distributed in any form or by any means, or
stored in a data base or retrieval system, without the prior written
permission of the publisher.

      10 QPD/ QPD 04

ISBN 0-07-049264-6

This book was set in Janson Text by Ann Eisner. The editors were
Pamela Wirt and Holly Gordon; the designer was Gayle Jaeger;
the production supervisor was Janelle S. Travers. New drawings were
done by Eric G. Hieber; Figures 50.1, 50.2, 51.1, 52.2, 53.1, 55.1,
and 55.2 were done by Arthur Ciccone. Semline, Inc., was printer
and binder.

Cover photographs courtesy of Becton Dickinson Microbiology
Systems, Cockeysville, Maryland. **Front cover:** (top left) *Moraxella
catarrhalis* on Trypticase soy agar with 5% sheep blood; (top right)
Shigella flexneri and *Salmonella typhimurium* on Modified SS agar;
(bottom left) *Trichosporon beigelii* on Sabouraud dextrose agar;
(bottom right) *Proteus mirabilis* on BTB (Bromthymol blue) agar.
Back cover: (top left) *Vibrio parahaemolyticus* on TCBS (Thiosulfate
citrate bile salts sucrose) agar; (top right) *Staphylococcus aureus* on
Mannitol salt agar; (bottom left) *Escherichia coli* and *Proteus vulgaris*
on CLED (Cystine-lactose-electrolyte-deficient) agar; (bottom right)
Clostridium difficile on *C. difficile* agar.

Color-insert photographs: **1, 2:** Courtesy of Centers for Disease
Control, Atlanta, Georgia; **3:** John D. Cunningham/Visuals
Unlimited; **4:** Dustman and Lukas/McGraw-Hill, Inc.; **5:** Biological
Photo Service; **6:** G. W. Willis, M.D./Biological Photo Service; **7–9,
12, 14, 16–21, 24–28, 30, 32:** Courtesy of James R. Milam,
University of Florida, Gainesville.

All text photographs courtesy of E. C. S. Chan unless credited otherwise.

CONTENTS

Preface vii
Introduction: Laboratory Protocol and Safety Procedures 1

PART I   **THE UBIQUITY OF MICROORGANISMS 5**
exercise 1   Microorganisms in the Environment 7
exercise 2   Mixed Microbial Flora 13
exercise 3   Aseptic Technique and the Transfer of Microorganisms 17

PART II   **MICROSCOPY IN MICROBIOLOGY 23**
exercise 4   Principles and Use of the Bright-Field Microscope 29
exercise 5   Microscopic Examination of Microorganisms in Wet Mounts 37
exercise 6   Microscopic Measurement of Microorganisms 41
exercise 7   Use of the Dark-Field Microscope 47
exercise 8   Use of the Phase-Contrast Microscope 51

PART III   **BACTERIAL MORPHOLOGY AND STAINING METHODS 59**
exercise 9   Staining the Whole Cell 61
     9.A   Simple Staining with Basic Dyes 61
     9.B   The Negative Stain 67
     9.C   The Gram Stain: A Differential Stain 71
     9.D   The Acid-Fast Stain: A Differential Stain 75
exercise 10   Staining for Cell Structures 79
     10.A   The Spore Stain 79
     10.B   The Capsule Stain 83
     10.C   The Flagella Stain and Detection of Motility 87
     10.D   The Cell-Wall Stain 91
exercise 11   Identification of a Morphological Unknown 95

PART IV   **CULTURE MEDIA PREPARATION AND THEIR STERILIZATION 99**
exercise 12   Preparation and Autoclaving of Nutrient Broth and Nutrient Agar (Plates, Deeps, and Slants) 101
exercise 13   Dry-Heat Sterilization 107

exercise 14 Preparation of a Chemically Defined Medium 111
exercise 15 Evaluation of Media to Support Growth of Bacteria 115
exercise 16 Selective, Differential, and Enriched Media 119

PART V   PURE-CULTURE TECHNIQUES 123
exercise 17 The Streak-Plate Method for Isolation of Pure Cultures 125
exercise 18 The Pour-Plate Method for Isolation of Pure Cultures 131
exercise 19 Enumeration of Bacteria by the Plate-Count Technique 135
exercise 20 Turbidity Measurement of Broth Cultures 143
exercise 21 Anaerobic Culture Methods 149
exercise 22 Cultural Characteristics 155
exercise 23 Maintenance and Preservation of Pure Cultures 163

PART VI   BIOCHEMICAL ACTIVITIES OF BACTERIA 169
exercise 24 Hydrolysis of Polysaccharide, Protein, and Lipid 171
exercise 25 Fermentation of Carbohydrates 175
exercise 26 Reactions in Litmus Milk 179
exercise 27 Additional Biochemical Characteristics 185
   27.A Hydrogen Sulfide Production 185
   27.B Production of Indole 189
   27.C Reduction of Nitrate 193
exercise 28 Morphological, Cultural, and Biochemical Characterization of an Unknown Culture 197

PART VII   CONTROL OF MICROBIAL POPULATIONS 203
exercise 29 The Effect of Temperature on Growth 205
exercise 30 Resistance of Bacteria to Heat 209
exercise 31 Bactericidal Effect of Ultraviolet Radiations 213
exercise 32 Effect of Osmotic Pressure on Microbial Growth 217
exercise 33 Comparative Evaluation of Antimicrobial Chemical Agents 221
exercise 34 Antibiotics: Agar-Diffusion Method 225
exercise 35 Effect of Enzyme on Bacteria: Spheroplast Formation by *Pseudomonas putida* 229

PART VIII   BACTERIAL GENETICS 233
exercise 36 Bacterial Variation Due to Environmental Change 235
exercise 37 Bacterial Variation Due to Genotypic Change 239
exercise 38 Nutritional Mutants 243
exercise 39 Isolation of Streptomycin-Resistant Mutants of *Escherichia coli* 247
exercise 40 Bacterial Conjugation 251

CONTENTS

exercise 41 Regulation of Enzyme Synthesis: Enzyme Induction and Catabolite Repression 255
exercise 42 The Ames Test: Using Bacteria to Detect Potential Carcinogens 263

PART IX VIRUSES 269
exercise 43 Bacterial Lysis by Bacteriophage: Phage Titer and the Plaque Assay 271
exercise 44 Phage-Typing 277
exercise 45 Propagation of Viruses by Tissue Culture (Demonstration) 281

PART X RECOMBINANT DNA TECHNIQUES 285
exercise 46 Isolation of DNA from Bacterial Cells 287
exercise 47 Restriction Enzyme Analysis of DNA 295
exercise 48 Transformation of *Escherichia coli* with Plasmid DNA Containing Genes for Antibiotic Resistance 301
exercise 49 Plasmid-Mediated Transformation of the Ampicillin-Resistance and the *lac* Genes 305

PART XI EUCARYOTIC MICROORGANISMS 307
exercise 50 Characteristics of Protozoa 309
exercise 51 Morphology and Cultivation of Algae and Cyanobacteria 315
exercise 52 Morphology and Cultural Characteristics of Molds 319
exercise 53 Sexual Reproduction of Molds 327
exercise 54 Dimorphism of *Mucor rouxii* 333
exercise 55 Morphology of Yeasts 337

PART XII MICROORGANISMS AND DISEASE 345
exercise 56 Normal Flora of the Human Body 347
exercise 57 Precautions and Methodology of Specimen Collection and Processing 351
exercise 58 Immunology and Serology: Immunoprecipitation 357
  58.A Precipitin Ring Test 359
  58.B Ouchterlony Immunodiffusion 363
exercise 59 Immunoagglutination: Bacterial Agglutination Tests 367
exercise 60 Streptolysin O Neutralization 373
exercise 61 Complement-Fixation 377
exercise 62 Complement-Mediated Lysis in Agar 387
exercise 63 Bacterial Infection of a Plant: Demonstration of Koch's Postulates 391
exercise 64 Airborne Infections 395
  64.A The Corynebacteria and the Gram-Positive Cocci 395
  64.B The Mycobacteria, Neisserias, and Gram-Negative Aerobic Rods 403
exercise 65 Foodborne and Waterborne Diseases 407

exercise **66** Contact Diseases 415
exercise **67** Bacteriological Analysis of Urine 419
exercise **68** Commercially Marketed Rapid and Convenient Bacterial Identification Kits 423

PART XIII ENVIRONMENTAL MICROBIOLOGY 433

exercise **69** Microbiology of Soil 435
    69.A Quantitative Enumeration of Microorganisms 435
    69.B Ammonification: The Liberation of Ammonia from Nitrogenous Compounds 439
    69.C Reduction of Nitrates and Denitrification 443
    69.D Symbiotic and Nonsymbiotic Nitrogen Fixation 447
exercise **70** Biological Succession in a Mixed Microbial Flora 451
exercise **71** Enrichment Culture Technique: Isolation of Phenol-Utilizing Microorganisms 455
exercise **72** Standard Method of Water Analysis: Multiple-Tube Fermentation 461
exercise **73** Standard Method of Water Analysis: Membrane-Filter Technique 469
exercise **74** Bioluminescence 475

PART XIV INDUSTRIAL AND APPLIED MICROBIOLOGY 479

exercise **75** The Enumeration of Bacteria in Raw and Pasteurized Milk 481
exercise **76** Microbiological Examination of Foods 487
exercise **77** Preparation of Fermented Foods: Sauerkraut and Yogurt 493
    77.A Preparation of Sauerkraut 495
    77.B Preparation of Yogurt 499
exercise **78** Fermentation of Grape Juice to Produce Wine 503
exercise **79** The Isolation of Antibiotic-Producing Microorganisms 507
exercise **80** The Production and Assay of a Microbial Enzyme (Penicillinase) 515

APPENDIXES 523

    A List of Cultures and Sources of Cultures 525
    B Stains, Staining Solutions, and Reagents 529
    C List of Media and Sources of Media 533
    D Cross Reference of Laboratory Exercises in This Manual to Chapters in Selected Introductory Microbiology Texts 541
    E Audiovisual Aids Sources 545
    F Selected Bibliography 547

    INDEX 549

PREFACE

Laboratory Exercises in Microbiology, in its new format and binding, is an extensive revision of the fifth edition. In spite of the many changes in physical appearance and content, it continues to provide the many loyal users of this manual (some of whom have used this manual since its first edition) the same basic organization of exercises that has withstood the test of time as a pedagogical "tool."

Each exercise is introduced by an **Objective** (to state the purpose of the exercise) and then is divided into the following sections: an **Overview** that serves to provide a concise background of the subject material; a **References** section that lists the essential reading from the companion textbook *Microbiology: Concepts and Applications* by Pelczar, Chan, and Krieg, as well as any particularly relevant literature source; a **Materials** section that lists in an orderly manner the supplies necessary for the exercise; a **Procedure** that outlines in numerical order each step a student follows to perform the work at hand; a **Results** section for the recording of data and their discussion and a **Questions** section that queries the student upon completion of the exercise. The final two sections are removable to facilitate submission to the instructor for evaluation and critique. Each exercise is designed for a student or a pair of students so that this manual can be used for very small or very large classes with equal facility.

The intent of these exercises is to introduce the student to the world of microorganisms and to provide a familiarity with basic techniques that are necessary for their study. Some techniques to be learned touch on the fundamental characteristics of microorganisms as creatures endowed with life in a holistic discipline. Other techniques are concerned with the surface mosaic of microorganisms as they interact immunologically with the vertebrate host; still others are involved with unravelling the genomic code of the microbial cell in all its molecular aspects. Additional techniques deal with the applied aspects of microbiology in medicine, agriculture, and the marketplace. In sum, this manual is a comprehensive introduction to microbiology at the practical level.

The exercises are organized into fourteen separate parts, beginning with experiments in Part I that demonstrate the ubiquity of microbes in the environment. Parts II and III provide the student with a solid foundation in the principles and proper use of the microscope, which is then used to expose the awesome diversity of microbial forms. Parts IV and V teach the preparation of culture media and their use in the cultivation of microbes. Part VI reveals the remarkable biochemical activities of such small forms. Part VII shows how the growth of microbial populations can be controlled. Part VIII provides knowledge on the study of microbial genetics and is a prerequisite for acquiring some recombinant DNA techniques in Part X. Part IX is on viruses, studied not only as infectious entities, but also as vectors or carriers of genetic information in genetic engineering. Eucaryotic microorganisms are not to be forgotten; they are the focus of Part XI. Part XII deals with microorganisms as the cause of disease. Part XIII is on microbial ecology; techniques to discover the important roles microbes play in the environment are found here. The manual ends with Part XIV on industrial and applied microbiology.

Appendixes provide the instructor with information about sources of laboratory supply, culture media and reagent composition, audiovisual supplements, and cross-references to other microbiology textbooks. A full-color insert is included to show actual results of many of the experiments carried out in this manual. Students are given the opportunity, through the use of references to specific color plates, to preview microbial reactions to various staining methods and to biochemical and physiological tests.

Science does not stand still. And so, microbiology as a science evolves with time. We have tried to incorporate all the essential basic advances that have taken place since this manual was last published in 1986. Even our outlook on working with microorganisms has changed. Not only must microbiologists use "aseptic technique" to work with pure cultures, they now have to practice "universal precautions" to avoid deadly microbial scourges, such as AIDS and antibiotic-resistant tuberculosis. This manual no longer recommends mouth pipetting, but insists on the use of Pipet-Aid and other mechanical volume dispensers. Throughout the manual, cautionary notes are set apart in boxes directing students to exercise particular care in performing certain procedures. In addition, special notes in boldface are provided within exercises to highlight supplementary information useful to instructors and students alike in carrying out the exercises. These notes replace the Appendix entitled Notes to Instructor by Exercise that appeared in earlier editions of the manual.

We thank Liliane Therrien for providing many of the laboratory setups of microorganisms for photography; Antonia Klitorinos and Yu-Shan Qiu for demonstrating in the photographs on laboratory procedures; J. R. Milam of the University of Florida for contributing a majority of the color slides appearing in the insert; and Richard W. Ikenberry of the University of Nebraska for offering valuable suggestions for updating the fifth edition. We also acknowledge with much appreciation the exercises and other material contributed generously by many of our colleagues. Their specific contribution is identified in the body of the manual.

With enthusiasm and appreciation, we wish to acknowledge the patience, professional guidance, and cooperation we have received from the truly competent editors at McGraw-Hill. "Thank you" to Pamela Wirt, Kathi Prancan, Holly Gordon, Gayle Jaeger, and Janelle Travers.

E. C. S. Chan
Michael J. Pelczar, Jr.
Noel R. Krieg

Introduction: Laboratory Protocol and Safety Procedures

Microbiology is the study of the biology of microorganisms and infectious entities. Microorganisms, or microbes, comprise many microscopic forms. Bacteria are microbes, but so are the single-celled protozoa and the multicellular members of the algae and fungi. Infectious entities, or agents, include the noncellular viruses that are not usually considered living microbes by many microbiologists. Microorganisms and viruses share the common properties of diminutiveness, usually rapid reproduction or replication, and wide geographical distribution. Microbiologists have developed unique methods and tools for studying their subjects. It has been said that microbiology is a science defined more by the techniques it uses than the subjects it covers!

Even the advancement of the discipline had to be tied to the development of techniques. For example, development of sterilization methods and pure-culture techniques were of paramount importance for the study of microbes and infectious entities because of their ubiquity and their natural occurrence in mixed populations. It is only from the pure culture consisting of just one kind of microbe or infectious agent that the microbiologist can study the particular behavior of a specific minute subject. This requirement at once highlights the difficulties to be encountered in attempting to study the interacting phenomena of mixed cultures in natural populations. Another example was the development of the electron microscope. The images obtained with it awed us with the differences between the procaryotic and eucaryotic cell as well as the fine structures of microorganisms and viruses. Thus the methods and techniques of microbiology are very unique and important to the science. This manual endeavors to introduce you to the basic methods and techniques of microbiology so that you can understand the behavior of the microscopic subjects and, in so doing, can handle them effectively and safely.

Microorganisms, viruses, and other infectious agents are studied because they cause disease, they can be harnessed for industrial applications, or they help us understand the natural world. They are usually studied in microbiology laboratories found in universities, hospitals, government agencies, and industrial institutions. Therefore the microbiology laboratory is a place where potentially infectious microscopic entities are handled, examined, and studied with safety and effectiveness.

Any culture of microorganism or infectious agent, even if it is not considered a pathogenic entity, should be considered potentially pathogenic, or as having the capacity to cause disease, in humans or other animals and plants. Risks of infection vary with each microbe or agent and the way it is handled. Safety standards and practices are instituted in order to reduce to an acceptable level the use of potentially pathogenic, and therefore hazardous, agents. Stringent standards and practices are set for the more dangerous agents and less strict ones for those less infectious. The aim of safety standards and practices in the microbiology laboratory is to allow work to proceed in this environment with the minimal risk of exposure, or of "catching" a disease, as a consequence of handling an infectious entity. Let us examine some of these safety standards and practices that satisfy this new awareness of laboratory safety when working with microorganisms and other infectious entities.

SAFETY STANDARDS AND PRACTICES

The attitudes and actions of those who work in the laboratory determine the level of safety. Workers (and students) must develop a positive and respectful attitude toward microorganisms and other infectious entities. Sloppy working habits, such as neglect of *aseptic technique* (a technique that prevents contamination and infection by infectious or extraneous agents) or failure to carry out disinfection, can cause contamination of the work at hand or transmission of a potential pathogen not only to yourself but also to others working with you. Also, laboratory equipment should be used properly in order to ensure safety; no short-cuts or expediency procedure should be taken at the risk of safety.

The microbiology laboratory should be a safe workplace. Legislation is in place, called the Workplace Hazardous Materials Information System (WHMIS), which requires that all hazardous substances, including microorganisms, be labeled in a specified manner and that there be a Material Safety Data Sheet (MSDS) available to accompany each hazardous substance. (A MSDS is now supplied with every chemical sold by a commercial company.) The professor in charge of teaching a laboratory course should ensure that adherence to this standard practice is observed.

BASIC REQUIREMENTS FOR A MICROBIOLOGY LABORATORY

1 Laboratory personnel must appreciate the potential hazards encountered in a microbiology laboratory. They should be trained in laboratory safety and be able to respond effectively when accidents happen.
2 The laboratory must be kept neat, orderly, and clean, and the benchtops should be free of nonessential materials.
3 Protective laboratory clothing (uniforms, coats, gowns) must be available and worn properly fastened by all personnel, including students, visitors, trainees, and others entering or working in the laboratory. Suitable footwear with closed heels and toes and with nonslip soles should be worn in the laboratory.
4 Safety face and eyewear (e.g., glasses, goggles, or face shields) should be worn when necessary to protect the face and eyes from splashes and sprays that may contain microbes and viruses, impacting objects, harmful substances, UV light, or other rays.
5 Eating, drinking, smoking, storing food or utensils, applying cosmetics, and inserting or removing contact lenses are not permitted in any laboratory area. Contact lenses should be worn only when other forms of corrective eyewear are not suitable.
6 Oral pipetting is prohibited. Use appropriate pipetting devices that bypass use of the mouth.
7 Long hair must be tied back or restrained.
8 Hands must be washed before leaving the laboratory and at any time after handling materials known or suspected to be contaminated, even when protective gloves have been worn.

9 Work surfaces must be cleaned and decontaminated with a suitable disinfectant (e.g., 5% Lysol solution; 1:5 dilution of household bleach, such as Javex or Clorox) at the end of the day and after any spill of potentially dangerous material. Loose or cracked work surfaces should be replaced.
10 All technical procedures should be performed in a manner that minimizes the creation of aerosols that may contain infectious material.
11 All contaminated or infectious liquid or solid materials must be decontaminated before disposal or reuse. Contaminated materials that are to be autoclaved or incinerated at a site away from the laboratory should have the outside disinfected chemically or be double-bagged and then transported to the autoclave or incinerator in durable leakproof containers that are closed and wiped on the outside with disinfectant before being removed from the laboratory.
12 Access to a Level 1 laboratory is at the discretion of the laboratory director or instructor.

There are generally four levels of containment with reference to laboratory classification. These levels are defined by the risk groups for infectious agents to be handled. Since the microbiology laboratory used for teaching (by design of the instructor for an *introductory* course) handles only Risk Group 1 agents (microorganisms, viruses, and parasites that are unlikely to cause disease in healthy workers, including students, or animals), it is classified as Level 1. A Level 1 laboratory requires no special design features beyond those suitable for a well-designed and functional laboratory. Containment cabinets are not required. Work may be done on an open benchtop, and containment is achieved through the use of practices normally employed in a basic microbiology laboratory.

However, if Risk Group 2 agents (those that incur moderate individual risk and limited community risk) are used, then containment Level 2 is required. Such agents are pathogens that can cause human or animal disease but under normal circumstances are unlikely to be a serious hazard to healthy laboratory workers, the community, livestock, or the environment. Laboratory exposures rarely cause infection leading to serious disease; effective treatment and preventive measures like vaccination are available, and the risk of spread is limited. A microbiology laboratory teaching medical microbiology and using pathogenic microorganisms and other infectious agents is generally classified as containment Level 2. For such a Level 2 laboratory, certain additional physical requirements are necessary as well as Class I or II biological safety cabinets to minimize the hazards of aerosols created. There are also containment Levels 3 and 4. (Discussion of these levels is obviously outside the scope of this manual.)

13 Hazard warning signs, indicating the risk level of the agents being used, should be posted outside each laboratory. The name of the laboratory supervisor or instructor as well as any special conditions for entry (e.g., entry denied to pregnant women or immunocompromised persons, or even children, for their own safeguard) should be listed on the hazard signs.

14 Protective gloves should be worn for all procedures that might involve direct skin contact with toxins, blood, infectious materials, or infected animals. Such gloves should be removed carefully and decontaminated with other laboratory wastes before disposal. It is preferable not to use reusable gloves.

15 Hypodermic needles and syringes should be used only for parenteral injection and aspiration of fluids from laboratory animals and diaphragm bottles. Extreme caution should be used when handling needles and syringes to avoid needle prick and the generation of aerosols during use and disposal. Needles should not be bent or sheared by hand. They should not ordinarily be replaced in the sheath or guard. They should be promptly placed in a puncture-resistant container and decontaminated, preferably by incineration or autoclaving, before disposal.

16 All spills, accidents, and overt or potential exposures must be reported to the laboratory supervisor or instructor as soon as possible. This person should file a report with the appropriate safety committee. Appropriate medical evaluation, surveillance, and treatment should be provided as required.

GENERAL LABORATORY INSTRUCTIONS

1 You must always read the assigned laboratory exercises before the start of the laboratory period.
2 Preliminary instructions and demonstrations will be given at the beginning of each laboratory period by the instructor. Ask any questions to clarify any of the procedures to be used. Then plan your work carefully. Do not attempt to start work before receiving instructions and seeing any demonstrations.
3 Accurate and detailed results are to be recorded at the completion of each exercise. Use the tables and other spaces provided for recording results and answering questions. Drawings of microscopic observations should be within the circular outlines. In making your drawing of a specimen, do not attempt to draw everything in the microscopic field. Simply select a few representative items, for example, cells, their shapes, arrangements, and structures.

SPECIFIC LABORATORY REGULATIONS

In addition to observing the preceding basic requirements, the following specific regulations for a teaching laboratory should be observed for the safety and convenience of everyone working in it.

1 Street outerwear should not be brought into the laboratory.
2 Begin every laboratory period by disinfecting, or washing down, your benchtop, and end it the same way.
3 If a culture is spilled, cover the area with disinfectant and notify the instructor. Report all accidents, however minor, to the instructor.
4 Never remove any cultures from the laboratory.
5 Inoculated media placed in the incubator must be properly labeled with your name, date, and nature of the specimen.
6 Pencils, labels, or any other materials should never be placed in your mouth.
7 Be very careful with gas burners. Long hair should be tied back neatly, away from the shoulders. Gas burners must be turned down or off when not in use during exercises. Be sure gas burners are turned off at the end of the lab period.
8 All reagents and equipment must be returned to their proper place at the end of the lab period.
9 All used cultures and contaminated glassware should be put into a designated container to be autoclaved.
10 Contaminated plastic and other disposables are to be discarded into a separate container also to be autoclaved.
11 Uncontaminated materials like scraps of paper and cotton must be placed into wastebaskets.
12 Never lay contaminated materials like pipettes on the benchtop.
13 Never discard contaminated liquids or liquid cultures into the sink for disposal. Put them into designated containers in the lab for later sterilization.
14 Personal conduct in a microbiology lab should always be courteous and professional. Personal attention to the principles of safety is always in order.

REFERENCES

Medical Research Council of Canada and Health and Welfare Canada: *Laboratory Biosafety Guidelines*, Office of Biosafety, Laboratory Centre for Disease Control, Health and Welfare Canada, Ottawa, Ontario, Canada K1A 0L2, 1990 (Cat. no. MR 21-1/1990E). (Aussi disponible en français sous le titre *Lignes directrices en Matière de Biosecurité en laboratoire*.)

U.S. Department of Health and Human Services, Public Health Service, Centers for Disease Control and National Institutes of Health: *Biosafety in Microbiological and Biomedical Laboratories*, 2d ed., Superintendent of Documents, U.S. Government Printing Office, Washington, D.C. 20402, 1988 [HHS Publication No. (CDC) 88-8395].

National Committee for Clinical Laboratory Standards (NCCLS): Document M29-T2, *Protection of Laboratory Workers from Infectious Disease Transmitted by Blood, Body Fluids, and Tissue*, 2d ed.; Document 117-P, *Protection of Laboratory Workers from Instrument Biohazards*, NCCLS, 771 East Lancaster Ave., Villanova, PA 19085, 1992.

GENERAL EQUIPMENT AND SUPPLIES NEEDED BY EACH STUDENT

Permanent Equipment Usually Furnished by Department

Microscope
Microbiology transfer needles
Forceps
Staining tray or trough
Bunsen burner
Staining rack complete with stains
Microscope slides
Cover slips
Microscopic slide labels (self-adhesive)
Lens and bibulous paper (usually in booklets)
Safety matches

Supplies Usually Furnished by the Student

Permanent marker for glass or plastic (such as Sanford's Sharpie marking pen)
Celluloid metric rule
Microscope slide box

REFERENCES KEYED TO EXERCISES

Where appropriate, exercises in the manual are keyed to the following sources for additional information:

"*MICROBIOLOGY*" refers to *Microbiology: Concepts and Applications* by M. J. Pelczar, Jr., E. C. S. Chan, and N. R. Krieg, McGraw-Hill, New York, 1993.

"*MM*" refers to *Methods for General and Molecular Bacteriology*, American Society for Microbiology, Washington, D.C., 1993.

Any other particularly relevant references for a specific exercise also will be cited.

THE UBIQUITY OF MICROORGANISMS

Microorganisms are numerous and ubiquitous in our environment. The exposure of a Petri dish containing sterile nutrient-agar medium to the air will demonstrate this. After the Petri dish has been incubated, colonies of microorganisms become visible showing that the surface of the medium had been inoculated with microorganisms carried by air currents. Microbes in the air come from different sources, such as soil, dust, and the upper respiratory tract of humans. Saliva, for example, teems with bacteria of all types. Pond water or a sample of hay infusion provides a menagerie of microbes—bacteria, yeasts, molds, protozoa, and algae. Microscopic examination of such specimens will reveal a diversity of morphological types among microorganisms. These microorganisms will exhibit considerable variation in shape, size, structure, and behavior. You will have discovered a whole new world: the world of microscopic living forms invisible to the naked eye!

EXERCISE 1

Microorganisms in the Environment

OBJECTIVE
To appreciate that microorganisms are found everywhere and to learn how to describe microbial colonies growing on agar medium.

OVERVIEW
The laboratory, like all other environments, has many microorganisms that hang suspended in the air or settle down with dust on various surfaces. This observation is especially important to workers in a microbiology laboratory. They must practice techniques that will prevent these microorganisms from contaminating materials they work with, such as sterile media, solutions, and equipment. Such techniques that keep out *unwanted* microorganisms are called *aseptic techniques*.

REFERENCES
MICROBIOLOGY, Chap. 2, "The Scope of Microbiology."
MM, Chap. 9, "Solid, Liquid/Solid, and Semisolid Culture."

MATERIALS
5 tubes each with 25 ml nutrient agar or
   1 flask with 125 ml nutrient agar
5 sterile Petri dishes (plates)
1 sterile cotton swab
1 boiling water-bath or microwave oven
1 45°C water-bath

NOTE: The term *tube of agar*, as used in the exercises in this manual, refers to a tube containing approximately 25 ml of a molten agar medium to be poured into one Petri dish. It is generally more convenient to dispense the agar medium in flasks, about 125 ml per flask, and to have students pour plates from this source.

PROCEDURE
1 Melt nutrient agar by placing the tubes or flask of agar in the boiling water-bath. Alternatively, place the tubes or flask into a microwave oven to melt the nutrient agar. (The use of the boiling water-bath shows that bacteriological agar melts or becomes molten at about 100°C. Use of the microwave oven is more convenient and speeds up the process of melting agar considerably.)
2 Cool or temper the melted agar to approximately 45°C by placing it in a water-bath set to this temperature.
3 One at a time, carefully remove the cover, cotton plug or cap, of a tube (flask) and flame the mouth in the Bunsen burner for about 5 s. Pour the contents of the tube (or about 25 ml medium from the flask) into each of five sterile Petri dishes [FIGURE 1.1] that have been labeled with your name, date, and any other identifying information, for example, *a*, *b*, *c*, *d*, and *e*, to identify treatment given each dish as described in step 4. **Always label your materials every time you carry out an exercise or do an experiment.**

NOTE: Reusable stainless steel caps are commonly used because they demand less labor in media preparation. **Be sure they are not too hot from the Bunsen flame before handling them.** However, cotton plugs wrapped in cheesecloth are easy enough to make, and they retain the shape of the tube or flask mouth after sterilization of the medium.

4 After the nutrient-agar medium has gelled, inoculate it in the following manner for each Petri dish:
  a Remove the cover from the first plate (*a*) and expose it to the air for 30 min.
  b With a cotton swab, wipe an area on your laboratory desk and streak the entire surface of the second plate (*b*) with this swab. Do not cut into the agar medium surface.

7

FIGURE 1.1
Petri dish preparation with nutrient-agar medium.

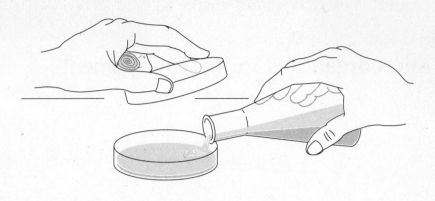

c Remove the cover from the third plate (c) and place a strand of hair on the agar medium surface.
d Touch the agar medium surface of the fourth plate (d) in several places with your fingers.
e Touch the agar medium surface of the fifth plate (e) in several places with your fingers after washing them with soap and water.

5 Incubate the Petri plates, inverted, at 25°C, or room temperature, for about a week.

6 Remove the Petri plates to your work area for observation of the colonies. A *colony* is a visible mass of microbial cells resulting from the growth of one cell or clump of cells that was deposited on the surface of the medium [FIGURE 1.2].

7 Probe the colonies on the nutrient-agar medium with a sterilized inoculating needle for consistency of colonies. An *inoculating needle* is a nichrome wire loop inserted into a metal holder with an insulated handle [FIGURE 1.3]. It is used for testing colony consistency, transfer of cells, and streaking of cells on media. The loop is sterilized by first heating it dry above the Bunsen flame (to avoid spattering of any material on it) and then by burning it to red-hot in the flame. Allow it to cool briefly before touching a colony to test consistency. After use, sterilize the inoculating needle again the same way.

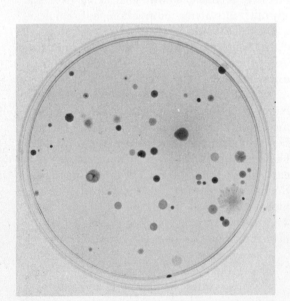

FIGURE 1.2
Nutrient-agar medium plate showing colonies after exposure to air.

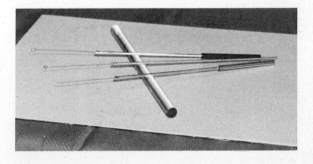

FIGURE 1.3
Inoculating needles with loop and straight wires.

NAME _____

Microorganisms in the Environment

RESULTS

1 Examine the plates and count the number of colonies on each. In the chart provided at the end of this section, record the number of colonies observed.

2 In the chart, describe the appearance of these colonies according to the descriptive terms for colonial morphology that are given below.

3 Draw representative colonies from each plate to illustrate colonial types. Use representative sketches, as shown below, to facilitate your drawings in the chart.

Descriptive Terms and Sketches for Colonial Morphology

Colony Surface:

Smooth, rough, rugose (wrinkled)
Contoured (an irregular, smoothly undulating surface, like that of a relief map)
Granular (fine, medium, coarse)
Dull
Glistening

Optical Characteristics:

Opaque—not allowing light to pass through
Translucent—allowing light to pass through without allowing complete visibility of objects seen through the colony
Opalescent—resembling the color of an opal
Iridescent—exhibiting changing rainbow colors in reflected light
Dull—not glassy or glistening
Glossy—not dull but shining, reflecting light

Consistency When Probed by Inoculating Needle with Loop:

Butyrous—growth of soft butterlike consistency
Viscid—growth follows the loop when touched or withdrawn
Membranous—growth thin, coherent, like a membrane
Brittle—growth dry, friable under the loop

Pigmentation of Growth:

White, buff, light yellow, straw yellow, pink, red, and so on
Soluble—color (pigment) diffuses into medium surrounding colonies
Nonsoluble—color (pigment) stays confined to colony

Form:

Circular   Irregular   Spindle   Filamentous   Rhizoid

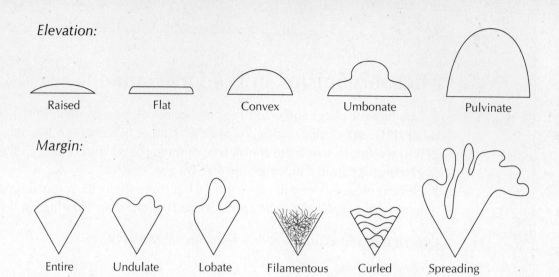

NUMBER AND APPEARANCE OF REPRESENTATIVE COLONIES ON NUTRIENT AGAR			
Specimen	No. of colonies on plate	Word description of predominant colonies	Sketch of some predominant colonies
a			
b			
c			
d			
e			

Temperature of incubation: Days of incubation:

NAME _____   1 MICROORGANISMS IN THE ENVIRONMENT *(Continued)*

QUESTIONS 1 Why is nutrient agar used as a medium?

2 Why is it necessary to hold the neck of the tubes or flask in the flame of the Bunsen burner?

3 How is a bacterial colony formed, and why is it sometimes called a *colony-forming unit?*

4 Why are Petri plates inverted during incubation?

5 What practical steps should be taken to keep down contamination in the laboratory?

6 What is the usual percent concentration of bacteriological agar used in making media?

7 At what temperature does bacteriological agar that is used in media liquefy? Solidify or gel?

8 Why was the agar medium tempered to 50°C before pouring?

9 Give some reasons why the Petri dish is so suitable for the aerobic cultivation of microorganisms.

EXERCISE 2

Mixed Microbial Flora

OBJECTIVE
To show and to understand that microorganisms in natural environments occur in mixed populations.

OVERVIEW
Even though microbiologists usually study *pure cultures* of microorganisms, that is, cultures containing only one kind of microbe, it must be realized that this is, in fact, an artificial situation. Most natural environments contain a mixture of microorganisms, and they must interact with one another in the struggle to grow and survive. In keeping with natural law, here too in the microscopic ecosystem, it is survival of the fittest.

The term *symbiosis* is now generally used in a broad sense to describe any type of interaction between different organisms, whether the association is beneficial or not. In this exercise, microorganisms from two natural sources will be allowed to grow to illustrate that myriad types of microbes inhabit an ecosystem [FIGURE 2.1]. Therefore, in addition to their ubiquity, microorganisms are in mixed culture in natural situations.

REFERENCES
MICROBIOLOGY, Chap. 2, "The Scope of Microbiology."
MM, Chap. 8, "Enrichment and Isolation."

MATERIALS
2 Petri plates of nutrient-agar medium
Sample of garden soil
1 sterile cotton swab

PROCEDURE
1 You are provided with two Petri plates of nutrient-agar medium.

2 Take a small pinch of garden soil and sprinkle it over the surface of the medium in one of the Petri plates.
3 Incubate the plate (inverted) at 25°C until the next laboratory period.
4 With the sterile cotton swab provided, wipe along the gum margin of several teeth.
5 Smear the swab gently over the surface of the medium in the remaining plate.
6 Incubate this plate (inverted) at 37°C in an incubator until the next laboratory period.

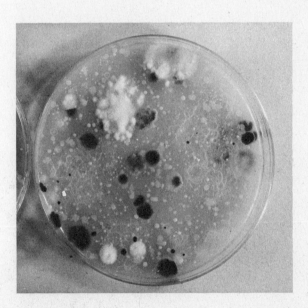

FIGURE 2.1
Nutrient-agar medium inoculated with soil sample showing mixed flora.

NAME _____

Mixed Microbial Flora

RESULTS 1 Examine the plates and estimate the number of *different* types of colonies on each plate. Record your observations in the chart provided.
2 Describe the appearance of several *predominant* colonies. Record your observations in the chart.

NUMBER AND DESCRIPTION OF DIFFERENT COLONY TYPES		
Sample	No. of different colonies	Word description of colonies
Soil		
Gum margin		
Incubation temperature of soil plate: °C. Gum margin plate: °C.		

QUESTIONS 1 Why were different temperatures used for the incubation of the two plates?

2 Indicate two types of bacteria that would not grow on the medium used.

3 Indicate one type of bacteria that would not grow due to the physical condition of incubation.

EXERCISE 3

Aseptic Technique and the Transfer of Microorganisms

OBJECTIVE
To acquire the skill of aseptic technique in the practice of microbiology.

OVERVIEW
Microorganisms are found everywhere. In the laboratory they are found on the surfaces of laboratory benches, equipment, and in the air. Such microorganisms always pose a threat to *contaminate* pure cultures. Pure cultures, each containing only one kind of microorganism, are used for laboratory study. Results from contaminated pure cultures are spurious results insofar as the study of the pure culture is concerned. A microbiologist worthy of his or her profession always keeps pure cultures *pure*, that is, free from extraneous or unwanted microorganisms. The methods of maintaining pure cultures, which also prevent such cultures from contaminating the worker and environment, are collectively called ***aseptic technique***. The acquisition of such a skill comes with practice and in time becomes second nature to every knowledgeable microbiologist.

Pure cultures of microorganisms are protected from the environment by a cover. For example, a Petri dish has a top cover; a culture tube has a cap or a cotton plug [FIGURE 3.1]. However, in order to transfer a culture of microorganisms from one medium to another, in a procedure called ***subculturing***, the cover must be removed briefly—but long enough to contaminate the pure culture by environmental microbes unless aseptic technique is observed. In this exercise you will learn how

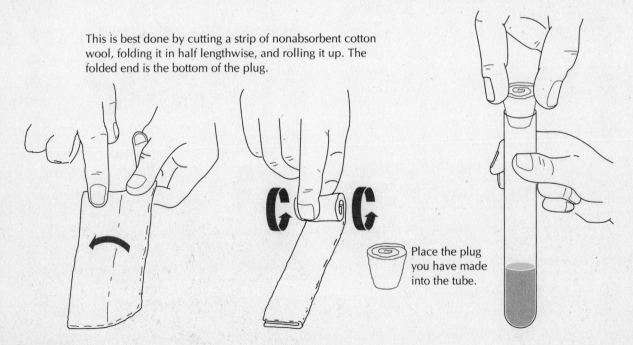

FIGURE 3.1
Procedure for making a cotton plug for a culture tube.

to transfer a pure culture of bacteria by aseptic technique.

REFERENCES
MICROBIOLOGY, Chap. 3, "Characterization of Microorganisms."
MM, Chap. 8, "Enrichment and Isolation."

MATERIALS
1 24-h nutrient-broth culture of *Serratia marcescens*
1 24-h nutrient-agar slant of *S. marcescens*
4 tubes of nutrient broth
4 nutrient-agar slants

PROCEDURE

> **CAUTION: This is the first time that you will be handling a culture of microorganisms. Remember all safety procedures and observe all instructions given by the instructor in order to carry out this exercise.**

To carry out a transfer of microorganisms aseptically, follow these general steps and refer to accompanying figures.

1 Sterilize the inoculating needle, straight wire or loop, by holding it in the Bunsen burner flame until the wire or loop gets red-hot. **Needle sterilization is always done before and after removal of microorganisms** [FIGURE 3.2].
2 Pick up a tube of culture and a tube of sterile broth or slant with your free hand. Hold them in the palm of your hand, secure them with your thumb, and separate them to form a V [FIGURE 3.3].
3 With the sterile loop or needle in hand, uncap or unplug the tubes in the slanting position with the free fingers and palm of the hand [FIGURE 3.3]. Never lay these caps or plugs down on the laboratory bench; they must be kept in the hand until transfer of the microbes is completed.
4 Flame the necks of the tubes by holding them in the Bunsen flame briefly [FIGURE 3.4]. This heats up the necks of the tubes so that convection currents carry airborne microbes away from the tube openings, thus preventing the entry of extraneous microorganisms.
5 Remove a small amount of the culture with the sterilized inoculating needle [FIGURE 3.5]. Cool the red-hot wire or loop by touching the inside of the tube before touching the cells. Either a loopful

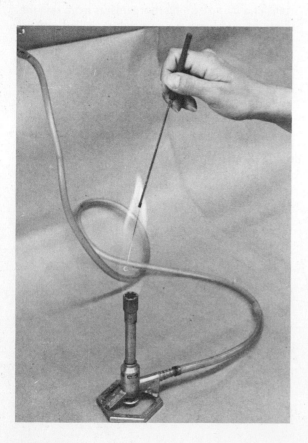

FIGURE 3.2
Sterilize inoculating needle.

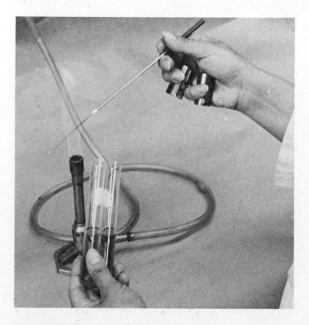

FIGURE 3.3
Secure and uncap culture and subculture tubes.

FIGURE 3.4
Flame tube necks to eliminate contamination.

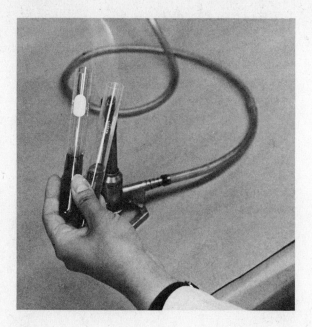

FIGURE 3.5
Transfer culture into subculture tube.

of broth culture or cells adhering to the loop, when it is touched gently to a mass of grown cells, is removed. The straight wire is usually used for inoculating into medium agar deeps, which is a tall butt of agar medium in a tube. The cell-laden loop is inserted into the subculture tube. With a broth medium, the loop is shaken gently in the broth to dislodge the cells. With an agar slant medium, the loop with cells is drawn gingerly over the slope surface in a straight line without breaking the agar surface [FIGURE 3.6].
6 Flame the neck of the tubes again. This second flaming kills the cells that may have fallen on the neck of the tubes during the transfer of cells. Replace the caps or plugs of the tubes. Flame your inoculating needle again to sterilize it, in order not to contaminate your work area; set it down in a safe place after completing each culture transfer.

The specific procedure for this exercise is as follows:

1 You are given a *broth* culture of *S. marcescens*.
2 Following the procedure outlined above, transfer the culture to two nutrient-broth tubes and to two nutrient-agar slants, transferring one at a time. Label one inoculated broth tube and one inoculated agar slant "25°C." Label the two remaining subcultures with "37°C." Also label all four tubes to indicate that they were transferred from a broth culture.
3 You are also given a *slant* culture of *S. marcescens*.
4 Transfer this culture to two nutrient-broth tubes and to two nutrient-agar slants. Similarly, label one inoculated broth tube and one inoculated agar slant "25°C"; the two remaining subcultures should be labeled with "37°C." Label all four tubes to indicate that they were transferred from a slant culture.
5 Incubate the appropriate cultures in the 25°C incubator (or at room temperature) and in the 37°C incubator for 24 to 48 h.

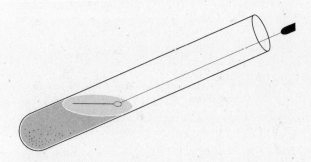

FIGURE 3.6
Streak an agar slant by drawing loop over surface.

NAME _____

Aseptic Technique and the Transfer of Microorganisms

RESULTS

1. Gently shake all the broth cultures and examine them for turbidity due to cell growth.
2. Examine the inoculated slants for cell growth.
3. Compare the cultures incubated at 25°C with their twin cultures incubated at 37°C. Note the appearance of orange-red pigmentation. (Normally pigmentation is only expressed with 25°C incubation.)
4. Record your observations in the chart provided.

GROWTH OF SUBCULTURES				
Subculture	Transferred from	Incubation temp., °C	Growth, + or −	Pigment, + or −
Nutrient broth	Nutrient broth	25		
Nutrient broth	Nutrient broth	37		
Nutrient broth	Nutrient-agar slant	25		
Nutrient broth	Nutrient-agar slant	37		
Nutrient-agar slant	Nutrient broth	25		
Nutrient-agar slant	Nutrient broth	37		
Nutrient-agar slant	Nutrient-agar slant	25		
Nutrient-agar slant	Nutrient-agar slant	37		

QUESTIONS 1 Why did you flame the necks of the tubes immediately after uncapping and before replacing the caps?

2 Do you see any difference in pigmentation between the cultures incubated at 25°C and those at 37°C?

3 If pigmentation is present, what is the color of this pigment? Is the pigment water-soluble?

4 How would you know if you contaminated the cultures during transfer?

MICROSCOPY IN MICROBIOLOGY

The microbe is so very small
  You cannot make him out at all,
  But many sanguine people hope
  To see him through a microscope.
      — HILLAIRE BELLOC

The microscope is one of the most important tools of the microbiologist because most microorganisms are too small to be seen with the naked eye. The unaided eye cannot see fine details smaller than 0.1 mm. This is because when we look at objects held as close to the eye as is possible for clear vision (about 10 in), their image on the retina of the eye is very small. However, when we use a simple magnifying glass or a microscope, the image of fine detail is made larger on the retina.

The microscope, as a precision instrument, has advanced tremendously in design since Antony van Leeuwenhoek (1632–1723), the father of microbiology, discovered microorganisms in water with the use of a simple lens, which was only a simple strong magnifier with a specimen holder. The microscope is used today in research not only for seeing fine detail but also for obtaining information on the composition of a specimen.

There are two main types of microscopes: the **light microscope** and the **electron microscope.** Light microscopes use visible light to form images, but the type of light used and its modification are the basis for the different types of light microscopy. The most common types of light microscopy are *bright-field, dark-field, phase-contrast,* and *fluorescent microscopy.* Electron microscopes use electrons instead of light rays, and magnets instead of lenses, to control the electron beams to form images. The two types of electron microscopes are the *transmission electron microscope (TEM)* and the *scanning electron microscope (SEM).*

ANATOMY OF THE LIGHT MICROSCOPE

To examine microorganisms by light microscopy, microbiologists use a *compound microscope* because compound microscopes give the highest magnifications. It is so called because it magnifies through two separate lens systems: the **eyepiece** and the **objective.** The latter lens system is nearest the object and forms an image of the object inside the microscope. This image (called the *first,* or *primary, image*) contains all the fine details and is magnified anywhere from 2× to 100×. It is then magnified further by the eyepiece to give an overall magnification, which is the magnification of the objective multiplied by the magnification of the eyepiece, for example, 100× of objective multiplied by 10× of eyepiece = 1000× the overall magnification. The main parts of the compound microscope are depicted in FIGURE II.1.

The **ocular lens** is found in the *eyepiece* which is slipped into the sleeves in the upper end of the draw, or eyepiece, tube. The number on the eyepiece indicates its magnification. The draw tube of a monocular microscope has a fixed length of 160 mm. In a binocular microscope, the draw tube is replaced by the **eyepiece,** or **binocular, tube,** the eyepiece sleeves of which can be displaced in relation to each other by a sliding action to adjust for the interpupillary distance of the user. (The spacing between the eyes varies from one individual to another.) The fields of view seen by both eyes should coincide exactly. The value indicated on the central wheel must then be set on each of the eyepiece sleeves to restore the correct mechanical tube length, which has been changed by the interpupillary distance setting. This also ensures that the objectives are *parfocal* (described below). However, if your vision is not the same in both eyes, you may adjust the ocular collar of the eyepiece on the poor side until both eyes see the same sharp image.

Each eyepiece may be equipped with eyecups. These should be extended when eyeglasses are not worn and retracted when eyeglasses are worn.

The **nosepiece** is used for quick change of objectives, the lens systems nearest the specimen, or object, to be magnified. Rotation of the nosepiece positions an objective above the stage opening. Certain designations on each objective should be noted. For example, consider

100/1.30
160/−

These designations mean the following:

100 = objective magnification is 100×
1.30 = numerical aperture (see below)
160 = mechanical tube length of 160 mm
− = objective is insensitive to larger than ±0.01-mm deviations in cover-slip thickness when cover slip is 0.17 mm thick. (One of two numbers may be shown. If the number here is 0.17, this means that the objective is to be used with a cover slip of 0.17 ± 0.01 mm thickness; 0 means that it has been corrected for zero cover-slip thickness, that is, for uncovered specimens.)

Other designations on an objective indicate the kind of objective, for example, whether it is an oil-immersion objective, a phase-contrast objective, or a *flat field objective,* that is, the whole field of view is in focus, not just the center.

Microscopes in most microbiological laboratories are equipped with three objectives: the low-power objective, the high-power objective, and the oil-immersion objective [TABLE II.1]. The *final* magnification of the microscope is the product of

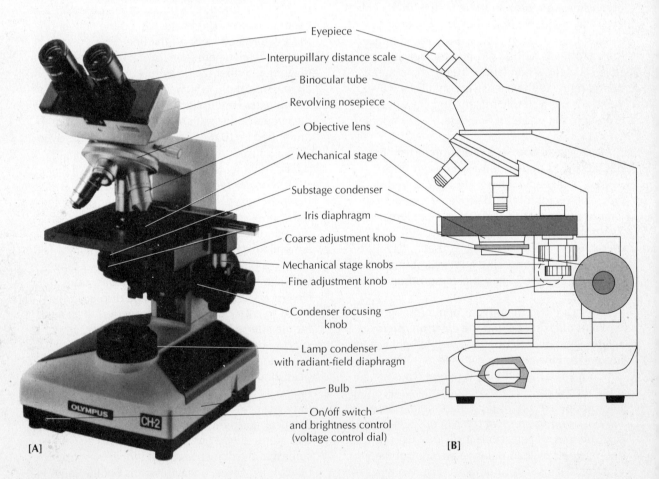

FIGURE II.1
The compound microscope.

TABLE II.1 Typical Characteristics of Objectives

Objective	Focal length, mm	Working distance, mm	Numerical aperture	Magnification	Resolution, μm
Low power	16	4–8	0.25	10×	1.10
High power	4	0.2–0.6	0.85	45×	0.32
Oil immersion	1.8	0.11–0.16	1.25	97×	0.22

the ocular and the objective lens used. For example, the 10× ocular used in conjunction with the oil-immersion objective will give a final magnification of 10× *times* 97× = 970×. The magnification, tube length, and focal length are interrelated in the following manner:

$$\text{Focal length} = \frac{\text{tube length}}{\text{magnification}}$$

For example, the focal length of a 10× objective is 160/10 = 16 mm. **Working distance** is the clearance between the lowest point of the objective and the object in focus. Therefore, it is obvious that there is very little working space when working with the oil-immersion objective, and great care must be exercised to avoid ramming the objective front lens onto the specimen.

Previously, changing objectives was rather inconvenient because the draw tube had to be raised before turning the nosepiece, and every objective had to be focused separately. Today the objectives are **parfocal,** which means that objectives can simply be swung into place and only slight focusing with the fine-adjustment knob is needed to bring the specimen into sharp focus.

There are two kinds of lenses in objectives, **achromats,** or without color, and **apochromats,** or with separated colors. A theoretically perfect achromat, though nonexistent, would bring all the wavelengths of light to a common focus so that all images formed from different wavelengths would be of the same size. But because of light dispersion and refraction by a glass lens (blue components bend more than the red components of the light spectrum), this is not possible in practice. Instead, an attempt is made to bring together those colors, red and green, to which the eye is particularly sensitive. The residual blue color fringe of the image is not readily detected by the eye and vanishes completely against a blue background. For this reason manufacturers usually provide a disk of blue (or green) glass to be placed in front of microscope lamps. Achromatic lenses are relatively cheap and are adequate for most purposes.

Apochromatic lenses are made on an entirely different principle and must be used with special eyepieces. The colors are intentionally widely separated, or overcorrected, and then brought back to a nearly common focus by a **compensating eyepiece.** An apochromatic objective used with a **regular eyepiece** gives a much poorer image than the much cheaper achromatic objective. Its use is justified only in color photography and in the most critical kind of research.

The **stage** is a platform with spring clips for holding the slide containing the specimen. It has an opening in the center that allows for the passage of light from an illuminating source below. To facilitate examination, the clips are usually replaced with a mechanical stage, which allows for smoother and better-controlled movements of the slide.

Beneath the stage is the **condenser** equipped with an **iris diaphragm.** The condenser is a system of lenses that forms light into a cone and focuses it on the specimen. The two main types in use in bright-field microscopy are the **achromatic-aplanatic** and the **Abbe.** The first type is used for research and color photomicrography because it is free from aberrations and its color correction closely matches that of the finest objectives, such as apochromatic objectives. The Abbe condenser is the most popular in laboratories because of its cost and suitability for routine work. Other special condensers are available for other kinds of microscopy, for example, dark-field and phase-contrast microscopy. With a numerical aperture of about 0.9, the Abbe condenser is perfectly adequate for most work. Every substage condenser is furnished with an *iris diaphragm,* the only function of which is to control the numerical aperture (see below) of the condenser. Optimum illumination of the specimen is achieved through proper use of the diaphragm. The iris diaphragm should never be used to decrease light *intensity* because this results in the loss of fine detail or even the introduction of detail that is not there. It is only used to adjust the *amount* of light admitted, for example, the shorter the working distance, the more the iris diaphragm must be opened.

The **light source** is positioned in the base of the microscope. Most microscopes are equipped with a built-in light source to provide direct illumination. However, some microscopes use an external light source, such as a lamp. These are provided with a mirror, one side flat and the other side concave, and the external light source is placed in front of the mirror to direct light upward into the condenser lens system. The flat side of the mirror is used for artificial light while the concave side is used for sunlight.

With an artificial light source, there is a **radiant-field diaphragm.** Just like the iris diaphragm, it should not be used to decrease light intensity. There are two ways to decrease the intensity of illumination: one method is lowering the voltage applied to the lamp, which has the disadvantage of making the light redder as the intensity is decreased; the other, and better, method is to place neutral-density filters in front of the illumination source.

SOME BASIC PRINCIPLES OF MICROSCOPY

Magnification

Magnification of a specimen is the function of a two-lens system in a compound microscope. The lens system in the objective magnifies the specimen and produces a **real image,** illuminated by the lamp and the condenser, that is projected up into the focal plane as shown in FIGURE II.2. The lens system in the ocular magnifies the real image, yielding a **virtual image,** which is seen by the eye. In general practice, the useful limit of magnification is at the most 1000 times the numerical aperture of the objective. However, in microscopy, magnification is secondary to resolution.

Resolution (Resolving Power)

The **resolving power** of a lens is its ability to show two closely adjacent objects as distinct, discrete, and separate entities. For example, we are interested in seeing whether a microscope can show two adjacent bacteria flagella as two separate entities instead of as one. Thus, the largest image a microscope produces by magnification is not actually the most useful.

The resolving power, or **resolution,** of a microscope is a function of the wavelength of light used and the numerical aperture (*NA*), a characteristic of the lens system that will be explained below. To determine whether a microscope has sufficient resolving power to distinguish between entities, it

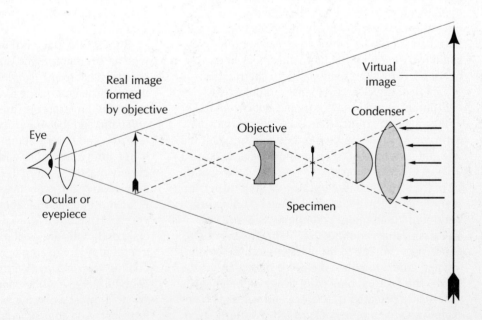

FIGURE II.2
Magnification in the two-lens system.

is necessary to determine the limit of resolution, or maximum resolving power. This limit is expressed in a formula which determines the smallest resolvable separation d, or the shortest distance between two entities that the microscope is able to differentiate. In the following formula, NA_{obj} and NA_{cond} are the numerical apertures of the objective and the condenser, respectively. The resolving power then, or the smallest resolvable separation between two entities d, expressed in micrometers, is:

$$d = \frac{\text{wavelength of light, } \mu m}{NA_{obj} + NA_{cond}}$$

Based on this formula, the shorter the wavelength of radiation of the electromagnetic spectrum, the smaller the distance between entities a given lens system can discern, and hence the greater the resolving power attainable. But the visible portion of the **electromagnetic spectrum** (i.e., the bands of color produced when light passes through a prism) is very narrow, and the very short wavelengths are found in the nonvisible ultraviolet portion of the spectrum.

To further illustrate the above equation with a typical light microscopy setup, assume the use of a green filter with the light source which radiates a wavelength of 0.55 μm. The oil-immersion objective used has an NA of 1.25, and the condenser has an NA of 0.9. Substituting these values into the equation, we have

$$d = \frac{0.55 \, \mu m}{1.25 + 0.9} = \frac{0.55 \, \mu m}{2.15} = 0.255 \, \mu m$$

Thus, the resolving power of this microscope, or the smallest resolvable separation d that this microscope can discern, is 0.255 μm. Because most bacterial cells are larger than this and are positioned with greater distance between them, they can be seen clearly by light microscopy.

Numerical Aperture

For proper resolution, it is necessary to have sufficient illumination of the specimen. Light must pass through the specimen and be caught by the objective. The **numerical aperture** is a mathematical expression that describes the way in which light is concentrated by the condenser and collected by the objective.

FIGURE II.3 shows an objective lens and specimen. As light rays pass from glass into air, rays are bent and some light is lost due to *refraction*. The greater the loss of light due to refraction, the lower will be the numerical aperture, and hence the resolution. In contrast, the numerical aperture is highest when the greatest amount of light successfully passes through the objective lenses. In other words, the more refracted light that an objective is capable of accepting, the greater its resolving power. The bending ability of light, as it passes through various substances, is expressed as a numerical value called the *refractive index*. When immersion oil is placed between the specimen and the oil-immersion lens, the light rays leaving the glass continue without refraction because there is no air interface and because immersion oil has the same refractive index as glass. The

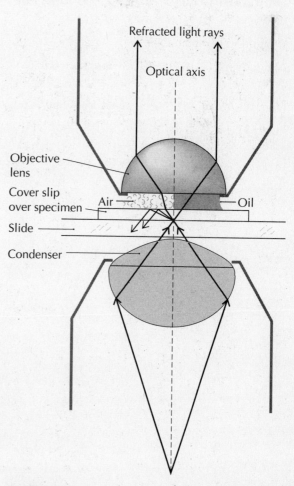

FIGURE II.3
Refraction of light through air and immersion oil.

angle θ subtended by the optical axis and the outermost rays still covered by the objective is a measure of the aperture of the objective; it is the *half-aperture angle*. The magnitude of this angle is expressed as a sine value. The sine value of the half-aperture angle multiplied by the refractive index n of the medium filling the space between the objective and the cover slip gives the numerical aperture:

$NA = n \sin \theta$

With dry objectives the value of n is 1, since 1 is the refractive index of air. When immersion oil is employed as the medium, n is 1.52. In the example given in FIGURE II.4, θ is 58°, and the numerical aperture can be determined as follows:

With dry objective,

$NA = n \sin \theta = 1.00 \times \sin 58° = 1.00 \times 0.85 = 0.85$

With oil-immersion objective,

$NA = n \sin \theta = 1.52 \times \sin 58° = 1.52 \times 0.85 = 1.29$

From this example it becomes obvious that an oil-immersion objective accomplishes a big gain in numerical aperture.

FIGURE II.4
Determining the numerical aperture.

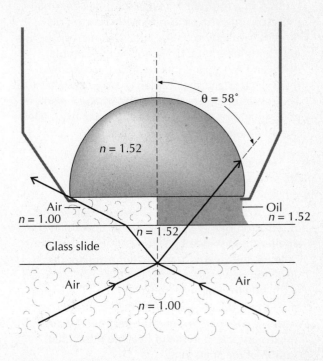

EXERCISE 4

Principles and Use of the Bright-Field Microscope

OBJECTIVE
To learn how to adjust a light microscope and to use it effectively.

OVERVIEW
Bright-field microscopy is the normal method of examining an object by transmitted light. Its major limitation is the absence of contrast between the object and the surrounding medium, which makes it difficult to examine living cells. Therefore, cells like microorganisms are stained with a color, which renders them nonviable, to facilitate their observation. This is like looking at stained-glass windows, but at the microscopic level. The stained cells absorb some of the white light passing through them. The rest of the light is then altered in color because some colors have been taken from it. The colors produced are seen as contrast by the eye. Thus, bright-field microscopy is best used if the specimen, or the background, is colored or has been artificially colored as by staining.

SETTING UP A BRIGHT-FIELD MICROSCOPE FOR MICROBIOLOGICAL EXAMINATION
There are essentially three general directions that should be followed by a microbiologist in setting up a microscope before making an examination of a specimen:

1 Align the illumination and set up a *Koehler illumination*, which is a procedure for aligning the optical components of the microscope.
2 Every time the objective is changed, rematch the condenser to it with the iris diaphragm.
3 Whenever the slide is changed, refocus the substage condenser, that is, refocus the image of the radiant-field iris.

Steps in Alignment of Microscope
An expanded procedure follows for setting up the bright-field microscope [FIGURE 4.1]. Refer to accompanying figures as you proceed through the steps.

Alignment of the illumination

1 For a microscope with an external light source, align the lamp with the flat mirror of the microscope so that an image of the spiral lamp filament is focused on the iris diaphragm of the substage condenser with the aid of a field condenser (on the lamp). With the radiant-field iris wide open, first focus the filament of the lamp on a piece of paper laid on the microscope mirror, and then, having moved the piece of paper up until it rests against the underside of the substage condenser, adjust the focus with the lamp field condenser until the filament is again sharp. (A small mirror held in the hand will allow one to see clearly what is going on underneath the stage.) The lamp should be tilted and moved backward and forward and the mirror again tilted until the image of the filament, when twice focused, is centered both on the mirror and on the substage iris diaphragm.

However, most microscopes today have a built-in light source in the base of the microscope. Behind the base, there are two centering screws; by adjusting them, the image of the filament may be centered on the iris diaphragm. At this point the filament may not be in sharp focus, but this is easily accomplished by loosening the knurled ring and moving the lamp socket backward and forward in the lamp receptacle until the image of the lamp filament is in sharp focus. Then tighten the ring until the socket is secure in this proper position. The lamp is now properly centered and the filament in focus on the iris diaphragm. In practice,

FIGURE 4.1
Numbered steps in the alignment of a bright-field microscope.

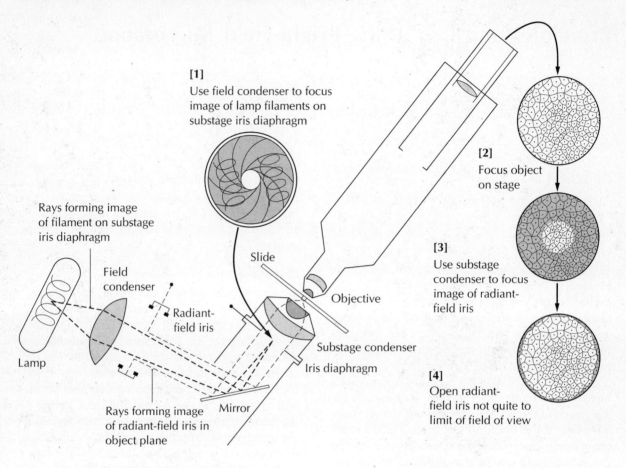

however, in most microscopes with a built-in illuminator, alignment of the lamp and focusing of the filament are not made because an image of the lamp diffuser is formed on the iris diaphragm. The centering is made by visually aligning the mount and the housing until a bright even-lighted disk is obtained.

For microscopes with an external light source, after aligning the illumination, rack the substage condenser up to the top. Remove the eyepiece from the tube. Adjust the mirror until the back lens of the objective is filled with light. It may be necessary to dim the light by a neutral-density filter.

Adjusting for Koehler illumination

2 Final adjustments for *Koehler illumination* may now be made. The first step is to place a slide on the stage and focus the specimen with a 10× objective. The lamp should be turned up to a comfortable brightness. At this point the radiant-field diaphragm and the iris diaphragm should be at their widest openings.

3 Now close the radiant-field diaphragm to a small opening. Focus the opening by slightly lowering the substage condenser [FIGURE 4.2]. Both the specimen and the leaves of the radiant-field diaphragm are now sharply defined; that is, the margin of the diaphragm is now focused in the plane of the specimen. If the small opening of the radiant-field diaphragm is off-center, center it by adjusting the condenser centering screws [FIGURE 4.3].

4 Open the radiant-field diaphragm until the entire field of view is just clear.

Matching the condenser to the objective

5 Remove the eyepiece and look down the barrel from at least 10 in away. Open or close the iris diaphragm until it cuts off about the outer tenth of the back lens of the objective [FIGURE 4.4]. Replace the eyepiece.

FIGURE 4.2
Substage condenser assembly and radiant-field diaphragm.

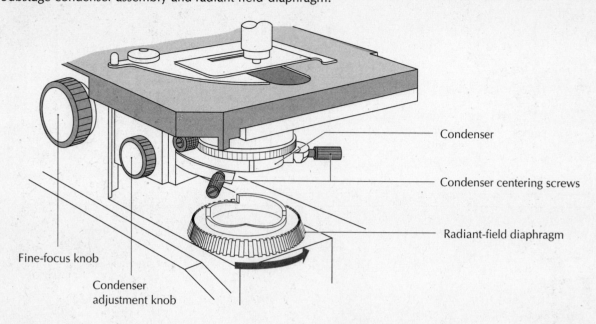

FIGURE 4.3
Focusing and centering a radiant-field diaphragm.

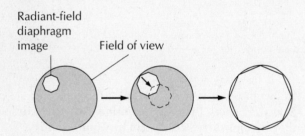

Refocusing the substage condenser

6 When another slide is put in place, swing back to the 10× objective and refocus the substage condenser to produce a sharp image of the radiant-field diaphragm. This is necessary because slides vary in thickness.

NOTE: Unfortunately, most student microscopes are of such simple construction that the adjustments just described cannot be carried out. However, the student should know that such alignments are necessary to use the microscope to its maximum potential.

Procedure for Using the Oil-Immersion Objective

The oil-immersion objective is used most often by microbiologists. Procedure for its use is as follows:

1 Place a drop of immersion oil on the top lens of the condenser.
2 Put the microscope slide on the stage carefully.
3 Rack the condenser up until the oil touches the slide.
4 Find and focus a suitable field on the slide with a low-power objective.
5 Focus the condenser as described previously.
6 Place a drop of oil on the cover slip or smear (specimen).
7 Rotate the oil-immersion lens into position.
8 Use the fine-adjustment knob to get the specimen into sharp focus after increasing the light intensity.
9 Match the condenser to the objective as described above.

STEREOMICROSCOPY

Sometimes the stereomicroscope is used in microbiology, such as in the examination of small bacterial colonies and mold structures. It gives a three-dimensional picture rather than a two-dimensional image. However, magnification is usually in the range of 5 to 100× only.

By definition, the stereomicroscope is binocular (hence some workers call it a "binocular microscope"), because it is really two microscopes in one; that is, each eyepiece has its own objective and eyepiece lenses. They are separated and inclined at an angle of 15° to each other so that the images seen by

FIGURE 4.4
Relationship between objective lenses and openings of the iris diaphragm.

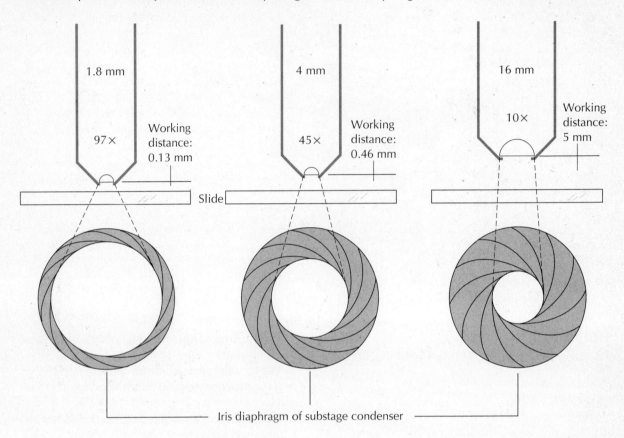

each eye differ as they do in normal vision. Thus, an object seen has breadth, length, and depth.

The stereomicroscope is used in the following manner:

1. Select either incident or transmitted lighting. Adjust for even illumination.
2. Adjust the eyepieces to the width of your eyes.
3. Place the specimen, such as a Petri dish culture, on the stage.
4. Focus with the focus-control knob, using the left eye only.
5. Using the right eye only, use the diopter ring adjustment located on the eyepiece to focus, if necessary.
6. Examine your specimen.

PRECAUTIONS TO BE OBSERVED IN USING THE MICROSCOPE

1. Do *not* touch the lenses. If they become dusty, blow them with an air bulb or with a stream of gas from a pressure can (e.g., DUST-OFF, Falcon Safety Products, Inc., Mountainside, NJ 07092). If they become dirty, clean them by wiping gently with lens paper.
2. With lens paper, wipe the oil off the oil-immersion objective after use to prevent the oil from drying or dust from collecting. Do not allow any oil to touch the other objectives.
3. Do not tilt the microscope when working with the oil-immersion objective. Otherwise the oil may flow.
4. Keep the instrument free of dust by keeping it covered when not in use.
5. Carry the instrument with one hand under the base and the other hand on the arm of the microscope.
6. Preliminary focusing to bring the object into view should always be done by raising the objective. Never allow the objective lens to touch the cover slip or the slide. All adjustable knobs should turn freely and easily; never force any movable part of the instrument.

7 Keep both eyes open when looking through the eyepiece of the monocular microscope. Closing one eye for long periods of time is tiring. Work with relaxed eyes. This can be done if you imagine that you are viewing the image as if it were at infinity. That is, look "through the microscope" but not "into the microscope." After some practice this should come easily.

REFERENCES
MICROBIOLOGY, Chap. 3, "Characterization of Microorganisms."
MM, Chap. 1, "Light Microscopy."
Smith, Robert F., *Microscopy and Photomicrography: A Working Manual*, CRC Press, Boca Raton, FL 33431, 1990.

MATERIALS
Microscope with radiant-field diaphragm
1 prepared slide with stained yeast cells

PROCEDURE
NOTE: Consult the preceding procedures for setting up a bright-field microscope.

1 Identify all parts of the microscope.
2 If necessary, align the illumination of the instrument.
3 Place the slide on the stage, specimen side up, and center the section to be examined as closely as possible in the center of the hole in the stage.
4 Rotate the low-power (10×) objective into position over the specimen and bring the cells into sharp focus. (The light intensity should be at a comfortable brightness.) This is done by looking through the eyepieces and then slowly raising the coarse-adjustment knob until the specimen comes into approximate focus. Remember to focus upward, never downward, with the focusing knobs. Bring the specimen into sharp focus with the fine-adjustment knob.
5 Raise the substage condenser all the way up.
6 Close the radiant-field diaphragm almost completely.
7 Focus the small opening of this diaphragm by slowly lowering the substage condenser. Both the yeast cells and the leaves of the radiant-field diaphragm should be in focus.
8 Center the radiant-field diaphragm opening by adjusting the centering screws on the condenser mount.
9 Open the radiant-field diaphragm until the leaves touch the periphery of the field and then a little beyond. Congratulations! You have adjusted the microscope properly with Koehler illumination.
10 Move the low-power objective to one side after examining your specimen. Place a drop of immersion oil on the specimen or cover slip mounted over the specimen in the prepared slide. Swing the oil-immersion objective into place over the specimen. Since the objectives are parfocal, only a slight adjustment of the fine-adjustment knob is necessary to bring the cells into sharp focus.
11 Increase the light intensity to see the cells clearly.
12 Match the condenser to the objective by removing the eyepiece, look down the barrel from at least 10 in away, and open the iris diaphragm until it cuts off about the outer tenth of the back lens of the objective. Use the iris diaphragm lever to do this. (Remember that as you go to objectives with higher numerical aperture, the diaphragm is opened more.) Replace the eyepiece.
13 Critically focus your specimen with the fine-adjustment knob.
14 Observe and examine your specimen. Record your results in the results section.
15 After use, clean the oil from the oil-immersion objective lens with lens paper. The oil on the slide should also be gently wiped off with lens paper.

NAME _____

Principles and Use of the Bright-Field Microscope

RESULTS Sketch a few typical cells giving an indication of relative size and morphology.

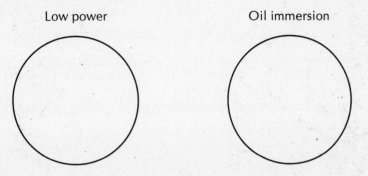

QUESTIONS

1. Explain why a *compound microscope* is so called.

2. How do you obtain the *final*, or *total*, magnification of a microscope?

3. Can you use the iris diaphragm to control light intensity? If not, what do you use it for?

4. Explain why magnification of a microscope is not as important as its resolution.

5. Describe in a sentence what is meant by the *numerical aperture* of a lens system.

6. Briefly explain what is meant by *Koehler illumination* by describing a few critical steps.

7. Why did you use immersion oil for your highest-power objective?

EXERCISE 5

Microscopic Examination of Microorganisms in Wet Mounts

OBJECTIVE
To learn how to make wet mounts and to examine several kinds of microorganisms under the brightfield microscope.

OVERVIEW
Now that you have become thoroughly acquainted with the proper use of the microscope (after having studied the material in Part II and having done Exercise 4), you will use it in this exercise to examine several kinds of microorganisms.

REFERENCES
MICROBIOLOGY, Chap. 3, "Characterization of Microorganisms."
MM, Chap. 1, "Light Microscopy."

MATERIALS
Microscope
Microscope slides and cover slips
Loop and thick straight inoculating (transfer) needles
Immersion oil
Cultures of a bacterium (*Bacillus cereus*), a yeast (*Saccharomyces cerevisiae*), a mold (*Aspergillus niger*), an alga (*Chlorella* sp.), and a protozoan (*Paramecium* sp.)

PROCEDURE
1 Prepare a wet mount of one of the pure cultures provided by placing a small drop of water—it may be from the tap—on a clean slide.

2 Transfer some of the culture by means of a loop transfer needle, or a thick straight transfer needle for the mold culture, and mix it in the drop of water to make a suspension (refer to Exercise 3).

3 Place a cover slip over the suspension. Complete examination of each culture before preparing the next wet mount to avoid drying of the specimen.

4 Place the preparation on the stage of the microscope, and set the instrument up for microbiological examination (refer to Exercise 4).

5 First, examine each wet mount under low-power (10×) magnification, observing the size and shape of the cells and any characteristics of motion.

6 Then swing the high-power objective into position over the slide. With parfocal objectives, only slight adjustment of the fine-adjustment knob is necessary to bring the cells into sharp focus. Again observe the size, shape, and motion, if any, of the cells.

7 Rotate the high-power objective partially out of the way, and place a drop of immersion oil on the center of the cover slip.

8 Bring the oil-immersion objective into position above the slide. Now look through the eyepieces and adjust the fine-adjustment knob until the cells are sharply focused.

9 Examine and observe the size and shape of the cells under the objective.

10 Examine the other microorganisms in the same manner.

11 Pay particular attention to comparative sizes of microorganisms, as well as to their shapes, structures, and movements.

NAME _____

Microscopic Examination of Microorganisms in Wet Mounts

RESULTS Sketch a few typical cells of each pure culture as seen under high-power magnification. Make sketches so that they reflect comparative sizes.

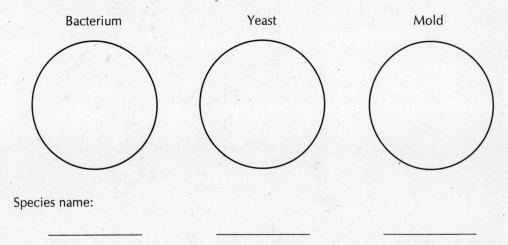

Species name:

_____ _____ _____

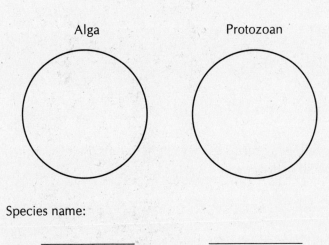

Species name:

_____ _____

QUESTIONS 1 What magnification was obtained in each of the examinations that you performed?

2 Why should preliminary focusing always be done upward?

3 What is the function of oil when used with the oil-immersion objective?

4 What is the limiting factor in obtaining useful magnifications through optical microscopes?

5 Identify the following terms as they apply to microscopy: *resolution*, *numerical aperture*, and *working distance*.

6 Did any of the microorganisms observed exhibit true motility?

EXERCISE 6

Microscopic Measurement of Microorganisms

OBJECTIVE
To learn how to calibrate an ocular micrometer and to measure the dimensions of several kinds of microorganisms.

OVERVIEW
The dimensions of microorganisms are usually expressed in micrometers (μm), a unit of measurement which is 1/1000 of a millimeter, or 1/25,400 of an inch. Microorganisms of various types range in size from a fraction of a micrometer to many micrometers. Their size can be measured using a microscope equipped with an *ocular micrometer* calibrated against a stage *micrometer*.

An ocular micrometer is a disk of glass with equal graduations etched on its surface [FIGURE 6.1]. When the disk is placed in the eyepiece, these graduations are superimposed on the microscopic field. The scale of these graduations is calibrated by superimposing them on the ruled lines of a stage micrometer in the microscopic field. The stage micrometer is placed on the microscope stage, and its ruled lines are focused on the focal plane of the microscopic field. The ruled lines are exactly 0.01 mm (10 μm) apart. You will have to determine how many units, or equal graduations, of the ocular micrometer superimpose a known length, determined by the number of ruled lines, on the stage micrometer. Once calibrated for each objective, the stage micrometer is removed, and the ocular micrometer can be used to determine the dimensions of various microorganisms on microscope slides.

REFERENCES
MICROBIOLOGY, Chap. 3, "Characterization of Microorganisms."
MM, Chap. 1, "Light Microscopy."

MATERIALS
Ocular micrometer
Stage micrometer
Prepared slides of a bacterium, yeast, mold, alga, and protozoan

PROCEDURE
Calibration of Ocular Micrometer
1 Center the stage micrometer on the stage of the microscope beneath the low-power objective and over the illumination source.
2 Remove the eyepiece from your microscope, and place the ocular micrometer on the metal shelf within it as shown in FIGURE 6.1.
3 Replace the eyepiece. Focus the eye lens, if present, to sharpen the graduations of the ocular micrometer by turning it, and rotate the ocular micrometer so as to superimpose its graduations upon the ruled lines of the stage micrometer when they are clearly focused. The appearance of the divisions of both the ocular and stage micrometers are shown in FIGURE 6.2.

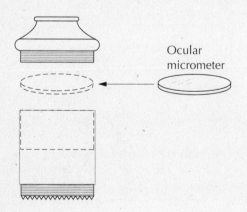

FIGURE 6.1
Insertion of ocular micrometer in eyepiece of microscope.

4 Focus on the ruled lines of the stage micrometer. The graduations of the ocular micrometer superimpose and are parallel with those of the stage micrometer.

5 By moving the mechanical stage, align a line of the stage micrometer with a line of the ocular micrometer at one end. Find another line on the stage micrometer that coincides with a line on the ocular micrometer at the other end. Count the number of divisions on the ocular micrometer within those coinciding lines to determine the distance of the stage micrometer divisions within those coinciding lines at both ends.

You can now calculate the distance between the ocular micrometer divisions. Because each division of the stage micrometer is $10\,\mu$m, you can determine how many ocular divisions are equivalent to each stage micrometer division to determine the number of micrometers in each division of the ocular scale. For example, 45 ocular micrometer divisions are equal to 10 stage micrometer divisions of $10\,\mu$m each. The distance then between the lines of the ocular micrometer is equal to:

$$\frac{10 \text{ stage micrometer divisions} \times 10\,\mu\text{m}}{45 \text{ ocular micrometer divisions}} = 2.2\,\mu\text{m}$$

6 In a similar manner, repeat the procedure with the high-power and oil-immersion objectives.

Measurement of Microorganisms

Replace the stage micrometer with one of the prepared slides of microorganisms and determine the dimensions of several cells. Repeat this with each of the different prepared specimens. The entire procedure for measuring microorganisms is illustrated in FIGURE 6.3.

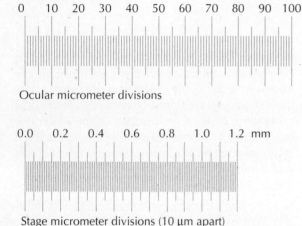

FIGURE 6.2
Divisions of the ocular micrometer and stage micrometer.

FIGURE 6.3
Measuring bacterial cells.

Stage micrometer placed on microscope stage just as any other slide is

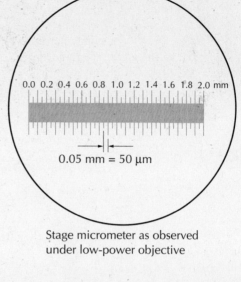

Stage micrometer as observed under low-power objective

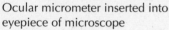

Ocular micrometer inserted into eyepiece of microscope

Stage micrometer on microscope stage

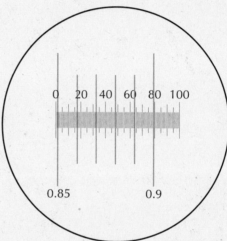

Ocular micrometer superimposed over stage micrometer as viewed under the high-power oil-immersion objective. In this case, 78 ocular micrometer divisions equal 50 μm. Therefore each of the smallest ocular micrometer divisions equals 0.64 μm. (Illustration slightly enlarged for clarification.)

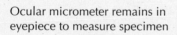

Ocular micrometer remains in eyepiece to measure specimen

Slide containing specimen replaces stage micrometer on microscope stage

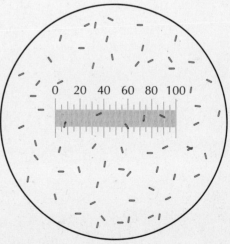

Cells of *Escherichia coli* observed under the oil-immersion objective with the ocular micrometer in place. The cells of *E. coli* ordinarily measure 0.6 by 2.0 to 3.0 μm.

NAME _____

6

Microscopic Measurement of Microorganisms

RESULTS

1 From your observations, complete the following:

Low-Power Objective:

_____ ocular divisions = _____ stage divisions

_____ ocular divisions = 1 stage division (10 μm)

1 ocular division = _____ μm

High-Power Objective:

_____ ocular divisions = _____ stage divisions

_____ ocular divisions = 1 stage division (10 μm)

1 ocular division = _____ μm

Oil-Immersion Objective:

_____ ocular divisions = _____ stage divisions

_____ ocular divisions = 1 stage division (10 μm)

1 ocular division = _____ μm

2 Record the measurements made of microorganisms in the chart provided.

MEASUREMENTS OF MICROORGANISMS				
Species (give specific name)	Length of several individual cells, μm	Average length, μm	Width of several individual cells, μm	Average width, μm
Bacterium:				
Yeast:				
Mold:				
Alga:				
Protozoan:				

45

QUESTIONS 1 How many micrometers are there in 1 in? 1 cm? 1 mm?

2 What range of variation did you observe between cells of the same species? (Give specific measurements.)

3 Would you expect unstained, living cells to be different in size from the same cells examined in a stained preparation? Explain.

EXERCISE 7

Use of the Dark-Field Microscope

OBJECTIVE

To understand how the dark-field microscope works and to use it for examining microorganisms.

OVERVIEW

The dark-field microscope increases contrast of unstained transparent specimens, without sacrificing resolving power, by using the light which the specimens scatter. The basic principle by which this is achieved is by means of a special condenser which prevents light rays from passing directly through the specimen as illustrated in FIGURE 7.1.

By use of the special condenser, the object is lit by a wide-angled hollow cone of light, obtained by total internal reflection. Because light rays are entering at wide angles, they will necessarily miss the objective unless the object bends, or diffracts, the light into the objective. The result of this is that the object then stands out brilliantly lighted against a black background, much as a bright star stands out against the dark sky, because only those rays passing through the object are seen while the undiffracted rays are not. Since bright objects against a dark background are seen easily by the eye, dark-field microscopy is useful for examining unstained living microorganisms suspended in fluid preparations. Microorganisms that are difficult to stain and that are very thin lend themselves very well to examination by dark-field microscopy. One such group of microbes is the spirochetes.

Technically, in order that the transmitted rays escape collection by the objective, the numerical aperture (NA) of the condenser must exceed that of the objective lens. This situation is easily obtained when one is using low-power or high-power dry objectives, since the NA of such objectives ranges from about 0.25 to 0.85 whereas that of most dark-field condensers is about 1.20. With oil-immersion objectives, however, the NA may equal or exceed that of the condenser, allowing the transmitted rays to be collected by the objective lens. In such a case, the NA of the objective must be decreased. The most convenient means of doing this is to use an oil-immersion objective with variable NA; by adjusting an iris diaphragm collar on the objective, the NA of the objective lens can be decreased below that of the dark-field condenser.

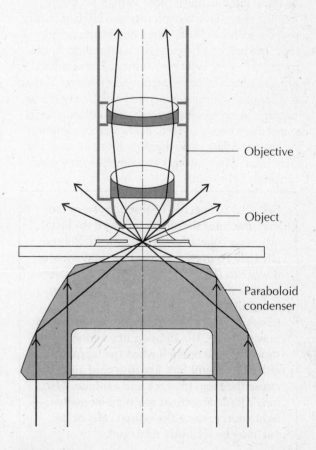

FIGURE 7.1

Path of light rays in the dark-field microscope.

REFERENCE

MICROBIOLOGY, Chap. 3, "Characterization of Microorganisms."

MATERIALS

Microscope equipped for dark-field illumination
Prepared slide of diatoms
Culture of spirochete such as *Treponema denticola*

PROCEDURE

1 Determine the *NA* of the dark-field condenser and the oil-immersion objective. The values are engraved on these pieces. If the objective has an *NA* equal to or higher than that of the condenser, this *NA* must be decreased below that of the condenser by turning the iris diaphragm collar on the objective. Even if the *NA* of the objective is decreased by only 0.1 *NA* below that of the condenser, this is sufficient. Rotate the collar to a value of 0.8.

2 Turn on the brightest light source available. The light source used for dark-field microscopy must be quite intense, much more so than for bright-field or phase-contrast microscopy. High-intensity sources such as a xenon lamp or a quartz-iodide lamp are preferred, although an incandescent filament (tungsten) lamp may suffice. The disadvantage in using any high-intensity light source is that a great deal of heat is generated which may kill or damage living specimens, so heat-absorbing filters may have to be employed. A 3% aqueous solution of copper sulfate in the light path can be used as a heat filter. Most workers are content with a good tungsten light source.

> **CAUTION:** Specific instructions from the manufacturer should be followed for installation and operation of artificial high-intensity light sources. *Improper use may involve danger to the operator.* For example, safety goggles must be worn when a xenon lamp is installed in its housing, because such lamps have enormous internal pressure and may explode. The housing of a xenon lamp should not be opened when the lamp is either on or still hot, because cold air may cause the lamp to crack and explode. Also, never look *directly* at any high-intensity light source, since the retinal cells of the eye may be seriously damaged.

3 Place the diatom test slide on the microscope stage. Place a drop of oil on the face of the condenser, and rack up the condenser until oil contact is made with the slide.

4 With a low-power objective in position, focus on the diatoms. Carefully raise or lower the condenser so that the smallest possible spot of light is produced [FIGURE 7.2].

If the spot of light is not in the exact center of the field, adjust using the centering screws on the condenser.

5 Swing aside the low-power objective, place a drop of oil on the slide, rotate the oil-immersion objective into position, and refocus sharply on the diatoms. If necessary, readjust the height of the condenser as before to obtain the brightest possible specimen image. Use the condenser centering screws to center the spot of light in the field.

How do the results compare with bright-field microscopy?

6 Replace the diatom test slide with a wet mount of *Treponema denticola*. This is done by putting a drop of the culture with a loop transfer needle on a microscope slide. Practice aseptic technique so that you do not contaminate your culture. Place a cover slip gently over the drop. Record your observations in the results section.

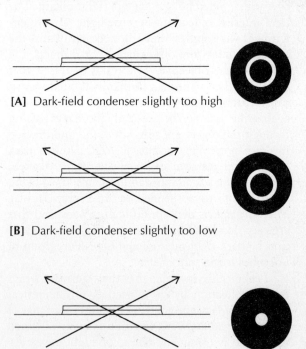

[A] Dark-field condenser slightly too high

[B] Dark-field condenser slightly too low

[C] Correct height of condenser

FIGURE 7.2
Correct positioning of the dark-field condenser.

NAME _____

Use of the Dark-Field Microscope

RESULTS

1 Describe the appearance of the diatoms under dark-field and under bright-field microscopy.

2 Sketch the appearance of spirochete cells seen under dark-field microscopy.

Magnification: _____ ×

49

QUESTIONS 1 What kind of microscopic image is obtained if the *NA* of the oil-immersion objective is not decreased below that of the dark-field condenser *NA?*

2 List *three* advantages of examining microorganisms using dark-field microscopy rather than bright-field microscopy.

EXERCISE 8

Use of the Phase-Contrast Microscope

OBJECTIVE
To understand how the phase-contrast microscope works and to use it for examining microorganisms.

OVERVIEW
Colorless, transparent, living microorganisms are difficult to see by bright-field microscopy because there is little contrast between them and their fluid suspending environment. There are specialized light microscopes that can increase this contrast and preserve the viability of the cells at the same time. Examples of such microscopes are the dark-field microscope (Exercise 7) and the phase-contrast microscope, which you will learn to use in this exercise.

Many transparent objects, such as bacterial cells, do not sufficiently alter the intensity, or brightness, of light passing through them by diffraction, refraction, absorption, or reflection. That is why they show little contrast with their surrounding medium under bright-field illumination. They do, however, change the *phase* of the light. Such change in phase is due to small differences in the thickness and chemical nature (refractive index) of the cells, especially their intracellular entities (e.g., nucleoids and inclusions). These differences retard part of the light as it passes through the specimen, that is, the light is shifted or changed in phase. Our eyes cannot detect phase shifts of light, but the phase-contrast microscope contains optical components that change phase differences into differences of light intensity which are visible to us. The specimen then becomes discernible. This is the basis of the phase-contrast microscope, first discovered in the 1930s.

Let us now examine what is meant by *change in the phase of light*. Light rays possess wavelike properties which form the basis for phase-contrast microscopy. In FIGURE 8.1A, two light rays are pictured as if they are waves; these two waves are said to be in the same phase because the positions of the wave crests coincide. This is in contrast to the two waves pictured in FIGURE 8.1B, where there is one-half wavelength difference in the position of the wave crests; these waves are said to be out of phase. If the two waves in FIGURE 8.1A could be superimposed, a phenomenon called *reinforcement* will occur as shown in FIGURE 8.1C. Here, the result of the superimposition can be considered as a single wave which has twice the wave amplitude, and hence twice the brightness or twice the intensity of either of the nonsuperimposed waves. The color of the light, however, would not be affected because the wavelength is not changed.

If the two waves in FIGURE 8.1B could be superimposed, a phenomenon called *cancellation* would occur as shown in FIGURE 8.1D. The result of this superimposition can be considered as a single wave which has an amplitude of zero, and hence no brightness at all, or "darkness."

Thus, if two waves were only one-quarter, or indeed any fraction, wavelength out of phase, the result of their superimposition would be partial reinforcement; the amplitude, and hence intensity or brightness, of the resulting wave has increased, but not doubled. In this way, increased contrast is obtained from phase-contrast microscopy.

The implementation of the principle of phase-contrast microscopy just outlined may be summarized as follows. Light passing directly through an object is separated from the light which is diffracted by the object using a special diaphragm in the condenser. A matching reversed diaphragm is used at the back of the objective. These diaphragms must be accurately aligned during use. The contrast in the image is much increased, although each small detail has a "halo" around it when viewed. These halos actually emphasize the appearance of the details of the object but make it difficult to measure them.

FIGURE 8.1

Light waves in phase-contrast microscopy. λ, wavelength; a, amplitude.

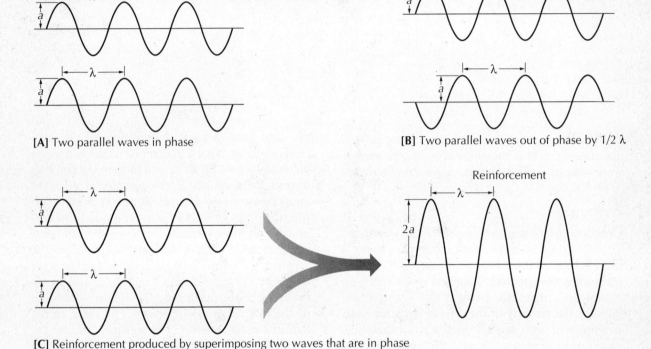

[A] Two parallel waves in phase

[B] Two parallel waves out of phase by 1/2 λ

[C] Reinforcement produced by superimposing two waves that are in phase

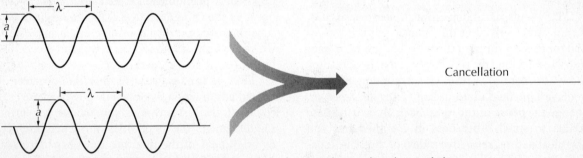

[D] Cancellation produced by superimposing two waves that are 1/2 wavelength out of phase

Now for the "nuts and bolts" of phase-contrast microscopy. For phase-contrast microscopy, two optical parts of the system exist which are not found in a bright-field microscope:

1. An annular diaphragm in the substage condenser
2. A phase-shifter disk located in the objective

These components are shown in FIGURE 8.2.

The annular diaphragm in the substage condenser blocks out all light rays except for a narrow ring; this restricts the condenser to transmission of only a hollow cone of rays to the object.

The phase-shifter disk in the objective is a disk of glass in which is etched a narrow ring. As FIGURE 8.2 shows, the construction of the optical system is such that *all of the transmitted rays* (heavy solid lines) that are delivered from the condenser in a hollow cone of light *pass through this etched ring* where the glass is thinner than the rest of the disk. On the other hand, *most of the rays diffracted* from the object (dashed lines) *pass through the thicker portion of the disk*, because the etched ring of thinner glass occupies only a small proportion of the area of the glass disk.

8 USE OF THE PHASE-CONTRAST MICROSCOPE

FIGURE 8.2

The annular diaphragm and phase-shifter disk in a phase-contrast system.

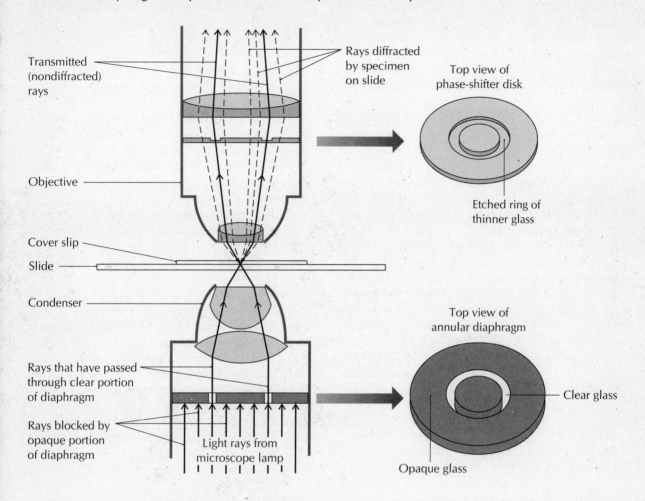

The effect of this system is to alter the phase of the transmitted rays as compared to the diffracted rays. As light rays pass through glass of varying thickness, they are retarded to varying degrees with respect to their phase. The thicker the glass, the greater the retardation. An example is given in FIGURE 8.3, where a ray passing through a certain thickness of glass is retarded by ¼ λ [FIGURE 8.3A] while a ray passing through a thicker portion of the glass is retarded by ½ λ [FIGURE 8.3B]. Let us assume that the ray pictured in FIGURE 8.3A is a transmitted ray that passes through the etched ring of the phase-shifter disk, while the ray pictured in FIGURE 8.3B is a diffracted ray that passes through the thicker regions of the phase-shifter disk. If the transmitted ray and the diffracted ray arose from the same location in the microscopic object, the objective lens would bring them to a common focus in the real image formed by the objective. Thus,

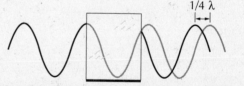

[A] Retardation of light ray by 1/4 wavelength (λ)

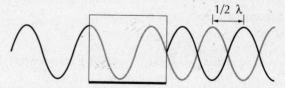

[B] Retardation of light ray by 1/2 wavelength (λ)

FIGURE 8.3

Retardation of the phase of light waves passing through glass of varying thickness.

the two rays would be superimposed in the real image; however, they would be out of phase by a difference of ¼ λ, and the resulting image would show a difference in intensity or brightness.

In the preceding example, the assumption was that both the transmitted and diffracted rays were in the same phase prior to reaching the phase-shifter disk. This is not accurate, however, if the specimen differs in refractive index from the medium. If the specimen has a slightly higher refractive index than the surrounding medium, the phase of the diffracted rays from the object will be retarded somewhat in comparison to the transmitted rays which bypass the specimen. Therefore, even before reaching the phase-shifter disk, the diffracted rays may have their phase retarded relative to the transmitted rays.

Because there are so many transmitted rays in comparison to diffracted rays, the darkness or brightness of the specimen still cannot be easily distinguished from the background. For this reason, the etched ring in the phase-shifter disk is coated with a very thin layer of evaporated metal to absorb many of the transmitted (background) rays. This coating acts in the same manner as a neutral-density filter. As a result, the specimen will appear distinctly darker or lighter than the background.

REFERENCE

MICROBIOLOGY, Chap. 3, "Characterization of Microorganisms."

MATERIALS

Microscope equipped for phase-contrast illumination
Prepared diatom test slide
Flat toothpicks (sterile)

PROCEDURE

NOTE: Proper setup for bright-field adjustment must precede phase-contrast adjustment.

1 In the usual type of phase-contrast condenser, there is a rotating disk that provides a series of separate annular diaphragms, each particular diaphragm being designed for a matching phase objective. If one changes objectives, one must also change the annular diaphragm by rotating the proper-size diaphragm into position. One of the positions on the rotating disk is usually left blank, without an annular diaphragm, for ordinary bright-field work.

Rotate the blank into position. Set up the diatom test slide for bright-field microscopy, using the oil-immersion phase objective (phase objectives can be used satisfactorily for bright-field work). Align the microscope properly (e.g., the condenser must be properly focused and centered—refer to Exercise 4, if necessary) *before* beginning the next step.
2 Open the iris diaphragm fully.
3 Rotate the appropriate annular diaphragm of the condenser for the oil-immersion objective into position.
4 Remove the eyepiece from the microscope, and replace it with the auxiliary focusing telescope provided by the manufacturer. (In some models, an aperture viewing prism is provided; put this into position instead of the telescope.)
5 Focus the telescope on the back of the objective lens. If the annular diaphragm is not properly centered, you will see an image similar to that illustrated in FIGURE 8.4A.
6 To properly center the annular diaphragm, turn the phase ring centering screws (if these are not present, use adjustment wrenches provided) until the hollow cone of transmitted light is exactly superimposed over the phase-shifter ring (gray rings seen in the objective). The final result should be as shown in FIGURE 8.4B.

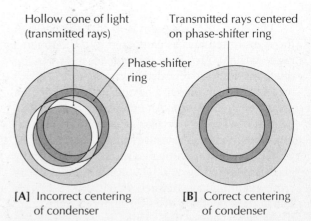

[A] Incorrect centering of condenser

[B] Correct centering of condenser

FIGURE 8.4

Appearance of back of objective lens as seen through auxiliary focusing telescope.

7 Replace the telescope with the eyepiece and examine the diatom. The intensity of the light source may be increased. Describe its appearance as seen under bright-field microscopy as compared to phase-contrast illumination.

NOTE: The above steps are outlined for the usual type of phase-contrast microscope. For other designs, consult the manufacturer's instruction manual.

8 Put a drop of saliva on a clean microscope slide. Use a toothpick and obtain a sample of material along the base of a tooth. Mix this plaque material with the drop of saliva. Put a cover slip over the drop. Examine the wet mount first by bright-field microscopy and then by phase-contrast illumination. Record your observations in the results section by sketching several of the microorganisms that you have seen.

NAME _____

8

Use of the Phase-Contrast Microscope

RESULTS

1 Describe the appearance of the diatoms under phase-contrast and under bright-field illumination. How do they compare?

2 Sketch several microorganisms seen in saliva under phase-contrast optics.

Magnification: _____ ×

57

QUESTIONS 1 What happens to the contrast when an incorrect annular diaphragm is rotated into position? (Answer this question only after testing with the microscope.)

2 Give one advantage of using phase-contrast microscopy rather than dark-field microscopy in the examination of microorganisms.

BACTERIAL MORPHOLOGY AND STAINING METHODS

One of the major characteristics of bacteria is **morphology**—size, shape, arrangement, and structure. These characteristics can be determined by examination of appropriately prepared specimens of microorganisms. When bright-field microscopy is used for the examination, it is desirable that the cells be stained to make them more readily visible by increasing contrast. Unstained microbial cells are practically transparent and are best observed by techniques that permit special and critical control of illumination, as in phase-contrast and dark-field illumination (Exercises 7 and 8).

However, dyes used in staining microorganisms do not readily penetrate living cells. Thus, dyes are applied on microorganisms only after they are killed. Killed cells soak up dyes and retain them in much the same way as a sponge soaks up water.

Staining of a bacterial film, called a smear, may be performed simply to reveal the shape, size, and arrangement of the cells. The cells are stained by the application of a single staining solution; the process is called a **simple stain.** However, it is possible to acquire additional information about the morphology and chemical composition of bacteria through the use of **differential stains.** Differential staining procedures usually involve treatment of the smear with a series of reagents. The appearance of the cells following this treatment may permit one to distinguish between two different bacterial types on the basis of the color they retain. For example, one type might appear blue while another may appear red after identical treatment. One may also distinguish between structural entities intracellularly or exterior to the cell wall.

Careful examination of appropriately stained bacteria provides invaluable information for the morphological characterization and identification of the specimen. Apart from differentiating and rendering visible the cell in whole or in part, staining also aids in the identification of microorganisms. For example, the reaction of positiveness or negativeness in the Gram stain is useful as a means of primary grouping.

Stains are made from dyes, which may be natural or synthetic. Natural dyes are mainly for histological purposes. Most bacterial stains are made from synthetic dyes, which are aniline or, more correctly, coal-tar dyes. They are all derivatives of benzene, a cyclic compound consisting of six carbon atoms.

There are three groups of dyes: **acidic, basic,** or **neutral.** The most commonly used dyes are salts. A salt is composed of a positively charged ion (cation) and a negatively charged ion (anion). Dyes are salts containing an organic ion and an inorganic ion. For example, methylene blue is actually the salt methylene blue chloride:

methylene blue$^+$ + chloride$^-$

Structurally, it may be represented as:

$$\left[(CH_3)_2N \text{—} \bigcirc \text{—} N \text{—} \bigcirc \text{—} S^+ \text{—} N(CH_3)_2 \right] Cl^-$$

Methylene blue chloride, like other salts, dissociates when dissolved in water into dissociation products: the positively charged organic ion and the negatively charged inorganic ion. It is the positively charged organic cation that gives this dye its blue color; this chemical grouping which exhibits color is called the *chromophore*. If the chromophore is positively charged, the dye is called a *basic dye* (like methylene blue). A basic dye has a chromophore that behaves like a base: it combines with an acid like HCl to form a dye salt, for example, methylene blue chloride.

Acidic stains have a reverse charge; the chromophore is in the negative ion. An example is eosin, which is really sodium⁺eosinate⁻. An acidic dye is so called because, like an acid, the chromophore combines with a base (like NaOH) to form the dye salt shown structurally as:

$$\left[\begin{array}{c} \text{structure of eosinate with NaO, Br substituents on xanthene ring, O bridge, carbonyl, phenyl-COO}^- \end{array} \right] \text{Na}^+$$

A summary of acidic and basic dyes may be viewed as follows:

Basic dye: Chloride or sulfate salts of colored bases to give a positively charged chromogen

Acidic dye: Sodium, potassium, calcium, or ammonium salts of colored acids to give a negatively charged chromogen

Neutral dyes are made by mixing together aqueous solutions of certain acid and basic dyes. The coloring matter (chromophores) in neutral stains is contained in both negatively and positively charged components.

All bacterial cells have a negatively charged cell surface when the pH of their environment is either near neutral or alkaline. Since the chromophores of basic dyes are cations, basic dyes stain microorganisms better under neutral or alkaline conditions; the negative charges on the cell surface attract the positively charged chromophores. To ensure the maximal surface charges on the cell, basic dyes are often made up as alkaline solutions. For instance, the base potassium hydroxide is incorporated to make the stain called *Loeffler's methylene blue.*

If the chromophore has a negative charge, it is repelled by the cell's negative surface charge. Thus, negatively charged chromophores do not stain bacterial cells. Instead, they surround the cells to give a colored background. Acidic dyes used in this way are called *negative stains.*

When the cells are dead, as in a smear, a basic dye, in addition to staining the surface of the cells, also penetrates into the cells and binds with negatively charged interior parts. Bacterial cells are rich in nucleic acids, which have negative charges. Thus, the cells become *entirely* stained as seen under the bright-field microscope.

EXERCISE 9.A

Staining the Whole Cell
9.A Simple Staining with Basic Dyes

OBJECTIVE
To perform simple staining on several bacterial species and to compare their morphological characteristics from such stained smears.

OVERVIEW
The use of a single stain to increase contrast between the bacteria and the background is referred to as *simple staining*. It is so called because stained preparations are made by the simple process of spreading a drop of a fluid suspension of cells on a glass slide, allowing it to dry, and applying gentle heat to fix, or stick, the cells to the slide; this is followed by the application of the staining solution. Simple staining is often employed when information about cell shape, size, and arrangement is required.

REFERENCE
MICROBIOLOGY, Chap. 3, "Characterization of Microorganisms."

MATERIALS
Pure cultures of *Bacillus subtilis*, *Staphylococcus aureus*, and *Aquaspirillum itersonii* in nutrient broth

Pure cultures of the preceding bacteria in isolated colonies on nutrient-agar medium in Petri plates

Staining solutions: Loeffler's methylene blue, Ziehl's carbol fuchsin, and modified Hucker's crystal violet

PROCEDURE
1 Wash and dry several microscope glass slides. Identify each slide with a marking pen with the name of the bacterium before use.
   NOTE: Clean microscope glass slides are essential for preparation of bacterial smears. Grease or oil must be removed by washing the slides with detergent, followed by generous water rinsing, and immersion in 95 percent alcohol. The slides are then removed, dried in air, and wiped with lintless paper before use. Even commercially "precleaned" slides should be prepared in this manner before use.

2 Refer to FIGURE 9.1 as you proceed through steps 2 through 6 to prepare your bacterial smear. Flame the transfer needle with an inoculating loop, and place one or two loopfuls of *cell culture in broth (liquid medium)* on a glass slide and spread evenly over an area about the size of a dime. Flame the transfer needle again as soon as the smear is made. Remember to practice aseptic technique (Exercise 3).

3 For making a smear from *bacterial growth on a solid medium*, place a small, clean drop of water (distilled or tap) on the slide by using the transfer loop.

4 Flame the transfer loop and aseptically remove a tiny bit of growth on the tip of the loop (some microbiologists prefer to use a transfer needle with a straight wire to do this) by raising the top of the Petri dish as little as possible on one side.

5 With a circular motion, emulsify this growth in the drop of water on the slide. Spread the drop out to approximately the size of a dime to obtain a thin smear. Flame the transfer needle again as soon as the smear is made.
   NOTE: Avoidance of thick, dense smears is necessary in order to see a monolayer of cells, not clumps or thick aggregates. A good smear is a thin whitish film when dried, not thick and turbid.

6 Allow the smear to air-dry; then *fix* the smear by passing the slide through the flame (right side up) two or three times. The slide should feel warm, but not hot, when placed against the back of your hand.
   NOTE: Unless fixed on the glass slide, the bacterial smear will wash away during the staining

FIGURE 9.1
Bacterial smear preparation.

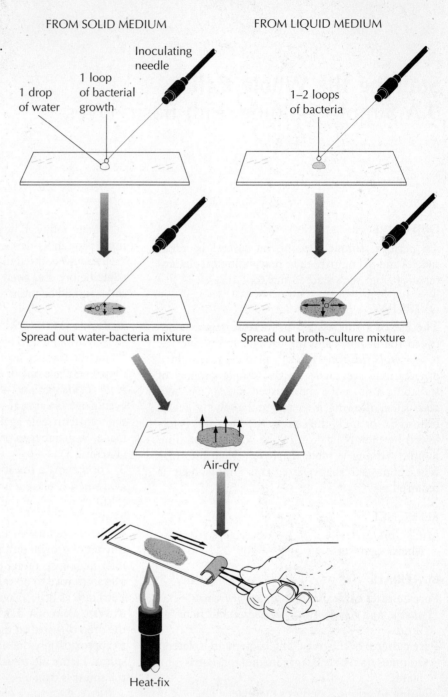

procedure. During heat fixation, the bacterial cells are killed, and the bacterial proteins are coagulated and "stick" to the glass surface, much like cooking an egg on a greaseless sticky pan. Heat fixation is the simplest way to fix cells to glass slides. If heat fixation is to be avoided, there are alternative chemical means of fixing cells to glass slides, for example, by the use of methanol.

7 Select one of the following stains, and cover the smear for the designated time: crystal violet, 2 to 6 s; carbol fuchsin, 15 to 30 s; methylene blue, 1 to 2 min. Staining should be done on a staining rack over a sink or other suitable receptacle.

8 After the smear has been exposed to the stain for the required time, remove excess stain by washing with a gentle stream of water; then blot

9.A SIMPLE STAINING WITH BASIC DYES

up excess water gently (do not wipe) with bibulous paper. Allow the smear to air-dry completely for several minutes.

FIGURE 9.2 shows the process of smear preparation and simple staining.

9 Examine the stained preparation under the microscope, using the oil-immersion objective. The preparation may be labeled more permanently with a self-adhesive microscope slide label using a fine-tip permanent-ink pen.

10 Prepare and examine simple stain preparations of each of the pure cultures of bacteria in liquid medium as well as on solid medium. Record your results in the results section.

NOTE: Instructors may wish to divide the work among students in various ways.

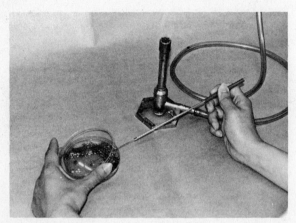
[A]

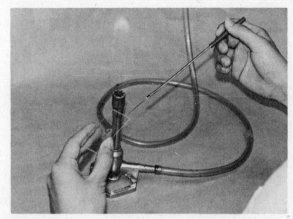
[B]

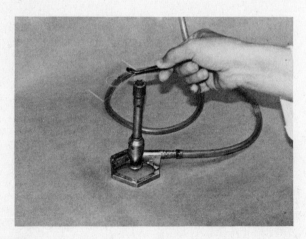
[C]

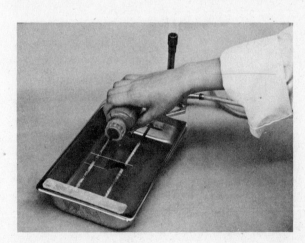
[D]

FIGURE 9.2
The process of smear preparation and simple staining. **[A]** Specimen is removed with transfer loop from colony. **[B]** Specimen is smeared (spread) on a clean glass slide in a drop of water to about the size of a dime. Smear is allowed to air-dry. **[C]** Smear is fixed by passing slide (smear side up) over the flame of a Bunsen burner several times. **[D]** Staining solution(s) is applied and then washed off. Smear is air-dried and examined under the microscope.

NAME _____

9.A

Simple Staining with Basic Dyes

RESULTS

NOTE: You may use the following terms to describe the cell morphology of the organisms observed:
Shape: Cocci, bacilli (rods), filaments, spirals (helicoidal)
Arrangement: Singly, pairs, chains, tetrads, or clusters

1 Make a drawing and give a word description of the organisms for each stained preparation made from a colony. Describe the type of colony from which each smear was made.

B. subtilis	S. aureus	A. itersonii
◯	◯	◯

Cell shape:

_____   _____   _____

Cell arrangement:

_____   _____   _____

Cell color with crystal violet:

_____   _____   _____

Cell color with carbol fuchsin:

_____   _____   _____

Cell color with methylene blue:

_____   _____   _____

Colony morphology:

_____   _____   _____

_____   _____   _____

_____   _____   _____

_____   _____   _____

2 Make a drawing and give a word description of the organisms for each stained preparation made from growth in liquid broth.

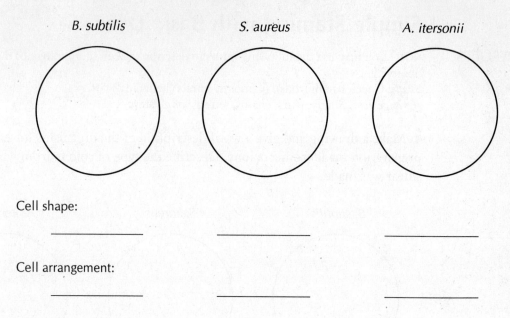

Cell shape:

_____   _____   _____

Cell arrangement:

_____   _____   _____

QUESTIONS **1** With the simple stains you used, were the dyes basic or acidic? Explain.

2 Were there any differences, under the microscope, between the cells grown in broth and those grown on solid medium?

EXERCISE 9.B

Staining the Whole Cell
9.B The Negative Stain

OBJECTIVE
To perform negative staining on several bacterial species and to compare their morphological characteristics from such stained smears.

OVERVIEW
Unstained bacteria can be made clearly visible when they are prepared as a film in nigrosin. The nigrosin provides a dark gray background; the cells are clear and unstained. The reason for this staining reaction is that the chromophore of nigrosin is in the negative ion, which will not combine with another negative ion. Therefore, an acidic dye, such as nigrosin or eosin, does not stain the negatively charged bacterial cell; instead it forms a deposit around the cell. Since this method of preparing a smear for microscopic examination does not subject the cells to strong chemical agents or physical treatment, for example, heating, there is little chance of distortion of their morphology. Thus the natural size and shape of the cells can be seen. Negative staining is also a method of choice for observing bacteria that are difficult to stain, such as some spirilli.

REFERENCES
MICROBIOLOGY, Chap. 3, "Characterization of Microorganisms."
Woeste, S., and P. Demchick, "New Version of the Negative Stain," *Applied and Environmental Microbiology* 57:1858–1859, 1991.

MATERIALS
Dorner's nigrosin solution
Nutrient-agar slant cultures of *Aquaspirillum itersonii*, *Staphylococcus aureus*, and *Bacillus subtilis*
Water-resistant black-ink marker (such as The Mighty Mark 7000, Faber-Castell Corp., Lewisburg, Tennessee)

PROCEDURE
1 Place a small drop of nigrosin near one end of a glass slide. Using the looped transfer needle, remove some organisms from one of the slant cultures and mix them in the drop. **Do not spread the drop out during this mixing.**
2 Following the technique illustrated in FIGURE 9.3, spread the drop out into a film. Using a clean slide held at an angle, draw one end just enough to the right to form contact with the droplet across its entire edge and the surface of the bottom slide. Then push the top slide across the surface of the bottom slide, drawing out the material in the droplet to form a wide film.
3 Allow the film to air-dry and examine it, using the oil-immersion objective of the microscope. The thickness of the film is not uniform, and hence the background in some areas will be darker

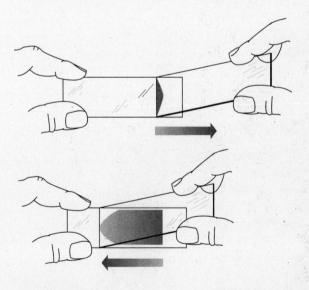

FIGURE 9.3
Preparing a smear for the negative stain.

than in others. Search several locations to select the best field for characterizing the morphology of the bacteria.

4 Repeat the procedure for the other two cultures.

ALTERNATE METHOD

1 Prepare a smear from one of the cultures provided. (See Exercise 9.A for making a smear.) Air-dry and heat-fix the smear.

2 Brush the blunt-tipped black-ink marker over the surface of the smear with one coat of ink.

3 Air-dry the smear and examine the "stained" (painted) smear under the oil-immersion objective.

4 Repeat the procedure for the other two cultures.

NOTE: The harsh heat-fixing used in this method may distort the size and shape of the cells. This disadvantage is countered, however, by the simplicity of the method.

NAME _____

9.B

The Negative Stain

RESULTS 1 Select a suitable microscopic field from each of the preparations and make drawings of several cells from a typical microscopic field.

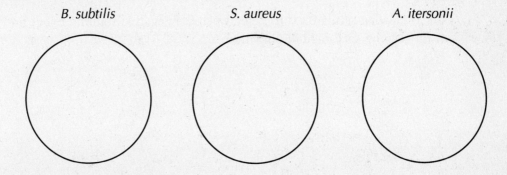

2 Describe the morphology of:

B. subtilis:

S. aureus:

A. itersonii:

69

QUESTIONS 1 Why is this technique called a *negative stain?*

2 What similarity is there between the negative-staining image and the image seen by dark-field illumination?

3 When is negative staining used?

4 What is the mechanism of action of a negative stain?

EXERCISE 9.C

Staining the Whole Cell
9.C The Gram Stain: A Differential Stain

OBJECTIVE

To understand the theoretical basis of the Gram stain for differential staining and to use the stain for comparing the Gram reaction and morphology of several species of bacteria.

OVERVIEW

The most important *differential stain* used in bacteriology for characterizing bacteria is the Gram stain. It was introduced in 1884 by the Danish physician and pathologist Hans Christian J. Gram (1853–1938). It has remained a common diagnostic staining reaction in bacteriology. In differential staining, a *primary stain* is applied to color all cells. A *decolorizing agent* is then introduced that removes color from specific cells or cell structures. And finally, a *counterstain* is applied which only those structures that were successfully decolorized will accept. In this fashion, the Gram stain divides bacterial cells into two groups based on whether they retain or lose the primary stain (crystal violet) after decolorization. Those organisms that retain the crystal violet appear dark blue or violet, and are designated **Gram-positive**; those that lose the crystal violet and that are subsequently stained by the counterstain safranin appear red and are designated **Gram-negative**. [See PLATES 1 and 2.]

The mechanism of the Gram stain may be explained on the basis of physical differences in the cell walls of these two groups of bacteria. The Gram-positive bacterium possesses a thick cell wall composed of a peptidoglycan network with associated matrix substances (teichoic and teichuronic acids). Gram-negative bacteria have a much thinner cell wall. Since thick walls always retain the primary dye while thinner walls do not, it is apparent that the physical mass and thickness of the cell wall are the important factors in determining the Gram reaction of a bacterium, and not the chemical composition of the cell wall. In brief, the Gram reaction may be viewed as a physical phenomenon brought about by the collapse of the cell wall upon dehydration by the decolorizing agent (ethanol, acetone), thus preventing the extraction of the initial crystal violet dye in the Gram-positive bacterium. The cell walls of typical Gram-positive and Gram-negative eubacteria are shown in FIGURE 9.4.

NOTE: The Gram stain does not always give clear-cut results. If Gram-positive cultures are too old, they stain Gram-negative. Gram stains should always be carried out with young, vigorous cultures. Some species of bacteria are inherently Gram-variable: some cells in the same culture stain

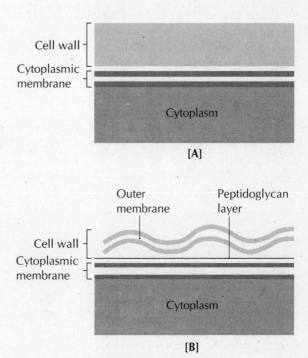

FIGURE 9.4

Cell surface of Gram-positive [A] and Gram-negative [B] eubacterium.

Gram-positive; others stain Gram-negative. To ascertain that a given test strain of bacteria truly stains Gram-positive or Gram-negative, use known cultures side by side on the same slide as the test organism and carry out the staining process.

REFERENCES
MICROBIOLOGY, Chap. 3, "Characterization of Microorganisms."
MICROBIOLOGY, Chap. 4, "Procaryotic and Eucaryotic Cell Structures."

MATERIALS
Nutrient-agar slant cultures of *Bacillus cereus*, *Escherichia coli*, *Staphylococcus aureus*, and *Neisseria subflava*
Staining solutions: Modified Hucker's crystal violet, Gram's iodine, 95% ethyl alcohol, safranin
Sterile toothpick

PROCEDURE
1 Prepare smears of *B. cereus* and *E. coli* side by side on the same slide. Fix the smears with gentle heating.
2 On a second slide, prepare and fix smears of *S. aureus* and *N. subflava*.
3 Stain smears with crystal violet for 1 min; then wash them off with Gram's iodine. Iodine is a *mordant* because it combines with a dye, such as crystal violet, to form an insoluble complex, in this case, an insoluble complex of crystal violet iodine.
4 Leave Gram's iodine solution on for 1 min; then wash with water and drain.
5 Decolorize with 95% ethanol for approximately 30 s until free color has been washed off. Wash slide with water and drain.
6 Counterstain smears for 30 s with safranin; then wash and blot-dry gently with bibulous paper.
7 Make microscope examinations of each stained preparation with the oil-immersion objective.
8 Using a sterile toothpick, take some scrapings from the base of your teeth and gums. Prepare a smear from this, and stain it by the Gram method. Examine several areas of this film using the oil-immersion objective.

The color reactions after each step of the Gram-staining procedure are outlined in FIGURE 9.5.

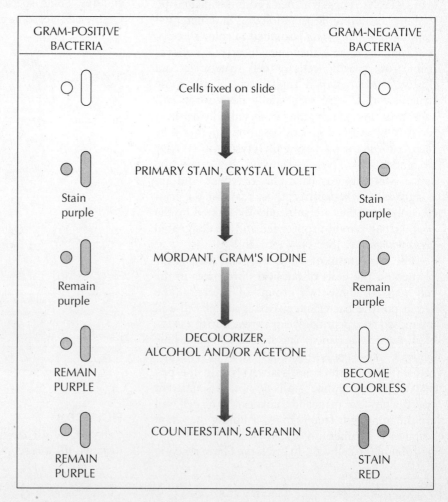

FIGURE 9.5
Gram-staining color reactions.

NAME _____

9.C

The Gram Stain: A Differential Stain

RESULTS

1 Identify the Gram reaction of each of the four species of bacteria used in this exercise.

2 Make a drawing from each preparation that will illustrate typical morphology of these microorganisms.

B. cereus   *E. coli*

○   ○

Cell color: _____   _____

Gram reaction: _____   _____

Cell morphology: _____   _____

S. aureus   *N. subflava*

○   ○

Cell color: _____   _____

Gram reaction: _____   _____

Cell morphology: _____   _____

73

3 Draw several morphological types of organisms, and determine their Gram reaction, as seen in the smear from the gumline scrapings.

Gumline smear showing
several cell types

Cell color: _____ _____ _____

Gram reaction: _____ _____ _____

Cell morphology: _____ _____ _____

QUESTIONS 1 Why should Gram staining be done on young cultures?

2 What are the advantages of differential staining over simple staining of bacterial cells?

3 Correlate the Gram-stain reaction with other properties of bacterial cells.

EXERCISE 9.D

Staining the Whole Cell
9.D The Acid-Fast Stain: A Differential Stain

OBJECTIVE
To perform an acid-fast stain on bacterial species and to use it to detect mycobacteria.

OVERVIEW
Some species of bacteria, particularly those in the genus *Mycobacterium*, do not stain readily by basic staining procedures, such as those of the Gram stain. These bacteria have waxy cell walls because the walls contain large amounts of lipoidal material. Such walls are hydrophobic, that is, they repel water, and are impermeable to stains and other chemicals in aqueous solution. In order to stain these bacteria, special staining procedures must be employed. Once the primary stain has been forced to penetrate the cell wall, it is retained within the cell, even if the cell is treated with a decolorizing agent such as acid alcohol. Because of this property, these cells are called *acid-fast* and the stain is called an *acid-fast stain*. [See PLATE 3.]

The acid-fast procedure employs three different reagents. It begins with a strong, red, basic dye called *basic fuchsin* which, when dissolved in an aqueous solution containing phenol, is called *carbol fuchsin*. Carbol fuchsin is soluble in the lipoidal material that constitutes the major portion of the cell wall of mycobacteria. Heat treatment enhances the penetration of this primary stain through the cell wall and into the cytoplasm. After this primary staining, the smear is decolorized with *acid alcohol*. All but acid-fast bacteria are decolorized by acid alcohol. Presumably, waxy walled bacteria are decolorized more slowly by acid alcohol than other bacteria. *Methylene blue* is used as the final reagent to stain the decolorized cells. Hence acid-fast bacteria appear *red* and non-acid-fast cells appear *blue*.

MATERIALS
Nutrient-agar (plus 18-ml glycerol/liter medium) slant cultures of *Mycobacterium smegmatis* and *Staphylococcus aureus*
Staining solutions: Ziehl's carbol fuchsin, Kinyoun carbol fuchsin, acid alcohol, methylene blue
Egg albumin or serum
Strip of blotting paper approximately ¾ by 1½ in
Prepared acid-fast stains of tuberculous sputum

PROCEDURE
NOTE: Smears of the bacteria should be prepared in a small drop of serum or egg albumin. The protein enhances the adherence of the bacteria to the glass surface and also provides material for light-background staining.

The Ziehl-Neelsen Method

1 Prepare and heat-fix smears of *M. smegmatis* and *S. aureus* side by side on a glass slide. On a second slide, prepare a single smear containing a mixture of the two species.
2 Cover the smears with a strip of blotting paper; the paper should not extend beyond the edges of the slide.
3 Saturate the paper with Ziehl's carbol fuchsin.
4 Heat the slide to steaming with a small flame or a hot plate. **Do not allow the slide to dry.** If necessary, add more stain. Allow the staining to continue for 3 to 5 min, and remove blotting paper with forceps.
5 Wash the slide with tap water, and then decolorize the smear for 10 to 30 s with acid alcohol. Exercise care so that the smear is not overdecolorized. Wash the slide with tap water.

6 Apply methylene blue for 30 to 45 s, wash, drain, blot-dry carefully, and air-dry. Examine the smears microscopically with the oil-immersion objective. Acid-fast organisms stain red; non-acid-fast organisms stain blue. The protein film, serum or albumin, will appear faintly blue.

7 Make a microscopic examination of the acid-fast stain of tuberculous sputum provided.

Kinyoun's Cold Procedure

An alternate procedure, a modification of the Ziehl-Neelsen method, circumvents the use of heat in staining and therefore is less messy.

1 Prepare and heat-fix smears of *M. smegmatis* and *S. aureus* as for the Ziehl-Neelsen method.

2 Cover the smears with a strip of blotting paper as in the Ziehl-Neelsen method.

3 Saturate the paper with Kinyoun carbol fuchsin, and let it stand for 15 to 20 min. Add more stain if the paper dries. Do *not* steam.

4 Remove paper with forceps, rinse with water, and allow to drain.

5 Decolorize with acid-alcohol for 10 to 30 s until no more stain appears in the washing.

6 Counterstain with methylene blue for 1 to 2 min.

7 Wash with tap water, drain, blot-dry carefully, and air-dry before examination under the oil-immersion objective.

NAME _____

9.D

The Acid-Fast Stain: A Differential Stain

RESULTS

1 Record the color of organisms in each preparation, and indicate their acid-fast reaction.

2 Draw a representative portion of a typical microscopic field from each stained preparation.

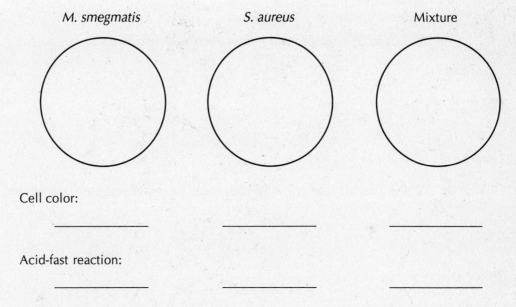

Cell color:

_____   _____   _____

Acid-fast reaction:

_____   _____   _____

3 Describe the appearance of the acid-fast stain of tuberculous sputum as seen under the oil-immersion objective.

QUESTIONS 1 What accounts for the acid-fast property of *Mycobacterium* species?

2 Why did you use *M. smegmatis* instead of *M. tuberculosis* in class?

3 For what diseases would you use an acid-fast stain?

4 Explain why the acid-fast stain is such a useful diagnostic technique.

EXERCISE 10.A

Staining for Cell Structures
10.A The Spore Stain

OBJECTIVE
To perform a spore stain on bacterial species and to be able to differentiate bacterial endospores from vegetative cells in such stained smears.

OVERVIEW
In response to unfavorable environmental conditions, certain bacteria, such as species in the genera *Bacillus* and *Clostridium*, develop a spore that possesses a notable resistance to physical and chemical agents [FIGURE 10.1]. For example, it is resistant to extreme desiccation, high temperatures, ionizing radiation, and many chemicals. The spore develops within the "mother" vegetative bacterial cell, called the *sporangium*, and hence is called an *endospore*. The endospore occupies a characteristic position within the cell: terminal, subterminal, or central. The location and size of the endospore vary with the bacterial species and are often of value in identification. As development proceeds, the sporangium eventually disintegrates, resulting in the *free spore*.

The nature of the spore necessitates a vigorous treatment for staining. Once stained, the spore is relatively difficult to decolorize. This property is the basis of staining endospores. The primary stain, malachite green, is driven into the cell with heat. The sporangium and any non-spore-forming vegetative cells are decolorized and counterstained to a light red with the secondary stain safranin.

REFERENCES
MICROBIOLOGY, Chap. 3, "Characterization of Microorganisms."
MICROBIOLOGY, Chap. 4, "Procaryotic and Eucaryotic Cell Structures."

MATERIALS
Nutrient-agar slant cultures of *Bacillus coagulans* and *Bacillus subtilis*
Blood-agar slant cultures of *Clostridium butyricum* and *Clostridium tetanomorphum*
Staining reagents: Malachite green, safranin
Gram-stain reagents

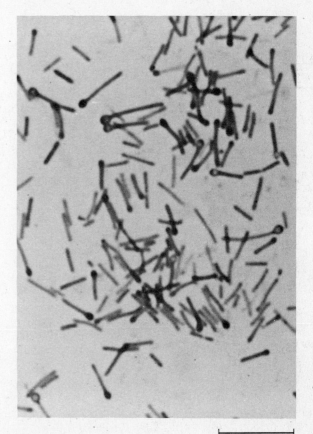

10 μm

FIGURE 10.1
Photomicrograph of spore-forming bacteria. Note the terminal position of the endospores within the sporangium (mother cell). *(Courtesy of Liliane Therrien and E. C. S. Chan, McGill University.)*

79

PROCEDURE

1 Prepare and heat-fix a smear from each culture. Two smears can be made side by side on a single slide. Remember to label them on the edge of the slide with a marking pen.

2 Cover the smear with malachite green staining reagent.

3 Heat the reagent gently by placing the slide on a warm hot plate or by exposing it to the flame of a Bunsen burner. Allow the reagent to steam for 2 to 3 min. More reagent may be added to prevent drying.

4 Wash the slide with tap water and drain.

5 Cover the smear with safranin for 30 s.

6 Wash the smear with tap water, drain, and blot-dry gently.

7 Examine the smear using the oil-immersion objective. The spores, both endospores and free spores, stain green; vegetative cells, including sporangia, stain red. [See PLATE 4.]

8 Prepare and examine a Gram stain of each species.

NAME _____

10.A

The Spore Stain

RESULTS From a typical microscopic field of each specimen, draw several of the bacterial vegetative cells with their endospores and free spores. Note the position and size of the endospores within individual sporangia. Note also the shape of free spores.

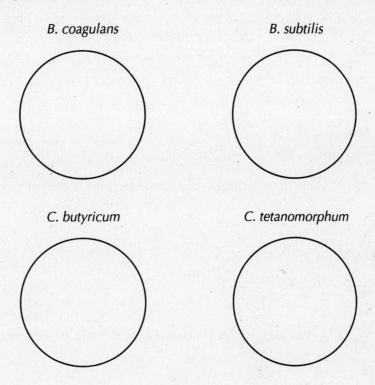

81

QUESTIONS **1** Why is a bacterial endospore more difficult to stain than a vegetative cell?

2 Were you able to see spores in the Gram-stained preparations of each species? Explain their appearance in the microscopic fields.

3 What do you think is the function of the endospore to a bacterial species?

4 At what stage of the bacterial growth cycle are endospores formed?

5 Give two species names of spore-forming bacterial pathogens and the diseases they cause.

EXERCISE 10.B

Staining for Cell Structures
10.B The Capsule Stain

OBJECTIVE
To perform a capsule stain on bacterial species and to be able to distinguish capsules from cells in the stained smears.

OVERVIEW
The cell wall of many species of bacteria is surrounded by a mucilaginous polymeric substance referred to as a *capsule* (discrete) or *slime layer* (amorphous). The size of the capsule varies with the species and among strains of the species. The composition of the medium may also influence the size of the capsule. Chemically, the capsular material is a polysaccharide, a glycoprotein, or a polypeptide.

Among disease-producing bacteria, the presence of a large capsule is generally an indication of a virulent form of the organism capable of causing disease. This is because the capsule protects the cell from being phagocytized by the host white blood cells. The identification of some bacteria, like the pneumococci that cause pneumonia, is facilitated by characterization of the capsule, for example, reaction with specific antibodies to detect capsular swelling (called the *Quellung phenomenon*). Capsules are also protective to the cell in other ways, such as retardation against drying, attraction and retention of nutrients, and adhesion to sites with favorable environments. [See PLATE 5.]

Capsule staining is difficult and is an "art." A successful capsule stain is a triumphant event! Most capsular materials are water-soluble and uncharged, so simple stains will not adhere to them. Thus, most capsule staining techniques stain the bacteria and the background more intensely than the capsule itself. FIGURE 10.2 shows the appearance of capsulated bacteria in a well-stained specimen.

MATERIALS
24-h skim-milk culture of *Enterobacter aerogenes*, *Flavobacterium capsulatum*, and *Proteus vulgaris*
24-h nutrient-broth culture of the same bacteria
Aqueous solution of nigrosin (5% w/v)
Absolute alcohol
Gram-stain reagents

PROCEDURE
1 Add a drop of culture to a drop of nigrosin solution on one end of a slide. Mix without spreading.

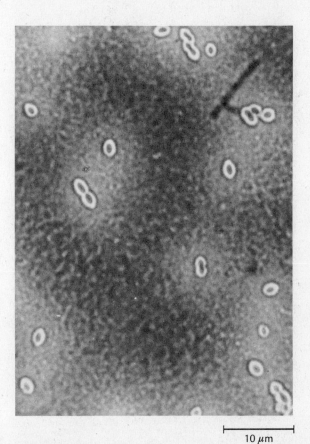

FIGURE 10.2
Photomicrograph of capsulated bacteria. *(Courtesy of Liliane Therrien and E. C. S. Chan, McGill University.)*

2 Spread the mixture over the slide using the end of a second slide to prepare a thin smear as in a negative stain (see Exercise 9.B).
3 Let the smear dry.
4 Fix the smear with absolute alcohol for 2 min; drain off the alcohol and let smear dry.
5 Stain with crystal violet for 2 min.
6 Wash the slide with tap water, and air-dry vertically before examination with the oil-immersion objective. The cells should stain blue while the capsules are colorless and surrounded by a black background of nigrosin stain.
7 Prepare and examine Gram stains of each species from the nutrient-broth cultures.

NAME _____

10.B

The Capsule Stain

RESULTS Make a drawing of a few cells from a typical microscopic field of the capsule stain of each species prepared from the skim-milk culture and the nutrient-broth culture. Record the Gram reaction of each species.

Skim-Milk Culture:

E. aerogenes   F. capsulatum   P. vulgaris

(○)   (○)   (○)

Gram reaction:
_____   _____   _____

Nutrient-Broth Culture:

E. aerogenes   F. capsulatum   P. vulgaris

(○)   (○)   (○)

Gram reaction:
_____   _____   _____

QUESTIONS **1** What is the chemical composition of bacterial capsules?

2 List three physiological (functional) roles ascribed to bacterial capsules.

3 What relationship exists between capsules of bacterial pathogens and the capacity of the bacteria to cause disease?

4 Assume that you have a nutrient-agar slant culture of some bacterial species. A capsule stain is prepared from this culture, and upon microscopic examination the organism is seen to have a very small capsule. How might capsule formation by this organism be enhanced?

EXERCISE 10.C

Staining Specific Cell Structures
10.C The Flagella Stain and Detection of Motility

OBJECTIVE
To perform the flagella stain on bacteria and to examine such stained smears for the presence of bacterial flagella; to detect motility in bacteria by several methods.

OVERVIEW
Motile bacteria, particularly those regarded as true bacteria, that is, *eubacteria*, possess one or more very fine, threadlike appendages called *flagella*. Flagella are *coiled* into rigid spirals that revolve around their points of attachment in the natural state. These are regarded as organs of locomotion. Unfortunately, because of the limitations of microscopic preparation procedures, whether for electron microscopy or light microscopy, the flagella are seen as "flat," wavelike organelles. Their length is many times that of the bacterial cell itself, but their diameters are much thinner, about 0.01 to 0.05 μm. Such diameters are about 100 times narrower than that of most bacterial cells and are 10 times smaller than the *limit of resolution* of the light microscope (0.2 μm). Therefore, bacterial flagella are seen *directly* only with the electron microscope. For light microscopy, they can be demonstrated only by special staining procedures: the thickness of the flagella is increased by coating them with mordants like tannic acid and potassium alum (in one method), and then staining them [FIGURE 10.3].

Flagella stains are difficult to do unless meticulous care is observed in carrying out the procedures. For example, rough handling will break off the flagella from the cell because flagella attachment is not very stable. Even though electron microscopy is the best direct method of determining flagellation in bacteria, the electron microscope is still not easily accessible. Thus, flagella staining remains an important procedure for the identification of bacterial species. [See PLATE 6.]

Motile bacteria can also be detected by examination of the organisms in wet preparation, where they are able to move about freely. Bacterial motility can also be determined indirectly by inoculating a special motility-agar medium; motile bacteria are able to migrate, or relocate, away from the line of inoculation.

MATERIALS
Gray's flagella mordant and Ziehl's carbol fuchsin
Specially cleaned slides
3 1-ml distilled-water blanks
Depression slide and cover slips
3 tubes of motility-test agar

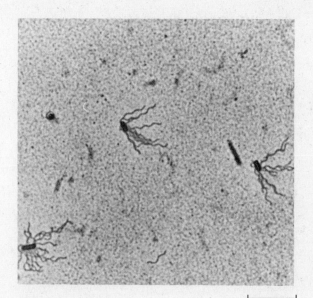

FIGURE 10.3
Peritrichous flagella seen on bacterial cells after staining. *(Courtesy of Liliane Therrien and E. C. S. Chan, McGill University.)*

18- to 22-h nutrient-agar slant cultures of
Pseudomonas fluorescens, Proteus vulgaris, and
Staphylococcus aureus
Small tube of petroleum jelly
95% ethanol (ethyl alcohol)
Lintless paper tissue

PROCEDURE
FLAGELLA STAIN

1 Obtain new slides and rinse them in 95% ethyl alcohol (ethanol). Wipe them dry with lintless paper tissue, and pass them through the flame of the Bunsen burner several times.
2 Prepare a light suspension of each culture in distilled water, and incubate 10 to 15 min to develop and extend the flagella fully. Handle suspension gently so that the flagella will not break off.
3 Transfer one loopful of the suspension of cells to one end of a clean slide. Tilt the slide, allowing the drop to run down. Air-dry this film. Do not heat-fix.
4 Flood the slide with flagella mordant, and allow to stand for approximately 10 min.
5 Rinse the stain off gently with distilled water.
6 Flood with Ziehl's carbol fuchsin for 5 min. Rinse off gently with water.
7 Air-dry (do not blot), and examine the slide, using the oil-immersion objective.

HANGING-DROP PREPARATION

1 Ring the outer edge of the concave well of a depression slide with a small amount of petroleum jelly. (Plain slides may also be used by building up a circle of petroleum jelly smaller than the cover slip.)
2 Place a small drop of the bacterial suspension on the cover slip, and then invert the slide with the concave area over the drop. In a properly made preparation, the drop will hang from the cover slip. The petroleum jelly forms a seal and prevents rapid evaporation of the drop.
3 Observe the specimen, using the *high-power objective*. Exercise care in focusing and adjusting the light. Proper illumination is critical since this is an unstained preparation. Concentrate on the edge of the droplet, where the organisms appear to be more active because of a greater oxygen supply.

ALTERNATIVE WET PREPARATION

In place of the hanging-drop procedure just described, the following technique can be used:

1 Smear a drop of immersion oil over a glass slide to cover an area approximately the size of a cover slip.
2 Place a small drop of the bacterial culture on a cover slip, and then invert and place this cover slip onto the slide with the immersion-oil film. Gently press the cover slip to the slide.
3 Observe the preparation using the oil-immersion objective and reduced illumination. Small droplets of the bacterial suspension will be observed to be trapped throughout the oil film. Motility of the organisms can be observed within these droplets.

INOCULATION OF MOTILITY-TEST AGAR

1 Make a stab inoculation from the agar slants of each culture into individual tubes of motility agar. Stab inoculation is done with the straight transfer needle. Some growth is removed aseptically from the slant culture and inoculated by stabbing straight down into the center of the motility-test agar medium.
2 Incubate these tubes at 35°C for 24 to 48 h.

NAME _____

10.C

The Flagella Stain and Detection of Motility

RESULTS

1 Make a drawing for each flagella-stain preparation. Note the number of flagella and the position of their attachments to the cell.

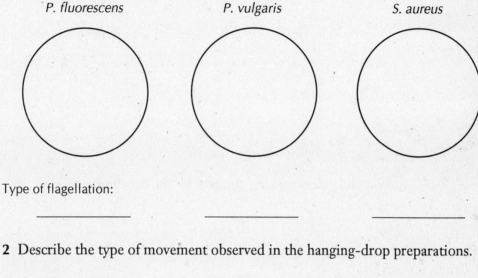

Type of flagellation:

_____   _____   _____

2 Describe the type of movement observed in the hanging-drop preparations.

Hanging-drop motility:

_____   _____   _____

3 Illustrate by a sketch the pattern of growth that developed in the motility-agar stab cultures. If the microbe is actively motile, you will see its growth radiating outward from the stabbed line. If it is not actively motile, the culture will grow confined within the stab line of inoculation.

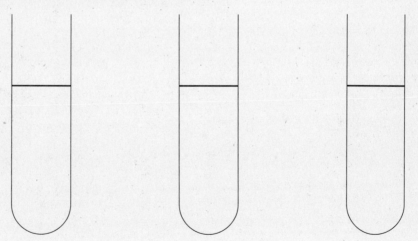

QUESTIONS

1 Do the results obtained by the three techniques performed correlate well with respect to motility and the presence of flagella?

2 Why must meticulous care be exercised in preparing flagella stains?

3 Why did you use young cultures for the flagella stain?

4 What is *Brownian motion?* Do bacteria exhibit this type of motion?

EXERCISE 10.D

Staining Specific Cell Structures
10.D The Cell-Wall Stain

OBJECTIVE
To perform the cell-wall stain on bacteria and to examine such stained smears for the presence of cell walls.

OVERVIEW
Most bacteria have rigid walls. The cell wall of a bacterium is easily seen in the electron microscope. However, in the light microscope it is difficult to distinguish the cell wall from the other components of the cell, but by using special reagents, the wall can be stained and observed [FIGURE 10.4]. Specifically, the cell wall is positively charged by treating it with a cationic surface agent such as cetylpyridinium chloride, which dissociates in water to give a positively charged cetylpyridinium cation and a negatively charged chloride ion. The cations neutralize the negative charges of the bacterial cell surface (wall), and the surface becomes positively charged because of the binding of the cations. It then takes on the acid dye with a negatively charged chromophore.

MATERIALS
24-h nutrient-agar slant cultures of *Escherichia coli* and *Bacillus cereus*
Microscope slide
$M/100$ cetylpyridinium chloride; saturated aqueous congo red; methylene blue

PROCEDURE
1 Make smears of the cultures provided, and fix them with heat.
2 Add 3 drops cetylpyridinium chloride solution to each smear.
3 Add 1 drop congo red to each smear. Mix the drops with a loop transfer needle. Do not scratch the smears.
4 Rotate the drops with a tilting motion by holding the slide in your hand.
5 Rinse off with tap water.
6 Counterstain for 10 s with methylene blue.
7 Rinse and blot-dry. Examine with the oil-immersion objective.

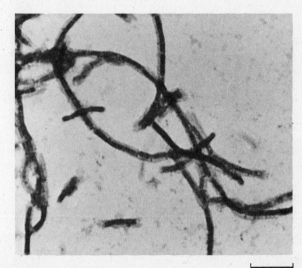

FIGURE 10.4
Photomicrograph of a cell-wall stain of a bacterium. *(Courtesy of Liliane Therrien and E. C. S. Chan, McGill University.)*

NAME _____

10.D

The Cell-Wall Stain

RESULTS Make drawings of each cell-wall preparation. Note the color of the cell wall and the cytoplasm.

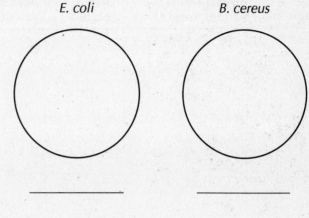

Cell wall color: _____   _____

Cytoplasm color: _____   _____

QUESTIONS 1 List three main differences between the cell walls of Gram-positive and Gram-negative bacteria.

2 Name one group of bacteria without cell walls.

3 Give four chemical substances found in the Gram-negative cell wall of bacteria.

Complete the following summary chart based on observations made in Exercises 9 and 10.

	SUMMARY OF MORPHOLOGICAL CHARACTERISTICS OF BACTERIA STUDIED IN EXERCISES 9.A–9.D AND 10.A–10.D					
Species	Shape and arrangement	Gram reaction	Acid-fast reaction	Spores	Capsules	Flagella

EXERCISE 11

Identification of a Morphological Unknown

OBJECTIVE
To determine the morphological characteristics of an unknown bacterial species.

OVERVIEW
Morphological characteristics, including staining reactions, represent one of the major properties of bacteria. In identifying an organism, it is customary to determine the characteristics of morphology and the staining reactions such as those obtained in the preceding exercises. In this exercise, you will be given pure cultures of bacteria identified by code rather than by name. You will determine the morphological properties of each of these unknown cultures.

MATERIALS
Staining solutions used in previous exercises
Agar slant cultures of two or more bacterial unknowns identified by code only

PROCEDURE
1 Perform the following techniques on each unknown culture: Gram stain, acid-fast stain, spore stain, capsule stain, and hanging-drop preparation. Use the procedures described in the preceding exercises.
2 Make microscopic examinations of each preparation.

NAME _____

11

Identification of a Morphological Unknown

RESULTS Draw a few typical cells from each Gram-stain preparation. Then record the results of the unknown culture below each drawing on the lines provided.

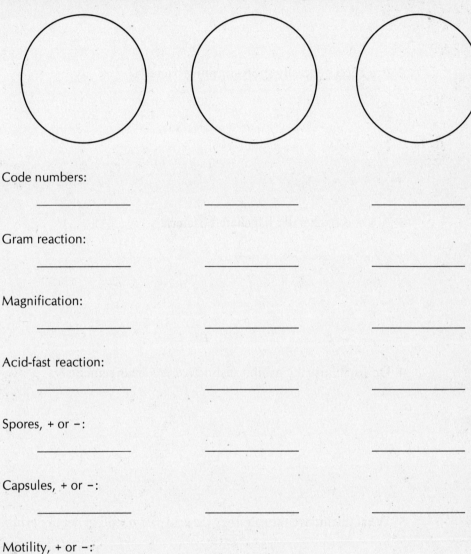

Code numbers:

_____   _____   _____

Gram reaction:

_____   _____   _____

Magnification:

_____   _____   _____

Acid-fast reaction:

_____   _____   _____

Spores, + or −:

_____   _____   _____

Capsules, + or −:

_____   _____   _____

Motility, + or −:

_____   _____   _____

QUESTIONS

1 Compare the characteristics of each of your unknown cultures with those of the pure cultures that you used in Exercises 9 and 10. Which pure cultures have morphological characteristics similar to your unknowns?

2 Are cocci generally spore-forming bacteria?

3 Are cocci generally flagellated bacteria?

4 Do bacilli usually exhibit distinctive cell arrangements?

5 What distinctive morphology do acid-fast bacilli generally exhibit?

CULTURE MEDIA PREPARATION AND THEIR STERILIZATION

In the laboratory, the composite of materials that supplies nutrients to grow microorganisms is called a **culture medium** (plural *media*). The medium also provides moisture and regulates the pH.

Bacteria exhibit great diversity in their requirements for nutrients; for example, some are capable of growing in a medium consisting of inorganic compounds only, while others require amino acids, vitamins, and even additional complex organic substances. The specific nutritional requirements have been determined for a great many species. However, the media usually used in the laboratory for routine cultivation of microorganisms consist of a mixture of *digests,* or extracts, from plant or animal tissues. The most common of these are meat extract, yeast extract, peptone, and agar. These ingredients, except for the agar, are used to prepare *broths,* or *liquid, media.* The addition of agar results in a *solid medium,* and, obviously, a reduction in the amount of agar results in a *semisolid medium.* Nutrient broth (peptone and meat extract) and nutrient agar (peptone, meat extract, and agar) may be considered basic liquid and solid media, respectively. They may be modified in a variety of ways by the addition of specific chemicals or complex supplements to produce a medium with some desired characteristics. For example, nutrient agar may be enriched by the addition of sterile blood; as such, it will support the growth of nutritionally fastidious bacteria. Either nutrient broth or nutrient agar may be supplemented with some substrate for use as a biochemical test medium.

Culture medium containing agar, while in the liquefied state, can be poured into either a Petri dish or a test tube. A Petri dish with agar culture medium is called an *agar plate.* If the agar culture medium in a tube is solidified in a slanted position, it is called an *agar slant;* if it is solidified in an upright position in a tube, it is called an *agar deep* [FIGURE IV.1].

There are innumerable media available for use in microbiology. Each formulation is designed to offer some advantage for the isolation, maintenance, characterization, or growth of certain groups of organisms. In general, culture media may be divided into two main types: **chemically defined** and **complex.** Chemically defined media are also called **synthetic media** because they can be synthesized easily from laboratory chemicals; such media are composed of known amounts of pure chemicals. Complex media are composed of complex organic materials that are rich in nutrients, such as the nutrient broth and nutrient agar mentioned previously.

A culture medium must be *sterilized* so that it will support the growth of only those cells inoculated into it. Microorganisms other than those inoculated into it and that grow in a sterilized medium are called **contaminants.** Generally, the *autoclave* is used for the sterilization of culture media. **Autoclaving** is a sterilization process that uses moist heat, or steam, under pressure so that *all parts* of the material to be sterilized reach 121°C for 15 min. In the microbiology laboratory, autoclaving is probably the most versatile and most often used sterilization process.

FIGURE IV.1
Various forms of culture media.

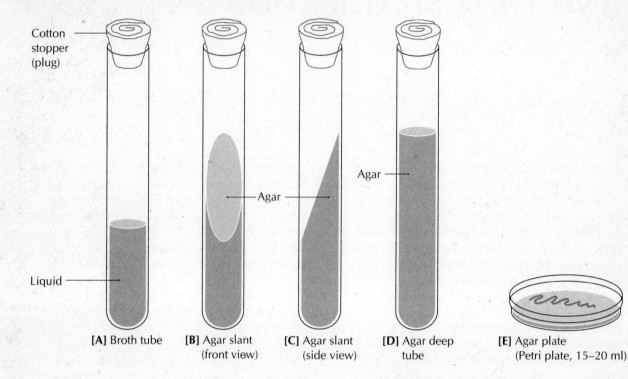

EXERCISE 12

Preparation and Autoclaving of Nutrient Broth and Nutrient Agar (Plates, Deeps, and Slants)

OBJECTIVE
To learn the preparation of complex culture media and the technique of autoclaving.

OVERVIEW
Media employed for the cultivation of bacteria, as well as for the cultivation of other microorganisms, may be liquid, solid, or semisolid. The ingredients range from pure chemical compounds to complex materials, such as extracts, or digests, of plant and animal tissues. The most common ingredients of bacteriological media used in routine laboratory work are beef extract and peptone for liquid media and beef extract, peptone, and agar for solid media. These basic ingredients may be supplemented with a variety of materials to provide a medium suitable for the cultivation or demonstration of a reaction for specific types or groups of bacteria.

When the medium is made, it is necessary to sterilize it by autoclaving [FIGURE 12.1]. The volume of the medium must be considered because heat transfer takes time, and effective sterilization of liquid requires that every part of the liquid be heated to 121°C for 15 min. When the liquid is in test tubes, heat transfer into the center of the tubes takes 10 to 15 min. Therefore, larger volumes of media, such as those in Erlenmeyer flasks, require proportionately longer autoclaving times.

NOTE: As a general rule in the preparation of media, each medium ingredient is dissolved in turn in a measured volume of distilled water. A medium containing agar should be boiled or put into the microwave oven for dissolving the agar before autoclaving.

REFERENCES
MICROBIOLOGY, Chap. 5, "Nutritional Requirements and Microbiological Media."
MICROBIOLOGY, Chap. 6, "Cultivation and Growth of Microorganisms."
MICROBIOLOGY, Chap. 7, "Control of Microorganisms: Principles and Physical Agents."

MATERIALS
Beef extract
Peptone
Agar
Yeast extract
Glucose
1 2000-ml Erlenmeyer flask or other vessel suitable for dissolving media ingredients
4 500-ml Erlenmeyer flasks
100 18- by 150-mm test tubes
Balance
1 500-ml graduated cylinder
Nonabsorbent cotton
pH meter
1 N HCl
1 N NaOH
Autoclave
Sterilizer indicator tape

PROCEDURE
1 Put a measured amount of water (approximately 500 ml) into a 2000-ml flask. Weigh out 5 g of peptone and 3 g of beef extract and add them carefully into the flask. Heat and stir frequently until the peptone and beef extract are dissolved. Adjust the total volume to 1000 ml by adding water.

NOTE: There is no need to heat to boiling unless agar is used. Heating and stirring are conveniently carried out by heater-stirrers (using magnetic bars).

2 Divide the beef extract-peptone (nutrient) solution prepared in step 1 into four equal parts (250 ml), placing each aliquot into a 500-ml Erlenmeyer flask, and treat as follows:

a Make no further additions. This is *nutrient broth*.

b Add 3.75 g of agar. Agitate and allow to soak for about 5 min; then heat to boiling with frequent stirring to dissolve the agar. This is *nutrient agar*.

c Add 1.25 g of glucose and stir to dissolve it. This is *glucose broth*.

d Add 1.25 g of yeast extract. Heat and stir to put it into solution. This is nutrient broth *enriched* with yeast extract (*yeast extract broth*).

3 Adjust the reaction of the four media to pH 7.0 using a pH meter and 1 N HCl or 1 N NaOH. Measure pH of molten agar at 60°C. Work rapidly and rinse electrodes well after use.

NOTE: The instructor or demonstrator will demonstrate to you the proper use of the pH meter. Remember that it has to be adjusted to the pH of a standard buffer before use.

4 Dispense each of the media into test tubes, 10 ml per tube for the broth media and 20 ml per tube for the nutrient agar (nutrient-agar deep).

5 Stopper the tubes with cotton; then sterilize them in an autoclave at 15-lb pressure (121°C) for 15 min. Use a short length of sterilizer indicator tape on a rack used for autoclaving.

NOTE: The laboratory instructor will demonstrate the operation of the autoclave. The sterilizer indicator tape is pressure-sensitive and shows the word *autoclaved* after the process.

6 After sterilization store at room temperature for use in Exercise 15.

12 PREPARATION AND AUTOCLAVING OF NUTRIENT BROTH AND NUTRIENT AGAR

FIGURE 12.1
[A] The laboratory autoclave, a pressure-steam sterilizer.
[B] Cross-sectional view of an autoclave illustrating operational parts and path of steam flow.

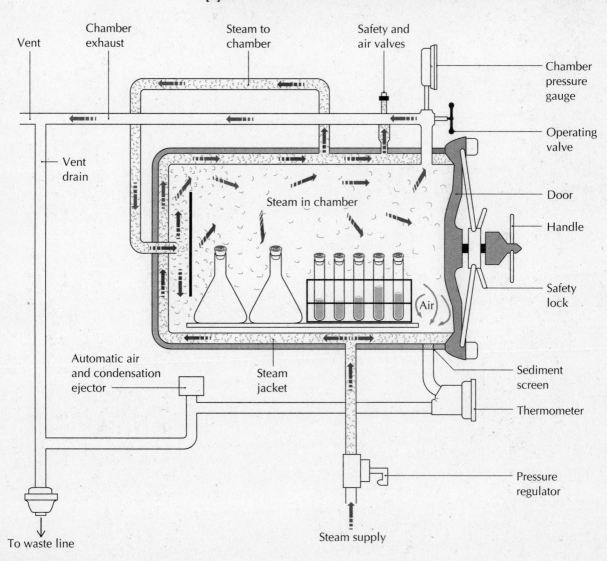

NAME _____

Preparation and Autoclaving of Nutrient Broth and Nutrient Agar (Plates, Deeps, and Slants)

RESULTS Describe the characteristics of the four media prepared.

Nutrient broth:

Nutrient agar:

Glucose broth:

Yeast-extract broth:

QUESTIONS 1 What is peptone, and how is it prepared from raw materials?

2 What kind of nutrients do yeast extract and beef extract provide?

3 What is the source and the chemical nature of agar?

4 What is the pH range for the optimum growth of most bacteria?

EXERCISE 13

Dry-Heat Sterilization

OBJECTIVE
To learn the technique of dry-heat sterilization.

OVERVIEW
In addition to moist-heat sterilization by autoclaving, there are other methods of sterilization. One of them is dry-heat sterilization; it is employed when it is neither convenient nor desirable to use moist heat, for example, in the sterilization of mineral oil, which is not miscible in water. Generally, the hot-air oven is used in the dry-heat sterilization of glassware and other heat-resistant equipment. This method of sterilization is not as commonly used in the microbiology laboratory today as it was in the past because of the availability of *presterilized* and *disposable* plastic ware (such as pipettes and Petri dishes). Nevertheless, glassware and metalware, wrapped in thick impervious paper (Manila paper) or aluminum foil, or in a closed container, can be sterilized with dry heat. Dry-heat sterilization, in the absence of moisture, does not contribute to rusting. Hence it still occupies a place in the sterilization of metallic instruments in dental and medical operatories.

As is the case with moist heat, time must be provided for heat transfer, that is, all components must be allowed to reach 160°C and held at that temperature for 2 h. Hot air must be able to circulate freely among the materials to be sterilized. In this respect, a convection oven (with forced air circulation) is more efficient. The oven temperature should not be allowed to rise above 180°C to prevent the charring of any paper or cotton.

This exercise will acquaint you with the operation of the hot-air oven and will also demonstrate that bacterial endospores are destroyed by dry-heat sterilization.

REFERENCE
MICROBIOLOGY, Chap. 7, "Control of Microorganisms: Principles and Physical Agents."

MATERIALS
4 tubes nutrient broth
4 sterile stoppered test tubes
3-ml spore suspension of *Bacillus subtilis*, containing approximately 10^{10} spores/ml, in a closed (capped or plugged) test tube with 5-mm glass beads
1 pair of forceps in 70% ethanol
Heat-insulated gloves

PROCEDURE
1 Flame the forceps briefly to burn off the ethanol.
2 Observing aseptic technique, remove one glass bead at a time from the spore suspension with a pair of forceps and place into each of four stoppered sterile test tubes.
3 Place two test tubes containing beads into the refrigerator.
4 Place the other two test tubes containing beads in the hot-air oven, adjusted to 160°C, for 2 h.
5 At the end of the exposure time, remove the tubes with beads from the oven with a pair of heat-insulated gloves, and permit them to cool to room temperature. At the same time, remove the test tubes with beads from the refrigerator, and let them equilibrate to room temperature.
6 Using aseptic technique, roll the bead from each tube into a nutrient-broth tube labeled "Heat exposed" or "Not heat exposed."
7 Incubate the tubes at room temperature for 24 to 48 h, and observe for growth, that is, *turbid* (with growth) or *clear* (no growth).

NAME _____

Dry-Heat Sterilization

RESULTS Observe each tube for the presence or absence of growth. In the chart that follows, record growth as + and no growth as 0.

Treatment	Tube number	Growth
Dry-heat exposure	1	
	2	
None	1	
	2	

QUESTIONS 1 What happened to the spores that enabled you to measure their viability by turbidity?

2 Describe how spore strips are used as indicators for sterilization processes under practical circumstances, for example, in a hospital.

EXERCISE 14

Preparation of a Chemically Defined Medium

OBJECTIVE
To learn the preparation of chemically defined media.

OVERVIEW
A medium that consists of known chemical compounds is designated as a *chemically defined*, or *synthetic, medium*. Nutrient agar, nutrient broth, and other such media that contain peptone and/or extracts of meat or other tissues are not chemically defined. These extracts, or digests, consist of a complex variety of substances, including amino acids, vitamins, carbohydrates, and inorganic salts. Chemically defined media for the growth of some bacteria, for example, *Escherichia coli*, are relatively simple in composition. The essential ingredients are glucose, an inorganic nitrogen source, and a few other inorganic salts. The composition of a chemically defined medium for lactobacilli, on the other hand, is much more elaborate; many amino acids, vitamins, inorganic salts, and other compounds must be provided. In this exercise, a relatively simple chemically defined medium (glucose-salts medium) will be prepared.

REFERENCES
MICROBIOLOGY, Chap. 5, "Nutritional Requirements and Microbiological Media."
MM, Chap. 7, "Nutrition and Media."

MATERIALS
0.5 g glucose
0.5 g sodium chloride
0.02 g magnesium sulphate
0.1 g ammonium dihydrogen phosphate
0.1 g dipotassium phosphate
1 500-ml Erlenmeyer flask
10 test tubes
Nonabsorbent cotton
1 N HCl
1 N NaOH
pH meter

PROCEDURE
1 Add 100 ml of distilled water to a clean 500-ml Erlenmeyer flask.
2 Weigh out the amounts of chemicals listed above. Add each chemical separately to the flask in the order listed, making sure that each chemical is dissolved by mixing it in, before adding the next one.
3 Determine the pH of a portion of the mixture with a pH meter. Adjust to pH 7.0 to 7.2 if necessary with acid or base.
4 Dispense the medium into test tubes, approximately 10 ml per tube.
5 Cotton-stopper the tubes, and sterilize the medium by autoclaving (15 min at 121°C). Store the medium for use in Exercise 15.

NAME _____

Preparation of a Chemically Defined Medium

RESULTS

1 Compare the appearance of the medium prepared in this exercise with that of the broths prepared in Exercise 12.

2 If the pH of the medium in the portion tested required adjustment, show the calculations made to determine the amount of NaOH or HCl needed for the total volume of the medium.

QUESTIONS 1 What are the sources of carbon and nitrogen in the chemically defined medium you prepared? In nutrient broth?

2 What are two desirable features of using chemically defined media?

3 How might you enrich the chemically defined medium prepared in this exercise and still have it remain a synthetic medium?

4 What is meant by the term *buffer?* What is the buffer system used in the medium you prepared?

EXERCISE 15

Evaluation of Media to Support Growth of Bacteria

OBJECTIVE
To determine the growth response of some bacterial species in several types of media.

OVERVIEW
Bacteria manifest a wide range of requirements for nutrients. At one extreme are the autotrophs, which require only a limited number of inorganic substances. At the other extreme are the fastidious heterotrophs, which must be supplied with a variety of complex organic substances. Media of various formulations have been developed for the satisfactory cultivation of each nutritional group of microorganisms. In this experiment you will determine the nutritional adequacy of several media for several species of bacteria.

REFERENCES
MICROBIOLOGY, Chap. 5, "Nutritional Requirements and Microbiological Media."
MICROBIOLOGY, Chap. 6, "Cultivation and Growth of Microorganisms."
MM, Chap. 7, "Nutrition and Media."

MATERIALS
Media prepared in Exercises 12 and 14
Nutrient-broth cultures of *Escherichia coli*, *Pseudomonas aeruginosa*, *Staphylococcus aureus*, *Branhamella catarrhalis*, and *Streptococcus lactis*

PROCEDURE
1 Transfer one loopful of each culture into one tube of each of the following media: nutrient broth, glucose broth, yeast-extract broth (Exercise 12), and glucose-salts medium (Exercise 14).
2 Make a stab inoculation of each culture into the nutrient-agar deep tubes (Exercise 12).
3 Incubate all inoculated tubes at 35°C for 48 h.

Evaluation of Media to Support Growth of Bacteria

RESULTS

1 Observe each of the tubes of liquid media for amount of growth (turbidity). Record the results in the chart provided. Use the following scheme for recording the amount of growth:

+++ = very turbid
++ = moderately turbid
+ = faintly turbid
0 = no turbidity

GROWTH OF VARIOUS BACTERIA ON DIFFERENT MEDIA				
Bacterial species	Nutrient broth	Glucose broth	Yeast-extract broth	Glucose-salts
Escherichia coli				
Pseudomonas aeroginosa				
Staphylococcus aureus				
Branhamella catarrhalis				
Streptococcus lactis				

2 Observe the agar deep-stab cultures, and make a sketch of the amount of growth along the line of inoculation.

E. coli   *P. aeruginosa*   *S. aureus*

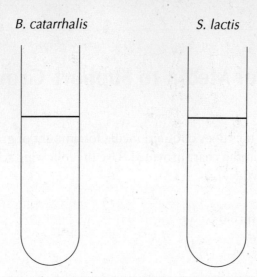

QUESTIONS 1 From the results you obtained, which of the species used in this exercise is the most fastidious or nutritionally demanding?

2 As this exercise was performed, would it be possible to determine whether any of the organisms used are autotrophs? Explain.

3 If a large amount of inoculum (several loopfuls) were used to inoculate each of the media, might this affect the results? Explain.

4 What type of instrument is available for measurement of bacterial growth on the basis of turbidity? What advantages are associated with this method of measuring growth?

EXERCISE 16

Selective, Differential, and Enriched Media

OBJECTIVE
To learn the advantages in the use of selective, differential, and enriched media in the laboratory cultivation of microorganisms.

OVERVIEW
Many special-purpose media are available to the microbiologist, the use of which facilitates the isolation of some specific group or type of microorganisms usually present with other species, that is, a mixed culture.

You will be using the following media:

Enriched media: The addition of blood or extracts of yeast or plant or animal tissues to nutrient broth or agar provides additional nutrients so that the medium will support the growth of fastidious heterotrophs, such as pathogens.

Selective media: Certain chemical substances will prevent growth of one particular group of bacteria without inhibiting others. When these chemicals are incorporated into the agar medium, the isolation of one particular group of bacteria is facilitated.

Differential media: The incorporation of certain reagents or chemicals into media may result in a kind of growth or change, after inoculation and incubation, which permits the observer to differentiate between types of bacteria.

BLOOD-AGAR PLATES
A variety of bacteria produce *hemolysins*, substances which bring about hemolysis of red blood cells of higher animals. Two general types of hemolytic reactions are observed: In α *hemolysis*, the bacterial colony is surrounded by a zone of greenish discoloration. Microscopic examination of the green zone reveals the presence of many discolored corpuscles due to the formation of a green substance, which is an iron-containing derivative of hemoglobin, possibly formed by reduction. In β *hemolysis* the zone around the colony is clear or uncolored. In β-hemolytic zones corpuscles cannot be found microscopically. The clearing around the colonies is the result of destruction of the red blood cells and the diffusion of the hemoglobin into the medium. *Streptococcus faecium* will give α hemolysis, and *S. equi* will give β hemolysis. The hemolytic reaction of *E. coli* is variable, depending on the strain. [See PLATES 7 and 8.]

DESOXYCHOLATE MEDIUM
This is both a selective and a differential medium which greatly facilitates the isolation of enteric bacilli (like *E. coli*) from fecal samples. Inhibition of Gram-positive bacteria is brought about by the pure chemicals sodium citrate and sodium desoxycholate (which occurs in bile). Sodium desoxycholate is a surface-active detergent and acts by destroying the integrity of the cell membrane by disrupting the interactions between membrane proteins and lipids. Citrate is required to potentiate the action of desoxycholate. The degree of inhibition can be more accurately controlled than when materials of unknown or variable composition, such as bile, are employed for this purpose. Differentiation of enteric bacilli is obtained by their formation of red colonies due to their ability to ferment lactose, that is, they produce acid which reacts with neutral red indicator.

PHENYLETHYL ALCOHOL MEDIUM
This is a selective medium for Gram-positive bacteria, such as streptococci and other micrococci. It is especially useful where specimens are contaminated with Gram-negative bacteria, since phenylethanol effectively inhibits this group. Phenylethanol interferes with both DNA replication and membrane integrity. The biosynthetic impairment may be secondary to the effect on the membrane to which the DNA-synthesizing machinery is attached. Evidently

the Gram-negative cell is more sensitive to the compound; the differential response probably resides in the cell-wall structures of Gram-positive and Gram-negative cells.

REFERENCES
MICROBIOLOGY, Chap. 5, "Nutritional Requirements and Microbiological Media."
MICROBIOLOGY, Chap. 6, "Cultivation and Growth of Microorganisms."
MM, Chap. 8, "Enrichment and Isolation."

MATERIALS
1 Petri plate of each of the following media: desoxycholate agar, phenylethyl alcohol agar, and blood agar
Nutrient-broth cultures of *Escherichia coli*, *Streptococcus faecium*, and *Streptococcus equi*

PROCEDURE
1 Label the name of the medium in each Petri dish clearly.

2 With a marking pen, mark off the bottom of each Petri dish of medium into thirds, and enter the name of each species in this area. Streak each sector with the appropriate culture as illustrated in FIGURE 16.1.

3 Label each plate with your name and date, and place the plates, inverted, in the 35°C incubator for 24 h.

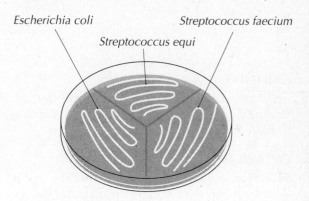

FIGURE 16.1
Streak inoculation of selective and differential media.

NAME _____

16

Selective, Differential, and Enriched Media

RESULTS

1 Examine carefully the growth on each plate, and fill in the results in the chart that follows. Record the relative amount of growth along the streak inoculation, that is, abundant, moderate, meager, or none.

2 Describe in the chart provided the appearance of the growth as well as the appearance of the medium adjacent to the growth.

Medium	Bacterial species	Amount of growth	Description of growth
GROWTH OF BACTERIA ON SELECTIVE, DIFFERENTIAL, AND ENRICHED MEDIA			
Desoxycholate agar			
Phenylethyl alcohol agar			
Blood agar			
+++ = abundant growth; ++ = moderate; + = meager; 0 = none.			

QUESTIONS 1 For what practical purposes, for example, in what routine laboratory procedures, would you use (a) desoxycholate medium, (b) phenylethyl alcohol medium, and (c) blood-agar medium?

2 Give one example of (a) an enriched medium, (b) a selective medium, and (c) a differential medium, other than those used in this exercise.

3 Might a single medium function for both differentiation and enrichment? Selection and differentiation? Explain.

PURE-CULTURE TECHNIQUES

A prerequisite to the characterization of a microbial species is that it be available for study as pure culture. Recall from Exercise 2 that although mixed cultures exist in natural environments, it is impossible to characterize individual species. The term **pure culture** denotes that all the cells in the culture had a common origin and are simply descendants of the same cell. It is possible to obtain a pure culture by transferring a single cell to a sterile medium. This can be accomplished by using a micromanipulator in conjunction with a microscope. However, indirect methods are almost always used to obtain a pure culture from a mixture of bacteria; for example, agar plate cultures are inoculated in such a way that isolated colonies develop.

The assumption is made that the microbial population of a colony develops from a single cell and hence represents a pure culture. This may not always be the case because some cells tend to clump together and cannot be separated easily. Therefore, it may be necessary to examine what is presumed to be a pure culture by additional cultural and microscopic tests. For example, one can restreak the colony to get isolated colonies that give the same colonial morphology; Gram-stained cells from one colony should look alike.

Actually the term **axenic culture** may be more appropriate than *pure culture* because an axenic culture is one in which an organism is grown in an environment free of any other living organism. Essentially, the term *pure culture* implies genetic purity, while *axenic culture* does not.

Once a pure-culture isolation has been made, it is desirable to maintain the culture, without change in its characteristics, in a viable condition for varying periods of time ranging from weeks to years. For short-term preservation of a culture (generally between 1 and 3 months), one simply makes periodic transfers to a fresh medium, for example, *nutrient-agar slants,* which are test tubes containing molten agar that have been left to solidify at an angle, or *cystine-trypticase-agar deep tubes.* Such cultures, with screw caps to minimize desiccation, are stored in the refrigerator at 4 to 10°C. Long-term preservation of a pure culture is best accomplished by **lyophilization,** that is, the culture species is dehydrated while in a frozen condition and then sealed under vacuum.

Transferring a pure culture of bacteria from one nutrient medium to another must be done without introducing unwanted microorganisms, or **contaminants.** This technique of transfer without the introduction of contaminants is carried out **aseptically.** The observance of aseptic technique must become second nature to any microbiologist. (Review Exercise 3.)

EXERCISE 17

The Streak-Plate Method for Isolation of Pure Cultures

OBJECTIVE
To learn how to streak bacteria on the surface of an agar medium to obtain isolated colonies.

OVERVIEW
Practically all specimens of material obtained from natural environments contain a mixed population of microorganisms. Before one can make a detailed study of the characteristics of the individual species constituting the mixture, it is imperative that each species be isolated in pure culture. The *streak-plate technique* provides a simple and practical procedure for this purpose [FIGURE 17.1]. It is essentially a dilution technique that spreads a loopful of culture over the surface of an agar plate. Although there are many ways to streak a plate to isolate colonies, the "four-way method" is described here [FIGURE 17.2].

REFERENCES
MICROBIOLOGY, Chap. 6, "Cultivation and Growth of Microorganisms."
MM, Chap. 8, "Enrichment and Isolation."

MATERIALS
4 sterile Petri dishes
4 tubes nutrient agar
Diluted broth cultures of *Serratia marcescens*, *Micrococcus luteus*, and *Arthrobacter globiformis*
Mixed culture of preceding bacteria

PROCEDURE
1 Pour each tube of melted and cooled (45 to 50°C) nutrient agar into separate Petri dishes.
2 Allow the agar to become firmly solidified. **Never attempt to streak a plate until the medium is firm.**

FIGURE 17.1
Technique for streaking an agar medium for isolating colonies.

3 Streak each of the three bacterial suspensions on a separate plate as shown in FIGURE 17.2A. To begin, streak the initial inoculum over an area corresponding to 1. Flame the loop, and then make a single streak through area 1 to the side of the plate, streaking several times through area 2. Do not allow the loop to reenter area 1. Again flame the loop, make a single streak through area 2 to the side of the plate, and continue making streaks in area 3, being careful not to reenter area 2. Repeat the process for area 4. In each step, after the loop has been flamed, it is advisable to cool the loop in the air briefly prior to the streaking procedure.

4 The fourth plate is to be streaked with a mixture of all three species. Place a small loopful of the mixed-culture suspension on this plate in position 1. Proceed to streak this plate as for other cultures.

5 Label the plates with your name and date, and incubate them at 25°C, or room temperature, for 48 h. **An inoculated plate is always incubated in an inverted position to prevent condensation from falling onto the surface and interfering with discrete colony formation.**

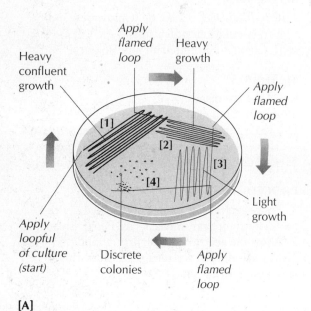

[A]

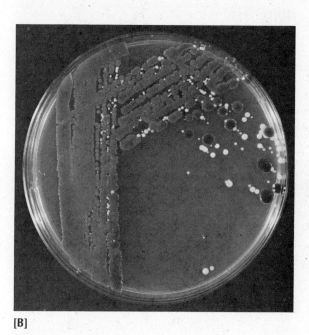

[B]

FIGURE 17.2

[A] Four-way streak-plate inoculation and [B] resultant isolated colonies.

NAME _____

The Streak-Plate Method for Isolation of Pure Cultures

RESULTS

1 Examine each of the streak plates, and make a sketch indicating the distribution of growth in the circles provided.

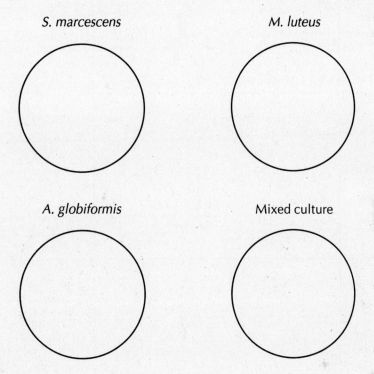

2 Describe the colonies that appear on the plates—that is, size, color, elevation—in the chart provided.

3 Prepare and examine Gram-stained smears from each of the colonial types. Enter your observations in the chart provided on the following page.

COLONIAL MORPHOLOGIES AND THEIR GRAM REACTIONS		
Bacterial species	Description of colonies	Cell morphology and Gram reaction of bacteria in each colony type
Serratia marcescens		
Micrococcus luteus		
Arthrobacter globiformis		
Mixture of species		

QUESTIONS   **1**   Define the following terms:

Colony:

Pure culture:

Axenic culture:

Mixed culture:

Normal flora:

Contaminant:

2   What is the rationale for assuming that a colony represents a pure culture?

3   List several points of technique that should be carefully observed when streaking a plate.

4 Why are Petri dishes incubated in an inverted position?

5 When does one use culture-medium slants, deeps, or broths?

6 When is a straight transfer needle preferable to a loop transfer needle for the inoculation of bacteria?

7 In streaking a pure culture for isolated colonies, the resulting colonies are of different sizes. How can this be explained?

EXERCISE 18

The Pour-Plate Method for Isolation of Pure Cultures

OBJECTIVE
To learn how to obtain isolated colonies by the pour-plate technique.

OVERVIEW
Isolated colonies can also be obtained from a mixed population of bacteria by diluting the specimen in a cooled (45 to 50°C) fluid-agar medium which is then plated. Thus, colonies develop throughout the medium and not just on the surface. Because the magnitude of the microbial population is generally not known beforehand, it is necessary to make several dilutions to ensure obtaining at least one plate with colonies that are distinctly separated on, or in, the agar medium. The plates with too many or too few colonies can be discarded. As in the streak-plate method, a colony is assumed to develop from a single cell, and the cells reproducing from it represent a pure culture. Since a colony may develop from a clump of cells, which may all be or may not be the same type of cells, the term "colony-forming unit" was coined to replace "colonies." Therefore, a suspension of bacterial cells may be said to contain 5×10^3 colony-forming units, or CFUs, per milliliter instead of colonies per milliliter.

In this exercise, the loop-dilution procedure is used because it is a simple and convenient way of obtaining isolated colonies.

NOTE: In mixed populations of microorganisms from natural sources, for example, a soil sample or a sample of plaque from teeth, some microbial species are present in much greater numbers than others. It is apparent that microbial species present in small numbers can be diluted out to extinction so that they are not detected on plates containing many colonies of the dominant species. This is an inherent limitation of dilution techniques in isolating for microorganisms. But this may be overcome by use of selective antibiotics that suppress the growth of the dominant species and permit those species with smaller numbers to develop into colonies.

REFERENCES
MICROBIOLOGY, Chap. 6, "Cultivation and Growth of Microorganisms."
MM, Chap. 8, "Enrichment and Isolation."

MATERIALS
1 10-ml tube of sterile saline solution
4 tubes nutrient agar
4 sterile Petri dishes
Mixed bacterial suspension: freshly prepared mixture of nutrient-broth cultures of *Staphylococcus aureus* and *Escherichia coli*

PROCEDURE
1 Melt the tubes of nutrient agar, cool them to between 45 and 50°C in a water-bath, and maintain them at this temperature during the dilution manipulations described below. Refer to FIGURE 18.1 as you proceed.
2 With a sterilized inoculating loop, transfer one loopful of the bacterial suspension into a tube of sterile saline. Rotate this tube back and forth between the hands (10 times) to ensure uniform distribution of bacteria.
3 Transfer two loopfuls of the suspension in saline solution to a tube of melted nutrient agar. Label this tube "1." Rotate this tube in the same manner as in step 2.
4 Transfer two loopfuls from tube 1 into another tube of nutrient agar labeled "2." Then pour the contents of tube 1 into a Petri dish labeled "1."
5 Rotate tube 2 to distribute the inoculum, and transfer two loopfuls from tube 2 to the tube of

agar labeled "3." Pour the contents of tube 2 into a Petri dish labeled "2."

6 Rotate tube 3 to distribute the inoculum, and pour its contents into a Petri dish labeled "3."

7 Pour a tube of sterile nutrient agar into a Petri dish labeled "Control" (Petri dish 4).

8 Label plates with your name and date, and incubate all plates (inverted) at 35°C for 48 h.

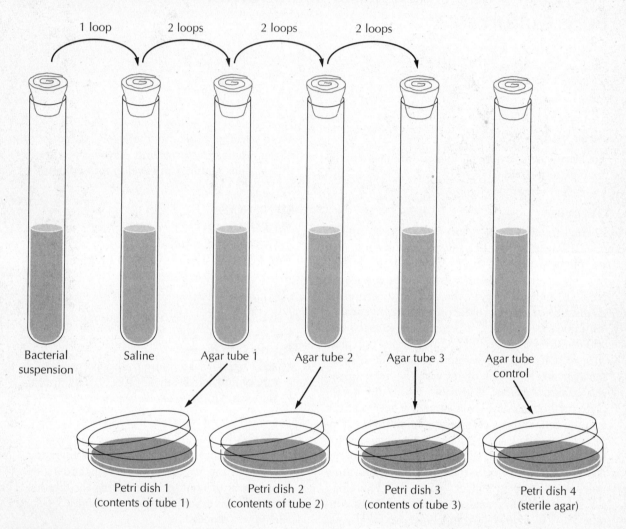

FIGURE 18.1
Dilution technique in pour-plate procedure.

NAME _____

The Pour-Plate Method for Isolation of Pure Cultures

RESULTS 1 Make a sketch from a selection of the three inoculated plates to illustrate the amount of growth, distribution of growth, and size of colonies.

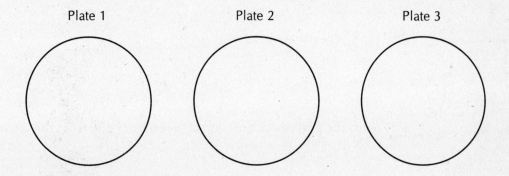

2 Prepare and examine Gram-stained smears from each of the two types of colonies that grew on the surface of the medium. Describe the appearance of surface colonies in the chart provided.

SURFACE COLONIES AND THEIR CELL CHARACTERISTICS		
Species	**Gram-stain reaction and morphology**	**Description of surface colonies**
Staphylococcus aureus		
Escherichia coli		

133

QUESTIONS 1 Why do both surface and subsurface colonies appear in a pour plate?

2 Would anaerobic bacteria grow in a pour plate as inoculated and incubated in this experiment? Explain.

3 What is the purpose of the control plate (Petri dish 4) in this exercise?

4 Why should the fluid-agar medium be cooled to between 45 and 50°C in the exercise?

5 How do the colonies on the surface of the pour plate differ in appearance from the colonies embedded in the agar?

6 Compare the advantages and disadvantages of the streak-plate method with the pour-plate method for obtaining pure cultures of microorganisms.

EXERCISE 19

Enumeration of Bacteria by the Plate-Count Technique

OBJECTIVE
To learn how to obtain a viable count of bacteria in a suspension by the plate-count technique.

OVERVIEW
One of the routine procedures for determining the bacteria content of many different materials is the *plate-count technique*. A typical plate-count procedure is a combination of a serial dilution and the use of a suitable growth medium to detect colonies from aliquots of the dilutions. This procedure is based upon the assumption that each viable cell will develop into a colony; hence, the number of colonies on the plate reveals the number of individual organisms contained in the sample that were capable of growing under the specific conditions of incubation and culture. However, a clump of cells would also give rise to one colony. This is the reason that it is necessary to mix dilutions well in order to break up any clumps present. Since this is difficult to do with some species of bacteria, microbiologists use the term *colony-forming units (CFU)* instead of *colonies* in a quantitative plate count. The plate count is an indirect measurement of cell concentration and provides results based on *viable* cells, that is, cells able to divide and form colonies. This is why the procedure is called the *viable plate-count technique*. When the number of viable bacteria in a sample is reported, it is expressed as the "number of bacteria (colonies)/ml" or "number of CFU/ml" of the sample.

GENERAL TECHNIQUE
Dilution of Specimen
The specimen is diluted in order that one of the final plates will have between 30 and 300 colonies. Numbers of colonies within this range per plate give the most accurate approximation of the microbial population for statistical purposes. Samples containing fewer than 30 cells per 0.1 ml diluent are subject to large fluctuations in numbers or sampling errors; plates containing more than 300 colonies have coincidence limitation, that is, overlapping colonies. Since the magnitude of the microbial population in the original specimen is not known beforehand, a range of dilutions must be prepared and plated to obtain one within the required colony range.

The initial dilution is usually prepared by placing 1 ml or 1 g into a 9- or 99-ml dilution blank. *Dilution blanks* are tubes or bottles containing a known volume of sterile diluent. The *9-* or *99-ml blank* refers to the amount of sterile diluent, usually a physiological saline or buffer solution, in the container. This dilution is shaken vigorously to obtain uniform distribution of organisms and to break up clumps of cells. Further dilutions are made by pipetting measured aliquots, usually 1 ml, into additional dilution blanks. Each dilution must be thoroughly shaken before removing an aliquot for subsequent dilution.

The procedure is illustrated in FIGURE 19.1.

Pipetting of solutions
There was a time when mouth-pipetting was standard practice in the microbiology laboratory. With the new awareness of the dangers of opportunistic infections, even normally innocuous microorganisms are treated with respect. There are now commercially available pipetting aids to bypass use of the mouth in pipetting [FIGURE 19.2]. These include Portapet pipettors, the pipette pump filler-dispenser, bulb-type safety pipette fillers, and the electrically powered or portable Pipet-Aid. The use of a Pipet-Aid is demonstrated in FIGURE 19.3.

NOTE ON PIPETTING ERRORS: When measuring fluid volumes, position your eyes so that they are horizontal with the top of the fluid column in the pipette. In this way you avoid parallax errors that come about from misalignment of the *meniscus*, the curved air-liquid interface at the top of the fluid in the pipette, with the graduated line on the pipette. The volume in the pipette is properly read when the bottom of the meniscus is positioned directly opposite the line on the pipette.

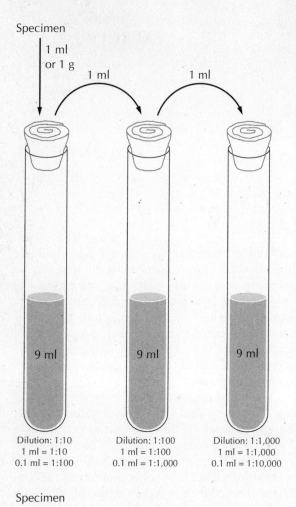

FIGURE 19.1
Use of 9- and 99-ml dilution blanks.

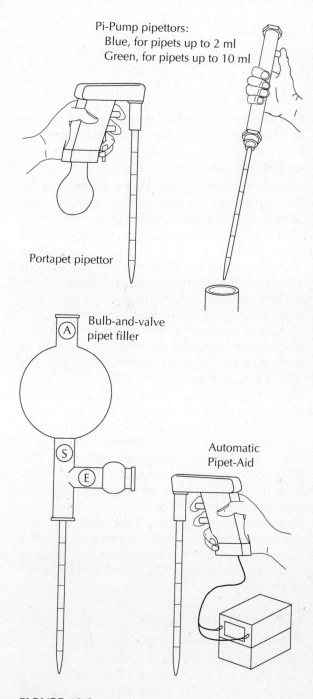

FIGURE 19.2
Pipette-filling and dispensing devices.

Plating of dilution

Sterile Petri dishes should first be labeled with your name, the specimen, and the dilution. From the appropriate dilution blank(s), 1 or 0.1 ml is pipetted into a Petri dish.

NOTE: If 0.1 is plated, the dilution factor is increased 10 times.

Samples from each dilution are plated in duplicate or triplicate. Statisticians tell us that, for greatest accuracy, triplicate plating of each dilution should be carried out, and then the average of these three counts should be used.

FIGURE 19.4 illustrates the manner of delivering the *inoculum,* material from dilution blank, into the Petri dish.

Addition of plating medium to Petri dish: The pour-plate method

The agar medium to be used must first be melted and then cooled in a water-bath adjusted to 45°C. Add approximately 25 ml of the cooled liquid medium to each Petri dish. Immediately thereafter, rotate the plate gently to distribute the inoculum throughout the medium. This is the *pour-plate method.*

Alternate spread-plate method

In the *spread-plate method,* the inoculum or dilution sample is placed on top of a prepoured nutrient-agar plate and is spread with a sterile glass spreader. This is accomplished by pipetting the appropriate dilution volume (1.0 or 0.1 ml), also called an *aliquot,* and placing it on the surface of the agar medium.

NOTE: It is best that these prepoured plates be left standing at room temperature for a few days, so that they dry out somewhat. If this is done, the medium can absorb the dilution volume of 1.0 ml quite rapidly; freshly prepared plates of medium should not be used with this volume of inoculum.

The delivered inoculum on the medium surface is spread uniformly by means of a sterilized glass spreader, any bent glass rod that has been first placed in ethanol and has had the excess alcohol flamed off [FIGURE 19.5]. A turntable is also helpful in ensuring uniform spreading.

> **CAUTION:** Be careful with burning ethanol. You will be supplied with 70% ethanol in a beaker. Do not allow the flaming alcohol to drip on your skin, your clothing, or back into the beaker, where it will burn into a big flame. If it drips back into the beaker, cover the beaker with a piece of wet cloth. Why will the flame go out?

Incubation of Plated Specimens

Upon solidification of the medium in the pour plates, *invert* the plates and place them in an incubator. In the case of spread plates, once the surface of the medium is dry, they too can be inverted and similarly incubated.

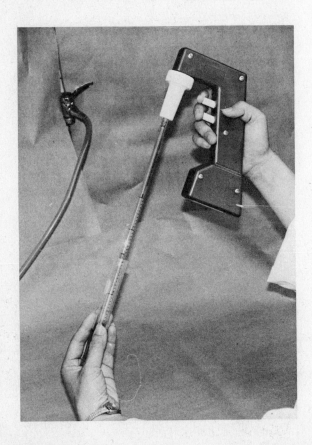

FIGURE 19.3
Demonstration of the use of a Pipet-Aid.

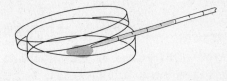

FIGURE 19.4
Technique for addition of inoculum into Petri dish.

FIGURE 19.5
Use of a glass spreader and turntable for the spread-plate technique.

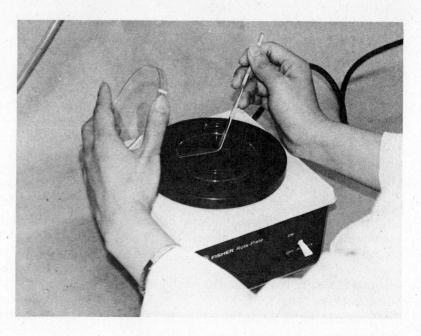

Counting the Colonies
After the prescribed period of incubation, select a plate that contains a number of colonies in the 30 to 300 range. Make an accurate count of these colonies by placing the plate on the platform of a colony counter, for example, a Quebec colony counter. This instrument facilitates the counting process, since the colonies are illuminated, magnified, and seen against a ruled background.

Calculation of Plate Count
The number of colonies counted on a plate multiplied by the dilution of the specimen contained on that plate equals the bacteria, or CFU, count per milliliter (or gram) of the specimen. For example, if 180 colonies were counted on the 1:1000 dilution of a milk sample, the calculation is:

180 × 1000 = bacteria, or CFU, per milliliter of milk sample

It is generally desirable to make triplicate platings of each dilution and to average the resulting counts.

REFERENCES
MICROBIOLOGY, Chap. 6, "Cultivation and Growth of Microorganisms."
MM, Chap. 11, "Growth Measurement."

MATERIALS
Bacterial suspension (faintly turbid suspension of *Escherichia coli*)
4 99-ml dilution blanks
9 sterile Petri dishes (or prepoured nutrient-agar plates)
2 bottles of nutrient agar
7 sterile 1-ml pipettes

PROCEDURE
1 Prepare the following dilutions from the bacterial suspension supplied:

1:1,000,000 (10^{-6}), 1:10,000,000 (10^{-7}), and 1:100,000,000 (10^{-8})
(Note the use of exponential notations for the dilutions.)

Plate each of these dilutions in triplicate. Either the pour-plate or the spread-plate technique may be used depending on what kind of Petri dishes are supplied, that is, whether or not they are prepoured nutrient-agar plates.
2 Following their incubation at 37°C for 24 to 48 h, observe the plates and describe the distribution of colonies from each dilution.
3 Select plates made from the appropriate dilution, make a plate count, and calculate the number of viable bacteria, or CFU, per milliliter of the original suspension.

NOTE: Diluents usually used are a physiological saline (0.85% NaCl) or a phosphate buffer saline (PBS). PBS has the following composition: 4.2 g NaCl, 3.17 g KH_2PO_4, 3.58 g $Na_2HPO_4 \cdot 12H_2O$, and 1000 ml distilled water; the pH will be 6.5.

The median of a group of measurements is the middle measurement, if there is one. Thus, the median for the group 1, 3, 4, 5, 7, 8, and 9 is 5. For the group 2, 6, 8, and 9, the median is not defined but may be taken as 7, which is halfway between the two middle numbers 6 and 8. Thus, if all the measurements of a series are numbered in order of magnitude, the magnitude of the measurement which is halfway up the series is called the *median*. The class median may be found conveniently by pairing off the largest and smallest measurements and repeating the process until only one or two are left. If only one is left, it is the median; if two are left, the median is usually taken as the measurement halfway between them.

The class *mean* in the exercise is the arithmetic average of the class results. Of course, if the class is very large, only a section of the class is considered by each student.

The average error between a series of plates, such as in the triplicate plates of a dilution, is calculated as follows:

Plate number	Plate count	Difference from average
1	120	9
2	130	1
3	137	8
Total	387	18
Average	129	6

Therefore, average error is ±6; the plate count of the particular dilution is 129 ± 6.

Where colonies per plate appreciably exceed 300, count colonies in portions of plate representative of colony distribution and estimate based on the total number per plate. Or else use the symbol TNTC, for "too numerous to count."

Numerous aids for plate counts are obtainable from commercial sources. For example, automatic colony registers (a probe contact with a surface registers a count) and Petri-plate rotators for surface spreading are interesting instruments to use or demonstrate.

Enumeration of Bacteria by the Plate-Count Technique

RESULTS

1 Describe the plates from each dilution in the chart provided.

DISTRIBUTION OF COLONIES IN DILUTIONS EMPLOYED	
Dilution	Description of colony distribution
10^{-6}	
10^{-7}	
10^{-8}	
TNTC = too numerous to count.	

2 Record below the colony count, or CFU, of each plate in the triplicate series counted. Calculate the average and from this, the viable population of the original suspension per milliliter, under the particular conditions of culture and incubation of the plate-count technique.

Plate counts of 10^{-6}: _____ 10^{-7}: _____ 10^{-8}: _____

Your CFU/ml:

3 Record below the results, that is, the viable population of the suspension, of other class members. From these data, determine the *mean* and *median results*.

CFU of class members:

Class mean: _____

Class median: _____

QUESTIONS

1 Did each of the plates in the triplicate series counted have the same number of colonies? How might differences be accounted for?

2 How might variations in the results obtained by various classmates be explained?

3 What dilutions would you consider appropriate if you were to perform a plate count on a 24-h nutrient-broth culture of *E. coli*?

4 Compute the average error in the triplicate plates of the dilution containing 30 to 300 colonies.

EXERCISE 20

Turbidity Measurement of Broth Cultures

OBJECTIVE
To estimate the amount of cells in a suspension of bacteria by turbidity measurement.

OVERVIEW
It is possible to follow the growth of a microbial culture by measuring a number of parameters, for example, growth can be followed by viable plate counts (Exercise 19). Other ways include direct microscopic counts of cell numbers and **turbidity**, ("cloudy") measurements. Turbidity is the effect of light scattering by a colloidal suspension. Many microbial suspensions are colloidal suspensions because the cells do not settle out of suspension quickly. These suspended cells scatter light, producing turbidity. As microbial cultures grow, the number of cells increases, accompanied by a proportional increase in turbidity.

Microbiologists follow changes in turbidity with an instrument called a **spectrophotometer,** which measures the amount of light that passes through a liquid medium. The amount of transmitted light decreases as the cell population and hence turbidity increases.

The working principle of a typical spectrophotometer is as follows and is shown also in FIGURE 20.1. A lamp produces light that is split into various wavelengths by a device called a *diffraction grating*. A wavelength of light is chosen that matches closely the color of the medium, for example, 620 nm is chosen for nutrient-broth medium because the color of the medium transmits most of the light at this wavelength. The light passes through a *cuvette*, which is an optically clear test tube that holds the liquid sample. The transmitted light that passes through the sample is collected by a *phototube*. The

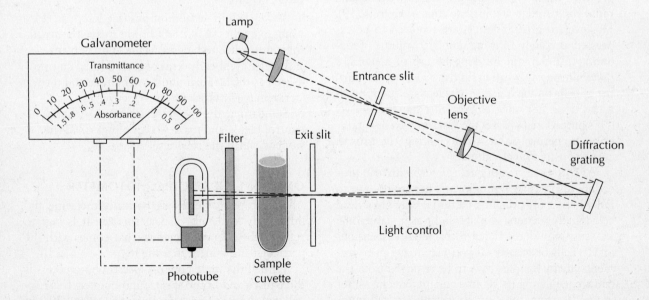

FIGURE 20.1
The working principle of a spectrophotometer.

quantity of light entering the phototube is converted to electrical energy and is indicated on a *galvanometer*. A turbid culture scatters more light than the sterile broth (control) and therefore transmits less light to the phototube, thus producing a lower reading on the galvanometer. The more turbid the culture, the lower the percent **transmittance.** Transmittance is measured as a percentage of the control cuvette.

However, turbidity changes as a result of cell reproduction are not usually measured as transmittance values. Instead, measurements in **optical density (OD)** are made. This is because OD is *directly* proportional to cell (biomass) concentration; as cells increase in number, so does the OD. Recall that bacterial cells multiplying by binary fission increase their numbers logarithmically. Optical density is also a logarithmic value as may be seen by its relationship to transmittance:

$$OD = 2 - \log (\%T)$$

OD is also referred to as *absorbance* (and may be so inscribed on the measuring scale) because spectrophotometers are often used to measure the concentration of color-absorbing solutes in solution. Absorbance values are exactly the same as OD values. When you plot OD values versus time in following the growth of a culture, you should get a straight line expressing the logarithmic nature of active unicellular growth.

Turbidimetric measurements to follow the growth of a microbial culture are convenient and rapid compared to viable plate counts (Exercise 19). However, if growth results are required to be expressed as colony-forming units per volume of medium, this can still be done by use of a *standard curve* plotting OD values on the Y axis versus number of cells on the X axis. A viable plate count needs to be done just once to plot this standard curve. This procedure bypasses the work involved in gathering numerous viable counts during the growth curve of a microbial culture.

As with every technique, the turbidimetric procedure also has its limitations. There is no way to tell whether the cells in a culture are dying; a dead cell usually scatters as much light as a viable one. Its sensitivity is restricted to bacterial suspensions of 10^7 cells or greater. If you take many measurements, many samples have to be removed from the culture, depleting its volume and introducing other inherent errors, for example, change in cell numbers. Besides, every time the culture is opened for sampling, there is a chance of contaminating it. Some microbiologists overcome this by using a flask with a cuvette side-arm that can be inserted directly into a spectrophotometer to read the turbidity of a culture without having to remove an aliquot. Others use multiple tubes of a culture containing the same amount of inoculum cells. All tubes are identical and are incubated in the same manner; tubes are removed at intervals as required for turbidity measurements and are discarded.

In this exercise you will learn how to make turbidity measurements of a bacterial culture diluted serially.

REFERENCES
MICROBIOLOGY, Chap. 6, "Cultivation and Growth of Microorganisms."
MM, Chap. 11, "Growth Measurement."

MATERIALS
1 18-h nutrient-broth culture of *Serratia marcescens*
6 capped tubes, each containing 9 ml sterile nutrient broth
5 1-ml sterile pipettes
7 5-ml sterile pipettes
1 spectrophotometer, such as Bausch and Lomb Spectronic 20
7 clean cuvettes

PROCEDURE
1 Make 10-fold serial dilutions (see Exercise 19) using five tubes of nutrient broth. Leave the sixth tube untouched to be used as the control blank.
2 Pipette aseptically 5 ml of each diluted culture as well as the original undiluted culture and the sterile nutrient broth into seven cuvettes.
3 Standardize the spectrophotometer with the control blank, and make OD readings of all the suspensions in the cuvettes.

ON USING THE SPECTROPHOTOMETER
All spectrophotometers are basically the same and differ only in details of construction. It is best to refer to the operation manual that comes with the particular instrument in using it.

One of the most popular spectrophotometers is the Bausch and Lomb spectrophotometer. It is rugged and simple to operate. Refer to FIGURE 20.2 in following the instructions of operation.

1 Turn on the instrument by rotating clockwise the power/zero control knob. Change the wavelength control knob to 620 nm. Allow the instrument to warm up for about 20 min.

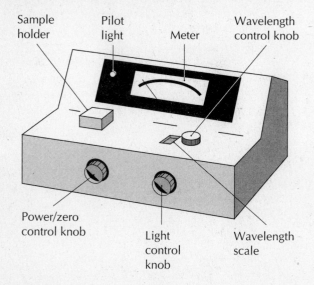

FIGURE 20.2
A typical spectrophotometer.

2 With the sample holder empty and closed, rotate the power/zero control knob to adjust the galvanometer needle to "0% Transmittance." (This is the dark-current adjustment; the galvanometer needle indicates that no light is being transmitted to the phototube.)

3 Align the index line of the sample holder with the vertical line etched on the cuvette holding sterile nutrient broth (the blank). Close the lid of the sample holder.

Adjust the galvanometer needle with light control knob until it reads "100% Transmittance." The instrument is now "standardized." Remove the blank cuvette.

4 Measure the OD (absorbance) of all the cuvettes containing bacteria by putting them into the sample holder individually and closing the sample holder lid. If you wish to ensure even suspension of the bacterial cells in the broth before measuring the OD, you may seal the cuvette with Parafilm (wax paper) and mix the suspension by gently inverting it several times. You may wish to standardize the instrument between readings in order to maintain accuracy. Record your results in the chart provided in the results section.

NAME _____

Turbidity Measurement of Broth Cultures

RESULTS 1 Record the OD of each sample in the chart provided.

Dilution	OD READINGS OF DILUTIONS
	OD 620 nm
0 (Original)	
1:10	
1:100	
1:1000	
1:10,000	
1:100,000	
Blank	

2 Plot the OD values (Y axis) versus bacterial dilution (X axis) and draw a line graph.

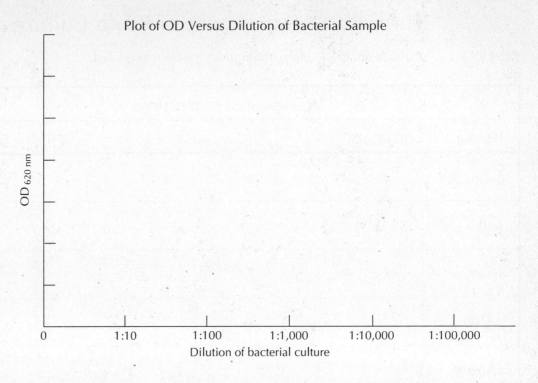

QUESTIONS 1 What kind of bacterial cells (culture) cannot be measured by turbidity?

2 Can you do turbidimetric measurements on a growing fungal culture in liquid medium? Explain.

3 What wavelength would you use to measure the growth of *E. coli* in a glucose-salts medium? Why?

EXERCISE 21

Anaerobic Culture Methods

OBJECTIVE
To learn how to grow anaerobic bacteria in the laboratory.

OVERVIEW
Molecular oxygen has a great influence on the growth of microorganisms. [See PLATE 9.] Thus, bacteria may be categorized with respect to oxygen in the following manner:

Strict aerobes: These bacteria have an obligate requirement for oxygen in their metabolism. Oxygen serves as the terminal electron or hydrogen acceptor in respiration.

Strict anaerobes: These bacteria cannot tolerate the presence of oxygen in their environment and will grow only in its absence. They grow by fermentation or anaerobic respiration mechanisms where organic compounds or inorganic ions serve as terminal-electron acceptors. Oxygen is toxic to the strict, or obligate, anaerobes probably because they lack or have low levels of enzymes like catalase and superoxide dismutase, which break down hydrogen peroxide and the superoxide radical, respectively.

Facultative microorganisms: When oxygen is present, these bacteria grow by respiration; but when oxygen is absent, they can utilize fermentation pathways or carry out anaerobic respiration.

Microaerophiles: These bacteria grow best when the oxygen concentration is less than the 21 percent usually found in the air. High oxygen concentrations are toxic to these organisms.

Not all anaerobes have the same requirement for anaerobiosis. Some anaerobic bacteria, such as *Clostridium perfringens* and *Bacteroides fragilis*, are moderately resistant to oxygen contact. But there are also anaerobic bacteria that cannot even tolerate a very brief exposure to air. Nor can some anaerobic bacteria tolerate toxic oxidation products, such as organic peroxides, formed when culture medium comes into contact with oxygen. The cultivation of anaerobic bacteria, therefore, requires special techniques.

Consequently, because of the properties of anaerobic bacteria, special **transport methods** have been designed to ensure survival of all anaerobes in a sample or specimen until it is processed in the laboratory. These methods essentially keep all the bacteria viable in an anaerobic system. The specimens should then be cultured promptly.

The *ideal* anaerobic culture method must meet the following three criteria:

1. Nonformation of toxic oxidation-reaction products when oxygen comes in contact with the culture medium during its preparation
2. Exclusion of air from the medium during incubation and cultivation
3. Maintenance of a low oxidation-reduction (O/R) potential in the culture system

The term ***O/R potential*** requires a brief explanation. In a medium, some organic molecules are in the oxidized state and others are in a reduced state. The proportion of oxidized to reduced molecules determines the relative state of oxidation or reduction of the medium. The overall oxidation-reduction state can be measured with an electrode, and its numerical value is the O/R potential. When organic compounds become more oxidized as a consequence of oxygen dissolving in the medium, the medium measures a *positive* O/R potential. When reducing agents dissipate dissolved oxygen and reduce medium components, the medium exhibits a *negative* O/R potential. Thus, the more negative the O/R reading, the more reduced is the medium. Strict anaerobes grow in a medium with low, or negative, O/R potential.

Reducing agents added to a medium to lower the O/R potential include cysteine or sodium thioglycollate. Cysteine is an amino acid that contains a sulfhydryl group (−SH) that easily donates its hydrogen to other organic compounds and reduces them in the process. Sodium thioglycollate also has a sulfhydryl group and is a good organic reducing agent; it also effectively removes dissolved oxygen.

Generally, culture in liquid media should never be used as the sole method for isolating anaerobes from samples. Recovery of anaerobes is generally poorer in liquid media than in solid ones. Furthermore, quantitation is not possible. However, for known pure cultures of some anaerobic bacteria, liquid media remain a convenient means of cultivation.

Two major methods of cultivating anaerobes are in common use in microbiology laboratories: the anaerobic chamber method and the anaerobic jar method.

THE ANAEROBIC GLOVE BOX OR CHAMBER

The *anaerobic chamber* or *glove box* (so called because the operator has to work through gloves) may be made of clear, flexible vinyl plastic. It is filled with oxygen-free gases (5% CO_2, 10% H_2, and 85% N_2) and is sufficiently large to house all the items necessary for carrying out anaerobic bacteriology, such as a small incubator or incubating shelves, a flameless inoculating-needle sterilizer, and even a microscope, if needed. In addition, the chamber contains thermostatically controlled fan-driven heatboxes, catalysts, and a desiccating agent. An anaerobic glove box is illustrated in FIGURE 21.1.

There are several advantages in using a glove box:

1 Specimens can be introduced into the glove box and processed under anaerobic conditions.
2 Media can be stored and reduced in the chamber so that their O/R potential is at a level which facilitates growth of anaerobes.

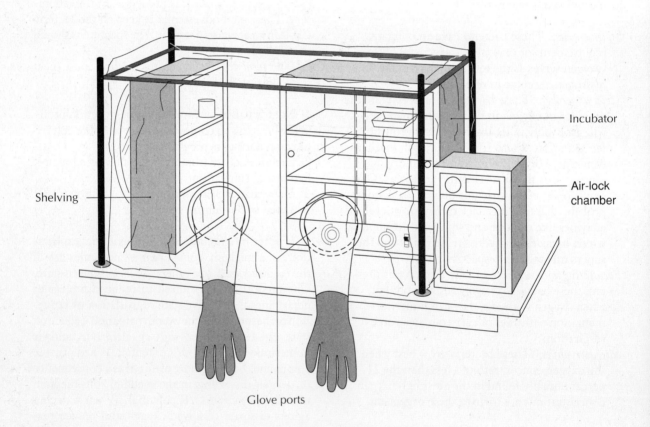

FIGURE 21.1
The anaerobic glove box.

3 During incubation cultures can be observed and subcultured at any time without exposing them to air.

THE ANAEROBIC JAR

The anaerobic jar method uses a polycarbonate jar that can be made anaerobic either by evacuating ambient air and replacing it with oxygen-free gas, or more commonly, by simply using disposable hydrogen plus carbon dioxide generator envelopes in the presence of cold palladium catalyst pellets. In replacing ambient air with a premixed combination of gases (5% CO_2, 10% H_2, and 85% N_2 for anaerobes; other mixtures for other kinds of bacteria), a special lid is required. One of the most convenient lids has a dual-valving system that facilitates evacuation of air from one valve and replaces it with any certified gas mixture [FIGURE 21.2]. In other words, with this type of jar, you can create your own choice of controlled, standardized atmospheres.

Laboratories engaged in research on anaerobes use the anaerobic chamber because it offers instant anaerobiosis and the capacity for large numbers of cultures. For small-scale work of individuals and where anaerobes are not so stringent, the anaerobic jar remains the method of choice.

REFERENCES

MICROBIOLOGY, Chap. 6, "Cultivation and Growth of Microorganisms."

MM, Chap. 6, "Physicochemical Factors in Growth."

MATERIALS

Anaerobic jar (BBL GasPak Anaerobic System, Almore Anaerobic System, or any other commercial source) or anaerobic glove box
2 nutrient-agar plates
3 nutrient-agar deep tubes
3 tubes nutrient broth
3 tubes cooked-meat medium
3 tubes thioglycollate broth with indicator
Cultures of *Bacillus subtilis* and *Staphylococcus aureus* in nutrient broth and *Clostridium sporogenes* in thioglycollate broth

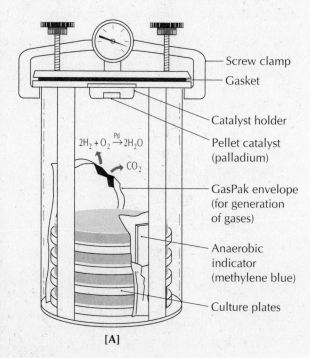
[A]

[B]

FIGURE 21.2
[A] Diagram of a typical anaerobic jar. [B] A novel anaerobic jar with the capacity for immediate gaseous exchange. *(Courtesy of Almore International, Inc., P.O. Box 25214, Portland, OR 97225-0214.)*

PROCEDURE

1 Melt the nutrient-agar deep tubes and cool them to between 45 and 50°C. Transfer one loopful of each culture into separate nutrient-agar deep tubes. Roll these tubes between the palms of your hands to distribute the inoculum throughout the medium before it solidifies.

2 Inoculate one tube of nutrient broth, cooked-meat medium, and thioglycollate broth with each of the three organisms.

3 Incubate all inoculated media aerobically at 35°C for 48 h.

4 With a marking pen, divide the bottom of the two nutrient-agar plates into three equal sections and write the name of each species in this area. Inoculate each sector with the appropriate cultures as illustrated in Exercise 16.

5 Label the plates with your name and date, and put one into a shared anaerobic jar or glove box. Incubate the other plate aerobically at 35°C for 48 h.

The instructor will show how the anaerobic jar is used after it is filled with plates. The conventional anaerobic jar, without a valving system, is designed to be used with a disposable hydrogen plus carbon dioxide generator envelope and eliminates the need for vacuum pumps, gas tanks, pressure regulators, or manometers. Each envelope contains one sodium borohydride tablet, one sodium bicarbonate tablet, and powdered citric acid. The jar employs a room-temperature catalyst system and therefore does not use any electrical connections or other means of heating the catalyst. The hydrogen released upon the addition of water to the envelope reacts with the oxygen in the presence of the catalyst to produce an anaerobic atmosphere. The carbon dioxide generated helps to support growth of organisms requiring it. To ascertain that anaerobic conditions have been created and maintained, a disposable anaerobic indicator, with methylene blue or resazurin, is often used.

In the same way, an anaerobic indicator with methylene blue may be used in the glove box. The instructor will show how to incubate your plate in the apparatus. After the plates are collected and positioned in a wire basket, they are placed in the air-lock chamber. The air-lock chamber is evacuated with a vacuum pump and filled with the nitrogen gas two times before introduction of the final gas mixture. The plates are then placed into the main chamber of the glove box where they may be set on a tabletop incubator or on the incubating shelves.

NAME _____

Anaerobic Culture Methods

RESULTS 1 Observe all media for the presence or absence of growth. Record results in the chart provided.

GROWTH OF THREE BACTERIAL SPECIES IN VARIOUS MEDIA			
Media	*Staphylococcus aureus*	*Bacillus subtilis*	*Clostridium sporogenes*
Nutrient-agar plate (aerobic)			
Nutrient-agar plate (anaerobic)			
Nutrient broth			
Cooked-meat medium			
Thioglycollate broth			
+++ = good growth; ++ = fair; + = poor; 0 = no growth.			

2 With a sketch, illustrate the distribution of growth of each organism in the nutrient-agar deep tubes.

S. aureus   *B. subtilis*   *C. sporogenes*

153

QUESTIONS

1 Describe the growth of each of the organisms used in this exercise according to its oxygen requirements.

2 Why will an anaerobe grow in thioglycollate broth even though the medium is exposed to atmospheric oxygen?

3 List three advantages and three disadvantages of the anaerobic glove box.

4 Explain why cooked-meat medium may be used to grow anaerobes.

EXERCISE 22

Cultural Characteristics

OBJECTIVE
To grow bacteria in the laboratory in order to identify their cultural characteristics; to learn how microbial cultural characteristics are recorded.

OVERVIEW
The appearance of the growth or mass of cells of bacteria that develops on various media comes under the heading of "cultural characteristics" [FIGURE 22.1]. Since various groups of bacteria manifest particular types of growth, this feature is useful as an adjunct in characterizing taxonomic groups.

REFERENCES
MICROBIOLOGY, Chap. 5, "Nutritional Requirements and Microbiological Media."
MM, Chap. 25, "Phenotypic Characterization."

MATERIALS
6 nutrient-agar plates
6 nutrient-agar slants
6 tubes of nutrient broth
6 nutrient-agar deeps
6 tubes of nutrient gelatin
Nutrient-agar slant cultures of *Streptomyces albus*, *Mycobacterium phlei*, *Bacillus subtilis*, *Pseudomonas aeruginosa*, *Micrococcus luteus*, and *Escherichia coli*.

PROCEDURE
1 Streak a plate from each culture so as to obtain isolated colonies (see Exercise 17).
2 Using the transfer needle, inoculate each organism into the remaining media in the following manner:
   a *Agar slants:* One streak up the middle of the slanted surface with a loop needle
   b *Broth:* Twirl the loop needle carrying the inoculum in the liquid
   c *Agar deeps:* Stab inoculation, puncture of the agar column from top to bottom with withdrawal of the straight needle through the same path
   d *Gelatin:* Stab inoculation as for agar deeps
   NOTE: The gelatin tubes must be maintained at a temperature near 20°C so that the medium will remain solid.
3 Incubate all the gelatin tubes at 20°C. The media other than gelatin inoculated with *S. albus* are to be incubated at room temperature (25°C). All other inoculated media are to be incubated at 35°C. The tubes should be observed after 48 h incubation and then further incubated for another observation between 4 and 7 days for any slow-growing culture. See FIGURE 22.2.

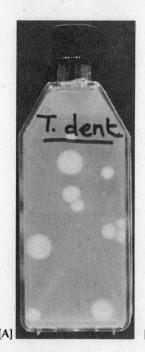

[A]   [B]

FIGURE 22.1
Colonial characteristics are sometimes important for the differentiation of species. [A] Discrete colonies of *Treponema denticola*. [B] Fluffy colonies of *Treponema socranskii*. (*Courtesy of Yu-Shan Qiu and E. C. S. Chan, McGill University.*)

FIGURE 22.2
Cultural characteristics of bacteria.

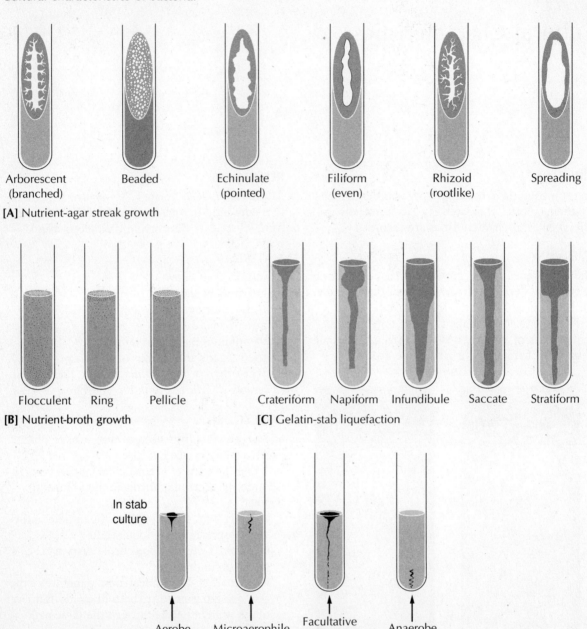

NAME _____

Cultural Characteristics

RESULTS Make drawings and record descriptions of growth for each species on each medium as called for in the charts provided on pages 158–162. See FIGURE 22.2 and PLATES 10 and 11 to facilitate the descriptions of the appearances of growth. (See also the description of colonial characteristics in Exercise 1.)

QUESTIONS 1 Why are cultural and biochemical characteristics so important for assigning bacteria to a taxonomic group?

2 Do pigment-producing bacteria always color the medium in which they are grown? Explain.

3 What cultural characteristics would enable you to distinguish between the genera *Clostridium* and *Bacillus*?

4 What is *Bergey's Manual of Systematic Bacteriology*?

CULTURAL CHARACTERISTICS OF BACTERIA: CHARACTERIZATION OF COLONIES

		Bacterial species					
		Streptomyces albus	*Mycobacterium phlei*	*Bacillus subtilis*	*Pseudomonas aeruginosa*	*Micrococcus luteus*	*Escherichia coli*
Sketch of colonies							
Size, mm							
Form							
Elevation							
Margin							
Consistency							
Chromogenesis							

NAME _____ 22 CULTURAL CHARACTERISTICS *(Continued)*

CULTURAL CHARACTERISTICS OF BACTERIA: CHARACTERIZATION OF GROWTH ON *NUTRIENT-AGAR SLANTS*

	Bacterial species					
	Streptomyces albus	*Mycobacterium phlei*	*Bacillus subtilis*	*Pseudomonas aeruginosa*	*Micrococcus luteus*	*Escherichia coli*
Sketch of growth						
Amount of growth						
Form of growth						
Chromogenesis						
Odor						
Consistency						

159

CULTURAL CHARACTERISTICS OF BACTERIA: CHARACTERIZATION OF GROWTH IN *NUTRIENT BROTH*

	Bacterial species					
	Streptomyces albus	*Mycobacterium phlei*	*Bacillus subtilis*	*Pseudomonas aeruginosa*	*Micrococcus luteus*	*Escherichia coli*
Sketch of growth						
Surface growth						
Degree of turbidity						
Odor						
Amount of sediment						
Nature of sediment						

22 CULTURAL CHARACTERISTICS *(Continued)*

CULTURAL CHARACTERISTICS OF BACTERIA: CHARACTERIZATION OF GROWTH IN *NUTRIENT-AGAR STAB CULTURE*

	Bacterial species					
	Streptomyces albus	*Mycobacterium phlei*	*Bacillus subtilis*	*Pseudomonas aeruginosa*	*Micrococcus luteus*	*Escherichia coli*
Sketch of growth						
Amount of growth						
Distribution of growth:						
Top						
Middle						
Bottom						

CULTURAL CHARACTERISTICS OF BACTERIA: CHARACTERIZATION OF GROWTH IN GELATIN STABS

	Bacterial species					
	Streptomyces albus	*Mycobacterium phlei*	*Bacillus subtilis*	*Pseudomonas aeruginosa*	*Micrococcus luteus*	*Escherichia coli*
Sketch of growth						
Amount of growth						
Liquefaction, + or −						
Degree of liquefaction						
Nature of liquefaction						

EXERCISE 23

Maintenance and Preservation of Pure Cultures

OBJECTIVE
To learn how microorganisms are maintained and preserved in the laboratory, especially for long-term storage.

OVERVIEW
A major aspect of microbiology involves techniques to keep cultures alive, in pure culture and in typical form without variation or mutation (*maintenance*), and methods to retain cultures in this status over a period of time (*preservation*). The choice of technique may be influenced by the length of time the cultures are to be maintained, for example, months or years, by the number of cultures in the collection, and by the availability of equipment, storage space, and skilled labor. In principle, all methods aim for *bacteriostasis*, that is, maintenance of viability without growth or reproduction.

In short-term maintenance and preservation, the conventional method is to transfer cultures to fresh media periodically. The time interval between transfer varies with the organism, the medium used, and storage temperatures. Some cultures may be transferred every few days, others only after weeks or months. Subculturing should be kept to a minimum in order to avoid the selection of variants or mutants.

Several methods are available for the long-term maintenance and preservation of bacterial cultures. They all employ refrigeration and/or desiccation. Many species can be preserved successfully for months or years by simply covering the growth with sterile mineral oil. (The oil is sterilized by heating in an oven at 170°C for 1 to 2 h.) The oil-covered cultures are usually stored in an upright position at refrigerator temperature (4°C).

Bacterial cultures can also be stored in the frozen state at the temperature of liquid nitrogen (−196°C) or in an ultra-low-temperature mechanical freezer (e.g., Revco, Inc.) at a temperature of −70°C. Heavy-walled borosilicate vials or polypropylene tubes with screw caps and silicon washers can be used. The preservation of bacteria at such cold temperatures is facilitated by the use of cryoprotective compounds, such as glycerol (10% v/v). Glycerol can be autoclaved for sterility at 121°C for 15 min. These agents pass readily through the cell membrane and appear to provide both intracellular and extracellular protection against freezing.

Freeze-drying, or *lyophilization,* is one of the most economical and effective methods for long-term preservation of microorganisms, including viruses. Lyophilization involves the removal of water from frozen microbial suspensions by sublimation under reduced pressure; that is, the water is evaporated without going through the liquid phase. Lyophilized cultures can be stored for long periods of time if properly prepared and stored. They can easily be rehydrated and restored to a growing state. Lyophilization can be carried out in several ways, and there are various types of apparatus for this purpose [FIGURE 23.1]. The simplest ones consist of a high-vacuum pump, a condenser, and a chamber or manifold. Successful lyophilization depends on using healthy cells and a cryoprotective suspending fluid, such as 20% skim milk, 10% horse serum, or other chemical agents. The general technique involves suspending the cells in a cryoprotective fluid in a glass vial which is quick frozen with a mixture of dry ice and alcohol. The vial is then thoroughly desiccated while frozen. The glass vial is finally sealed with a hot flame.

REFERENCES
MICROBIOLOGY, Chap. 7, "Control of Microorganisms: Principles and Physical Agents."
MM, Chap. 12, "Culture Preservation."

FIGURE 23.1
The lyophilization apparatus.

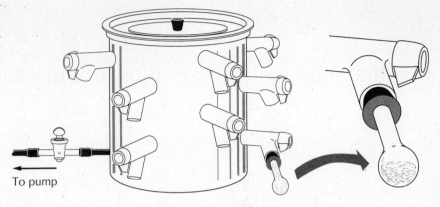

MATERIALS
6 trypticase-agar slants in screw-cap tubes
3 cystine-trypticase-agar deep tubes with screw caps
30-ml sterile white mineral oil
Nutrient-agar slant cultures of *Bacillus subtilis*, *Neisseria subflava*, and *Chromobacterium violaceum*
Lyophilized specimen of bacterial culture

PROCEDURE
1 Transfer each culture to two agar slants and one tube of cystine-trypticase agar. Incubate *C. violaceum* at room temperature and the other two species at 35°C for 24 to 48 h.
2 Following this incubation period, overlay one set of the agar-slant cultures with sterile mineral oil. Be sure that the oil extends approximately ¼ in above the top of the agar slant as shown in FIGURE 23.2.
3 Store all the cultures at room temperature for future testing for viability. Be sure that they are properly labeled, that is, with your name, date, and name of bacterial species.
4 At future times, as designated by the instructor (intervals of several weeks), make subcultures from each of the stored cultures to determine their viability.
4 Steps for opening and culturing a typical lyophilized specimen are illustrated in FIGURE 23.3.

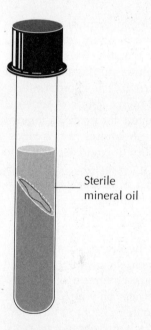

FIGURE 23.2
Agar slant with mineral-oil overlay.

FIGURE 23.3
Procedure for opening and culturing lyophilized specimen (double-vial preparations).

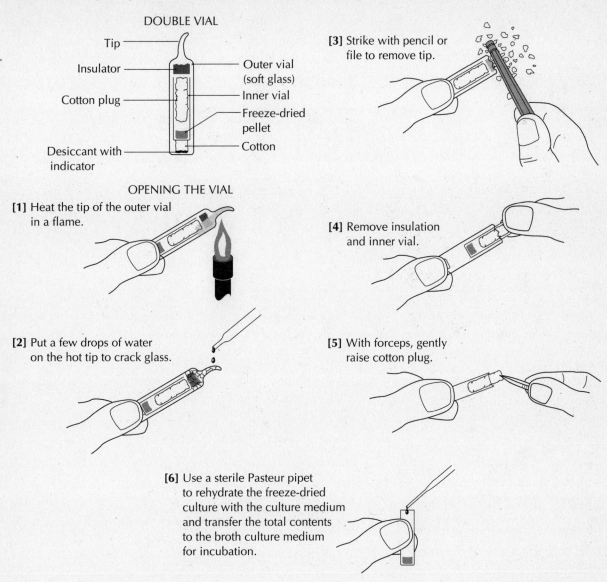

Maintenance and Preservation of Pure Cultures

RESULTS Following each subculture of the stored cultures, record whether they are viable in the chart provided.

VIABILITY CHECK OF STORED CULTURES				
Medium	Date tested	*Bacillus subtilis*	*Neisseria subflava*	*Chromobacterium violaceum*
Nutrient-agar slant				
Nutrient-agar slant with oil				
Cystine-trypticase agar				

QUESTIONS 1 What are the primary aims of culture maintenance and preservation?

2 What is the function of the American Type Culture Collection?

BIOCHEMICAL ACTIVITIES OF BACTERIA

Knowledge of the ability of bacteria to dissimilate certain substrates and to synthesize various products is indispensable for the adequate characterization of a species. It is not uncommon for two different bacterial cultures to be very similar in their morphological and cultural characteristics but to exhibit very striking differences in their metabolic reactions.

Routine qualitative tests designed to permit convenient detection of important biochemical features generally consist of a nutrient medium plus substrate in which the organism is cultured. An "indicator" may also be included to reveal the accumulation of a product. In other instances, changes in the physical nature of the medium may be adequate evidence to conclude dissimilation of a substrate, for example, liquefaction of gelatin, or synthesis of a substance such as polysaccharides responsible for "stringy" milk.

More definitive measurements of metabolic products are often desired. Such data enable one to establish better criteria for a taxonomic group. For example, in the routine test to determine the ability of two cultures to ferment glucose, both are found to produce acid; the result for each is the same. However, if an experiment is performed to identify the acidic constituent, the results could quite possibly be very different. One culture may produce only a single acid, the other a different acid or even several different acids.

Knowledge of the biochemical activities or potential biochemical activities of a microbial culture has many applications in biology beyond that of characterizing a species. The biochemical activities of microorganisms make them either beneficial or harmful. Benefits from microbial biochemical activities include the degradation of waste such as bodies of dead animals and plants back to minerals in the recycling of the elements, the production of consumable products like bread, cheese, wine, and antibiotics, the manufacture of commercially useful materials like solvents and organic acids, and the breakdown of compounds from oil spillage.

Harmful biochemical activities of microorganisms include the causation of myriad diseases and the destruction of property as in the decay of wooden structures and accelerated corrosion of metals.

EXERCISE 24

Hydrolysis of Polysaccharide, Protein, and Lipid

OBJECTIVE
To use special media to detect hydrolytic properties of some enzymes excreted by bacteria.

OVERVIEW
Some microorganisms produce enzymes capable of splitting the large complex molecules of polysaccharides, proteins, and lipids. These enzymes are extracellular, being exoenzymes, and accomplish the breakdown of their respective substrates via hydrolysis. *Polysaccharidases* hydrolyze polysaccharides to sugars; *proteases* hydrolyze proteins to peptides and amino acids; *lipases* hydrolyze lipids to glycerol and fatty acids. When organisms are grown in a medium that contains one of these substrates, evidence of its degradation can be obtained. [See PLATES 12, 13, and 14.] In this exercise, the presence of an exoenzyme is determined by looking for a change in the substrate around a bacterial colony.

REFERENCES
MICROBIOLOGY, Chap. 1, "Essential Biochemistry for Microbiology."
MICROBIOLOGY, Chap. 3, "Characterization of Microorganisms."
MM, Chap. 23, "Enzymatic Activity."

MATERIALS
1 poured plate each of starch agar, tributyrin (butterfat) agar, and milk agar
50 ml of Gram's iodine solution in dropping bottle
Broth cultures of *Bacillus cereus*, *Escherichia coli*, and *Pseudomonas fluorescens*

PROCEDURE
1 With your glass marking pencil, divide the outside bottom half of each Petri dish into three equal sections.
2 Streak each culture on the surface of each medium [FIGURE 24.1].
3 Incubate the inoculated plates at 35°C for 48 to 72 h.

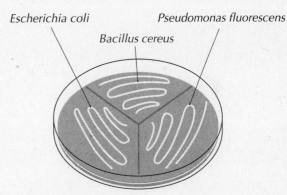

FIGURE 24.1
Streak inoculation of medium for hydrolysis by organisms.

Hydrolysis of Polysaccharide, Protein, and Lipid

RESULTS

1 *Evidence of starch hydrolysis:* Flood the surface of the starch-agar plate with Gram's iodine solution. Iodine reacts with starch resulting in the formation of a dark blue color; in the absence of starch (where the starch was hydrolyzed by the enzyme amylase), the blue color does not develop. Note the reaction surrounding each streak of growth. Sketch the appearance of growth and the surrounding medium for each organism, and record results in the chart provided on page 174.

2 *Evidence of lipid hydrolysis:* The tributyrin, as you noted, is dispersed throughout the agar medium as an emulsion; the medium appears opaque. Hydrolysis of the lipid tributyrin results in products (glycerol and fatty acid) that are soluble, and hence there is a clearing of the medium. Observe the appearance of the medium around each streak of growth, and record the results in the chart on page 174.

3 *Evidence of protein hydrolysis:* Casein is the principal nitrogenous constituent of milk; it is a protein. It exists in a colloidal state and, as such, is responsible for the white color of milk. If the casein is hydrolyzed, the products of this hydrolysis are soluble; a clear zone surrounds the streak of bacterial growth. Record the results of your observations in the chart on page 174.

QUESTIONS

1 Write a general equation for the enzymatic hydrolysis of each substrate used in this exercise.

2 Define the following terms:
Lipolytic, proteolytic, and saccharolytic:

Amylase, lipase, and protease:

3 Explain whether the enzymes you detected by your procedure are exoenzymes or endoenzymes.

HYDROLYTIC REACTIONS BY BACTERIAL SPECIES			
Bacterial species	Sketch of reaction on starch agar *(label blue areas in medium)*	Sketch of reaction on tributyrin agar *(label clear areas in medium)*	Sketch of reaction on milk agar *(label clear areas in medium)*
Escherichia coli			
Bacillus cereus			
Pseudomonas fluorescens			

EXERCISE 25

Fermentation of Carbohydrates

OBJECTIVE
To use fermentation tubes to study the catabolic activities of bacteria on carbohydrates; to learn that carbohydrate reactions are important in the identification of bacterial species.

OVERVIEW
A wide variety of carbohydrates are fermented by bacteria, and the pattern of fermentation is characteristic of certain species, genera, or other taxonomic groups of organisms. Thus, knowledge of the carbohydrates fermented by a particular organism aids in identification of the organism. Reactions of carbohydrates employed for identification are usually degradative catabolic reactions used by the bacteria as part of their energy-producing metabolism. These reactions can be determined by inoculating the organism into a tube of carbohydrate broth containing a Durham tube. The medium consists essentially of nutrient broth plus 0.5 percent of the particular carbohydrate plus an indicator. The Durham tube is a small inverted vial in the broth. Since the usual end products of carbohydrate fermentation are acid or acid and gas, the indicator will reveal production of acid and the inverted vial will trap gas if it is produced [FIGURE 25.1].

In addition to the preceding routine method for testing the fermentation of carbohydrates, sometimes it is useful to determine whether organisms utilize carbohydrates by *oxidative metabolism*, which requires the presence of molecular oxygen, rather than by *fermentative metabolism*, which does not require oxygen but may occur in the presence of oxygen. The oxidation-fermentation test medium of Hugh and Leifson is useful for this purpose. It is helpful in the identification of bacteria because some bacteria are both oxidative and fermentative whereas others are either oxidative or fermentative. One of the medium tubes is covered with a layer of sterile melted petrolatum or sterile paraffin oil; the other is not. After incubation, acid formation only in the open tube indicates oxidative utilization of the sugar. Acid formation in both the open and sealed tubes is indicative of a fermentation reaction. Lack of acid production in either tube indicates that the organism does not utilize the sugar by either method.

REFERENCES
MICROBIOLOGY, Chap. 3, "Characterization of Microorganisms."
MICROBIOLOGY, Chap. 11, "Microbial Metabolism: Energy-Yielding Biochemical Processes."

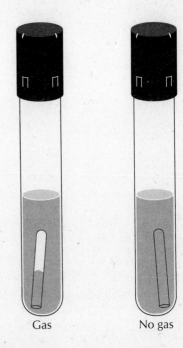

FIGURE 25.1
Fermentation tubes containing inner Durham tubes.

MATERIALS

6 each of fermentation tubes containing phenol red-glucose broth, phenol red-lactose broth, and phenol red-sucrose broth (Each fermentation tube has an inner Durham tube.)

Broth cultures of *Bacillus subtilis*, *Escherichia coli*, *Proteus vulgaris*, *Alcaligenes faecalis*, and *Staphylococcus aureus*

NOTE: Durham tubes are simply inserted into each test tube containing liquid medium. Upon autoclaving, the air is expelled from the Durham tubes, which become filled with liquid medium.

PROCEDURE

1 Inoculate a series of the three different carbohydrate broths with *B. subtilis*. Use one loopful of the broth culture as inoculum.
2 Repeat this procedure with each of the remaining cultures.
3 Select one tube of each carbohydrate broth and keep it uninoculated as a control.
4 Be sure that each tube is labeled for carbohydrate and the organism inoculated. Label the uninoculated tube of carbohydrate as control.
5 Incubate all tubes at 35°C for 48 h.

Fermentation of Carbohydrates

RESULTS After the incubation period, compare each of the inoculated tubes with the control tube of the same medium to determine whether growth occurred and whether acid or gas was produced. Record results in the chart below. The pH indicator phenol red is red in color at a neutral pH of 7 and changes to yellow at a slight acid pH of 6.8. Thus, slight amounts of acid can cause a color change.

FIGURE 25.1 of the fermentation tubes shows one with gas and another without gas. Also see PLATE 15.

	FERMENTATION OF CARBOHYDRATES		
	Phenol red-glucose broth	**Phenol red-lactose broth**	**Phenol red-sucrose broth**
Bacillus subtilis			
Escherichia coli			
Proteus vulgaris			
Alcaligenes faecalis			
Staphylococcus aureus			
A = acid only; AG = acid and gas; a = slight acidity; 0 = no carbohydrate fermentation.			

QUESTIONS 1 Name the three essential components of a carbohydrate fermentation tube.

2 List two (a) acids, (b) neutral products, and (c) gases that may be produced during bacterial fermentation of glucose.

3 What is the pH range of the indicator used in this experiment, and what color denotes a neutral, an acid, or an alkaline reaction?

4 If you use glucose-broth medium without a pH indicator, how can you determine whether a particular bacterium fermented glucose?

5 If the carbohydrate broth does not change color after it has been inoculated and incubated, how can one tell whether the unchanged color is due to failure of the organism to grow or failure of the organism to ferment the carbohydrate?

6 What is the difference between fermentation and respiration?

7 If an organism grows only in the glucose medium of Hugh and Leifson that is exposed to air, explain whether the organism is oxidizing or fermenting the glucose.

EXERCISE 26

Reactions in Litmus Milk

OBJECTIVE
To learn the use of litmus milk in the characterization of bacteria.

OVERVIEW
Milk, particularly skim milk, not only serves as an excellent culture medium but is also used extensively for biochemical characterization and identification of bacteria. Protein and carbohydrate substrates, primarily casein and lactose, respectively, are contained in dried skim milk (butterfat is removed before drying), and microbial degradation of either or both can be detected. The addition of litmus to the milk provides an indicator of the organism's ability to produce acid or alkali as well as its oxidation-reduction activities. Therefore, litmus milk can provide much information on microbial reactions and is used to characterize bacteria.

Litmus is a substance extracted from certain lichens, e.g., the genera *Lecanora* or *Rocella*, and consists of phenol-containing compounds (litmus may be more than one compound and its exact chemical composition is unknown). It has been used since the late 1800s as a pH indicator. Litmus is light blue to gray around pH 7, pink below pH 5, and blue-violet above pH 8. After autoclaving and cooling, litmus milk is about pH 6.8 and is light blue to gray in color. Litmus is also an oxidation-reduction (O/R) indicator. When the O/R potential of the medium is lowered by oxygen removal, litmus becomes colorless (*leuco form*) and is reduced. However, in its leuco form, litmus can no longer function as a pH indicator.

Characteristic reactions of microorganisms in litmus milk are described in the following paragraphs.

LACTOSE FERMENTATION
Organisms capable of using this sugar as a carbon source for energy production employ the enzyme β-galactosidase to break down lactose in order to metabolize it further in the following manner:

$$\text{Lactose} \xrightarrow{\beta\text{-galactosidase}} \text{glucose} + \text{galactose}$$

$$\text{Glucose} \xrightarrow{\text{glycolysis}} \text{pyruvic acid} \longrightarrow \text{lactic acid} + \text{butyric acid} + CO_2 + H_2$$

Acids, excreted by bacteria in fermentation, affect casein that exists as large molecules in colloidal suspension, giving milk its white turbid appearance. In the presence of sufficient acid, casein becomes a curd and litmus milk becomes one semisolid gel. The end products of lactose fermentation may include the gases carbon dioxide and hydrogen as shown in the preceding equation. In litmus milk the production of such gases may be evidenced by breaks or bubbles within the curd. Some bacteria, such as the clostridia, produce so much gas that the curd is broken to pieces so that this type of fermentation is known as "stormy fermentation."

RENNIN PRODUCTION
Curd formation can also come about by the action of an enzyme called **rennin**, or **rennet**, produced by some bacteria. Rennin acts on casein to form paracasein, which is converted to calcium paracaseinate and forms a curd through a process called *sweet curdling*. The rennet curd is a soft semisolid clot that will flow slowly when the tube is slanted; it is different from acid curd which is firm and does not flow.

PROTEOLYSIS
Proteolysis is also termed *peptonization*. In this process, the bacteria produce proteolytic enzymes which hydrolyze the milk proteins, primarily casein, to their component amino acids. The protein is solubilized, and the amino acids are decarboxylated and/or deaminated, resulting in an alkaline pH in

the medium. The litmus milk turns deep purple in the top portion and becomes liquefied, giving a translucent brownish (wheylike) appearance in the bottom portion.

LITMUS REDUCTION

Fermentation is an anaerobic process in which biooxidations occur without the participation of molecular oxygen. These oxidations are due to the removal of hydrogen atoms or electrons from a substrate. In order that such oxidations may occur, there must be available electron acceptors. In litmus milk, litmus acts as such an acceptor so that when it accepts hydrogen or an electron from a substrate, it becomes reduced and turns white. Litmus reduction can also occur in litmus milk when bacteria rapidly consume oxygen during growth. As the oxygen supply is depleted, facultative bacteria can shift from a respiratory to a fermentative metabolism and produce sufficient acid to form a curd. The curd hinders oxygen diffusion into the medium, resulting in a litmus-milk tube filled with curd, having a pink top (acid litmus in the presence of oxygen) and a white bottom (reduced litmus in the absence of oxygen).

Following are the types of reactions that may occur in the presence of litmus milk. Refer also to FIGURE 26.1 and PLATE 16.

Lactose fermentation: Acid, acid curd, and gas, evidenced by breaks or bubbles in curd, may or may not be produced; litmus milk turns pink.

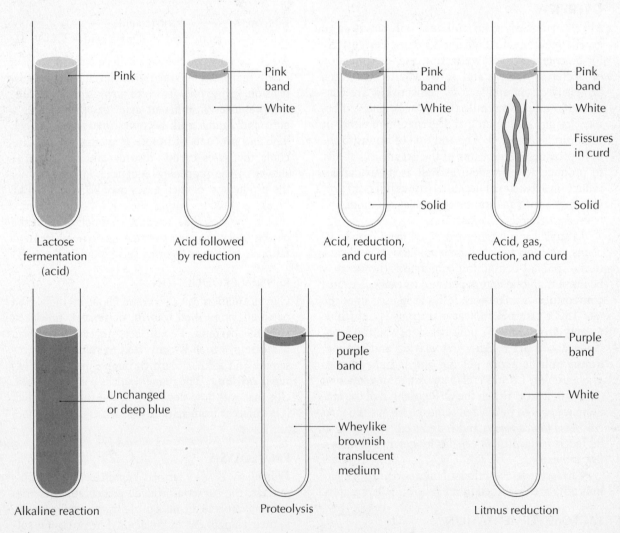

FIGURE 26.1
Some litmus-milk reactions.

Proteolysis: Casein solubilized; top portion of medium resembles a colored broth; litmus turns blue (alkaline reaction).

Litmus reduction: Litmus decolorized (white) except for colored ring at top.

Ropiness: Abundant production of capsules or slime causes milk to become viscous.

Rennet coagulation (sweet curdling): Coagulation of milk with little acid formation; action is due to a rennetlike enzyme.

REFERENCE
MICROBIOLOGY, Chap. 3, "Characterization of Microorganisms."

MATERIALS
6 tubes litmus milk
Nutrient-broth cultures of *Streptococcus lactis*, *Streptococcus faecalis* subsp. *liquefaceins*, *Escherichia coli*, *Alcaligenes faecalis*, and *Pseudomonas aeruginosa*

PROCEDURE
1 Inoculate tubes of litmus milk with each of the cultures listed. Use an uninoculated tube as a control.

2 Incubate the tubes at 35°C for 1 week. Observe at 48 h and at the end of incubation.

NAME _____

Reactions in Litmus Milk

RESULTS Observe and record changes that appear at 48 h and again at the end of 1 week's incubation. It is not a simple task to interpret results properly from litmus-milk reactions. However, the list of reactions appearing in the Overview section, FIGURE 26.1, and PLATE 16 should help in their interpretation.

	REACTIONS IN LITMUS MILK			
Species	Description, 48 h	Reaction,* 48 h	Description, 1 week	Reaction,* 1 week
Streptococcus lactis				
Streptococcus faecalis				
Escherichia coli				
Alcaligenes faecalis				
Pseudomonas aeruginosa				
Control				

*A = acid; AC = acid curd; AG = acid and gas; P = proteolysis; ALK = alkaline reaction; ACR = acid curd reduction; ACP = acid curd proteolysis; I = inert (no reaction).

QUESTIONS

1 Why is milk a good culture medium?

2 List two products of lactose fermentation that might occur in milk.

3 What are the products of proteolysis?

4 What products are responsible for the development of an alkaline reaction in milk? For a viscous consistency?

5 When a curd develops in a sample of milk, how can you determine whether it is a rennet or an acid curd?

1 **Gram stain.** Gram-positive rods. Exercise 9.C.

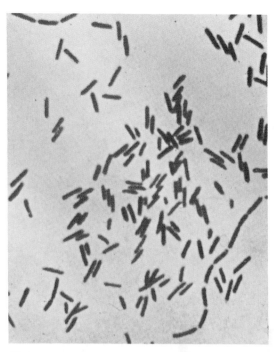

2 **Gram stain.** Gram-negative rods. Exercise 9.C.

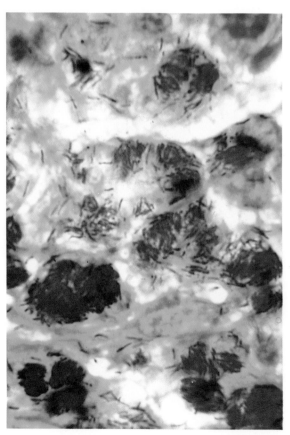

3 **Acid-fast stain.** Acid-fast mycobacteria. Exercise 9.D.

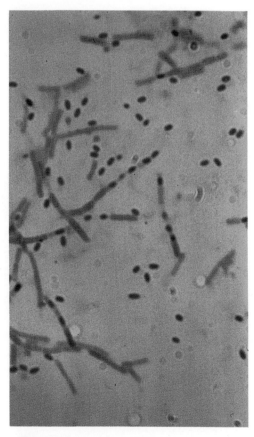

4 **Spore stain.** Spore-forming *Bacillus*. Exercise 10.A.

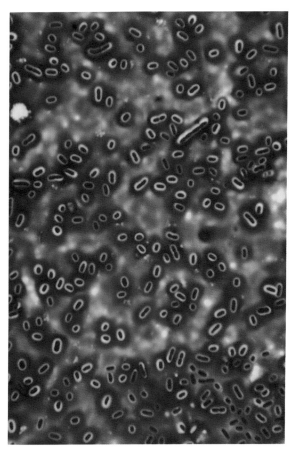

5 **Capsule stain.** Capsulated bacilli. Exercise 10.B.

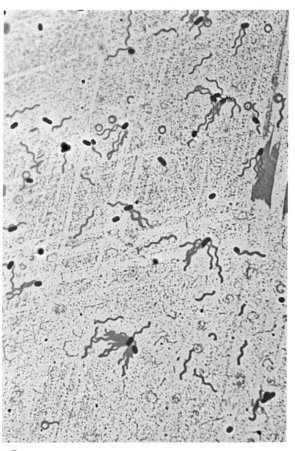

6 **Flagella stain.** Bacilli with mordant-thickened flagella. Exercise 10.C.

7 **Hemolysis.** Alpha hemolysis on blood-agar plate. Exercise 16.

8 **Hemolysis.** Beta hemolysis on blood-agar plate. Exercise 16.

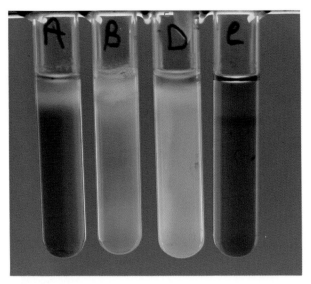

9 **Relation of oxygen to growth.** Left to right: aerobic; facultative; anaerobic; control. Exercise 21.

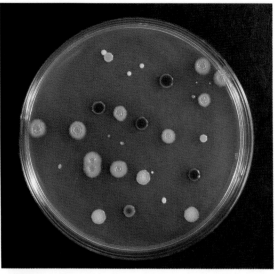

10 **Cultural characteristics.** Pigmented and nonpigmented bacterial colonies. Exercise 22.

11 **Cultural characteristics.** Growth in broth culture. Left to right: control; flocculent; ring; pellicle. Exercise 22.

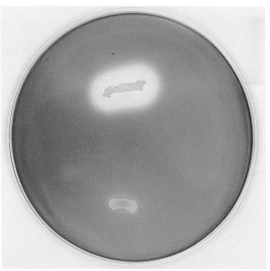

12 **Starch hydrolysis.** Positive on top; negative on bottom. Exercise 24.

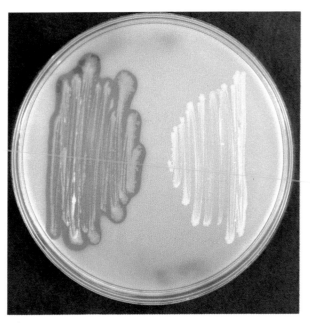

13 **Lipid hydrolysis.** Positive (left); negative (right). Exercise 24.

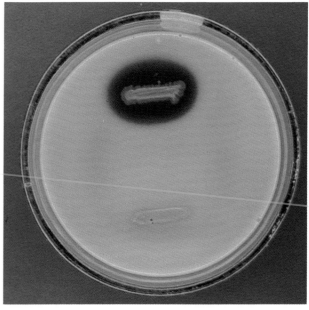

14 **Casein hydrolysis.** Positive on top; negative on bottom. Exercise 24.

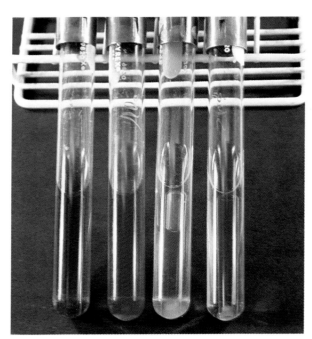

15 **Carbohydrate fermentation.** Left to right: control; negative; acid and gas; acid. Exercise 25.

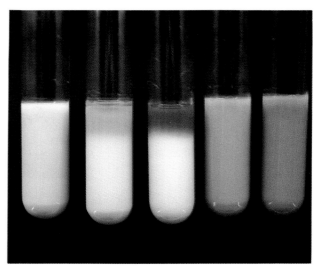

16 **Litmus milk reactions.** Left to right: acid; acid, curd, reduction; alkaline, proteolysis, reduction; alkaline; control. Exercise 26.

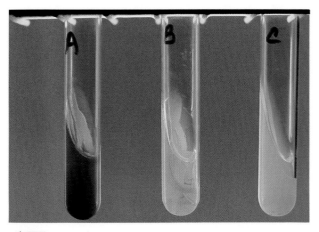

17 **Hydrogen sulfide production.** Left to right: positive; negative; control. Exercise 27.A.

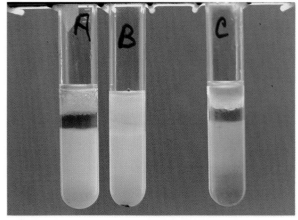

18 **Indole production.** Left to right: positive; negative; control. Exercise 27.B.

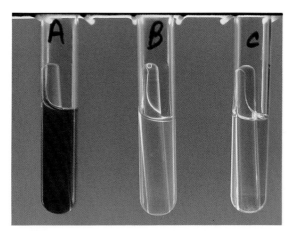

19 **Nitrate reduction.** Left to right: positive; negative; control. Exercise 27.C.

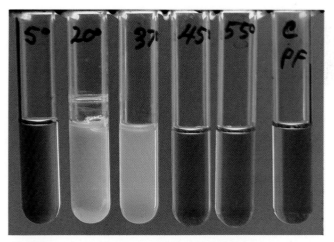

20 **Effect of temperature on growth.** Left to right: 5°C, no growth; 20°C, growth; 37°C, growth; 45°C, no growth; 55°C, no growth; control. Growth at temperatures of 20 to 37°C indicates that this organism is a mesophile. Exercise 29.

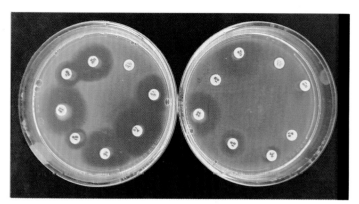

21 **Kirby-Bauer antibiotic sensitivity.** The organism growing in the plate on the right is resistant to a broad range of antibiotics. The clear zone indicates growth inhibition. Exercise 34.

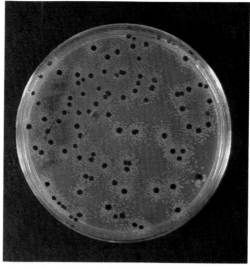

22 **Plasmid transformation.** *E. coli* cells transformed by pBLU plasmid show characteristic blue colonies on LB/Amp/X-gal medium. Exercise 49.

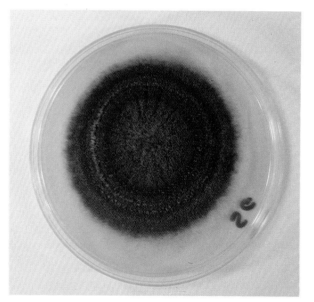

23 **Cultural characteristics of mold.** Growth on Sabouraud agar plate. Exercise 52.

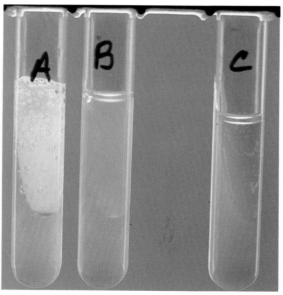

24 **Catalase test.** Left to right: positive; negative; control. Exercise 56.

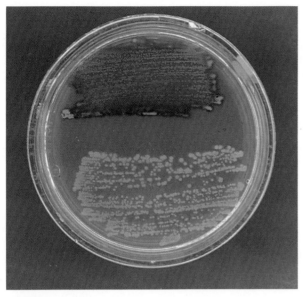

25 Oxidase test. Positive on top; negative on bottom. Exercise 64.B.

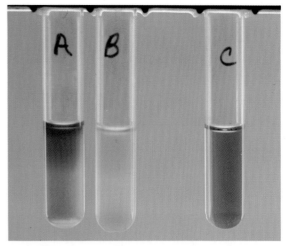

26 Voges-Proskauer test. Left to right: positive; negative; control. Exercise 65.

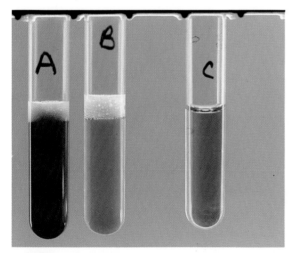

27 Methyl-red test. Left to right: positive; negative; control. Exercise 65.

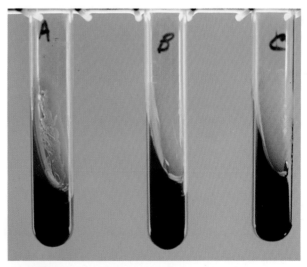

28 Citrate utilization test. Left to right: positive; negative; control. Exercise 65.

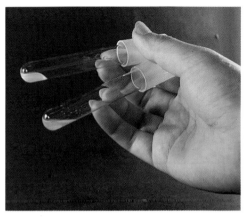

29 **Coagulase test.** Positive on top; negative on bottom. Exercise 65.

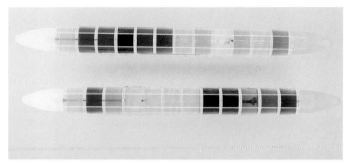

30 **Rapid test characterization.** Rapid identification of enterics can be made from single inoculation using Enterotube II System. Exercise 68.

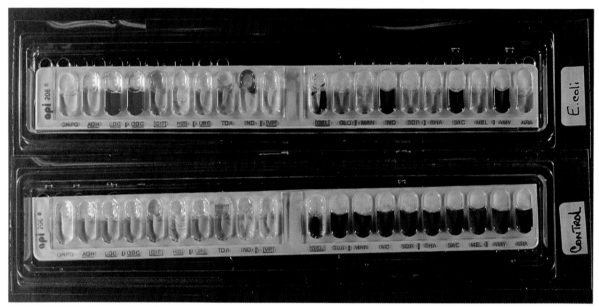

31 **Rapid test characterization.** Results of an API 20E System identification of *Escherichia coli*. Exercise 68.

32 **Membrane filtration technique.** *E. coli* growing on a membrane filter saturated with Endo medium and demonstrating its characteristic green sheen. Exercise 73.

EXERCISE 27.A

Additional Biochemical Characteristics
27.A Hydrogen Sulfide Production

OBJECTIVE
To perform laboratory tests to detect hydrogen sulfide production in the characterization of bacteria.

OVERVIEW
Hydrogen sulfide is a metabolic by-product of certain bacteria. It is produced primarily in two ways:

1 Some bacteria *oxidize* reduced organic sulfur-containing compounds in order to obtain energy or for growth; in this process they remove the sulfur and excrete it as hydrogen sulfide. For example, the sulfur-containing amino acid cysteine is assimilated by the organism with the enzyme cysteine desulfurase in the following manner:

$$\begin{array}{c}CH_2-SH \\ | \\ HC-NH_2 \\ | \\ COOH\end{array} + H_2O \longrightarrow \begin{array}{c}CH_3 \\ | \\ C=O \\ | \\ COOH\end{array} + H_2S\uparrow + NH_3$$

Cysteine　　Water　　Pyruvic acid　Hydrogen sulfide　Ammonia

2 Some bacteria produce hydrogen sulfide by the *reduction* of inorganic sulfur compounds such as the thiosulfates ($S_2O_3^{2-}$), sulfates (SO_4^{2-}), or sulfites (SO_3^{2-}). These compounds may be used by certain bacteria as electron acceptors in anaerobic respiration, producing hydrogen sulfide as an end product, in much the same way as oxygen is used to produce water in aerobic respiration. In effect, the sulfur atoms act as hydrogen acceptors during oxidation of the inorganic compound. This process may be illustrated as follows:

$$2S_2O_3^{3-} + 4H^+ + 4e^- \longrightarrow 2SO_3^{2-} + 2H_2S$$

Thiosulfate　　　　　　Sulfite　Hydrogen sulfide

Hydrogen sulfide is a colorless gas and therefore is invisible. However, it has the odor of rotten eggs and is repulsive even at low concentrations. This is fortunate for detection because the gas is very poisonous. In the microbiology laboratory, the hydrogen sulfide gas liberated is detected by incorporating a heavy-metal salt, for example, ferrous ammonium sulfate, into the medium; hydrogen sulfide reacts with it to form an insoluble black metal sulfide precipitate. For example, hydrogen gas reacts with ferrous or lead ions to form an insoluble black ferrous or lead sulfide.

Culture media used for the detection of hydrogen sulfide production by bacteria generally contain two ingredients: peptone and thiosulfate (in addition to the heavy-metal salt mentioned above). Even though peptone contains sulfur-containing amino acids, thiosulfate is added as a supplemental substrate from which the bacteria can produce hydrogen sulfide.

Several types of media are commercially available. Some of them test for sugar fermentation reactions, indole production, and motility as well, for example, SIM agar. Others, like peptone-iron agar, are designed to detect only hydrogen sulfide production.

NOTE: Media containing ferric salts also show a black precipitate upon the formation of H_2S because, after autoclaving of the medium and growth of the organism, both ferrous and ferric ions are present due to changes in oxidation-reduction potential.

REFERENCES
MICROBIOLOGY, Chap. 3, "Characterization of Microorganisms."

Blazevic, D. J., and G. M. Ederer, *Principles of Biochemical Tests in Diagnostic Microbiology*, Wiley, New York, 1975.

MM, Chap. 25, "Phenotypic Characterization."

MATERIALS
3 tubes peptone-iron agar
Nutrient-agar slant cultures of *Proteus vulgaris*, *Escherichia coli*, and *Serratia marcescens*

PROCEDURE
1 Inoculate each of the organisms into separate tubes of peptone-iron agar. Use the transfer needle, and make a stab inoculation of the medium.
2 Incubate tubes at 35°C for 7 days.

NAME _____

27.A

Hydrogen Sulfide Production

RESULTS Observe tubes for evidence of hydrogen sulfide production (blackening along the line of inoculation). Sketch each tube, illustrating the type of reaction exhibited. [See PLATE 17.]

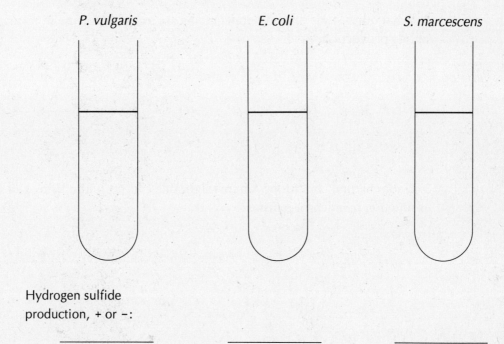

Hydrogen sulfide
production, + or −:

_____   _____   _____

187

QUESTIONS

1 What are the constituents in peptone-iron agar that make it possible for use in the detection of hydrogen sulfide production?

2 What other kinds of commercial media are available for use in testing hydrogen sulfide production?

3 Under natural conditions, where might one expect to find hydrogen sulfide production by microorganisms?

4 What is the chemical basis for the formation of a black precipitate to indicate the production of hydrogen sulfide?

EXERCISE 27.B

Additional Biochemical Characteristics
27.B Production of Indole

OBJECTIVE
To perform laboratory tests to detect indole production in the characterization of bacteria.

OVERVIEW
Amino acids are produced as a result of hydrolytic degradation of proteins, peptones, and peptides. One of these amino acids, tryptophan, serves as a substrate for biochemical differentiation or characterization, since some species are capable of hydrolyzing this amino acid. One of the end products of this reaction is indole, which can be detected by a colorimetric test. The indole test is considered to be a very important test in the identification of a wide variety of microorganisms commonly encountered in the clinical laboratory. The tryptophan in the test medium is supplied by the peptone. The peptone generally used is a commercial preparation called *tryptone*, since it is only partially hydrolyzed protein and contains a high concentration of tryptophan, which is degraded by the preparation procedures of some other kinds of peptone.

Tryptophan is catalyzed within the cells by the enzyme tryptophanase into three end products: indole, ammonia, and pyruvate. Pyruvate is further oxidized to provide energy for the cell; indole and ammonia are excreted as waste products. This catalysis is shown by the following chemical equation:

$$1/2 O_2 + \text{Tryptophan} \xrightarrow{\text{tryptophanase}} \text{Indole} + H_3C-\underset{O}{\overset{\|}{C}}-COOH + NH_3$$

(Pyruvic acid)

Indole production is detected by Kovacs' reagent, which contains a compound called *para-dimethylaminobenzaldehyde (DMAB)* that reacts with indole to give a color reaction. The bright red color formed is due to the compound rosindole dye produced by the reaction depicted in the following equation:

$$2 \text{ Indole} + \text{Kovacs' reagent (}p\text{-dimethylaminobenzaldehyde)} \xrightarrow{\text{acid condensation}} \text{Rosindole dye (bright red)} + H_2O$$

In Kovacs' reagent, the acidified DMAB is dissolved in amyl alcohol. When mixed thoroughly with the culture medium, the amyl alcohol extracts only the indole but not the tryptophan. The tryptophan is left in the aqueous portion; otherwise it would interfere with the test because it also reacts with DMAB to give a red color. In the amyl alcohol, the DMAB reacts with the indole according to the reaction illustrated above. Thus, after standing a few minutes, the immiscible amyl alcohol will rise to the top of the culture medium containing rosindole dye. Therefore, a bright red layer on top of the culture medium indicates a *positive* test for indole.

REFERENCES

MICROBIOLOGY, Chap. 3, "Characterization of Microorganisms."

MM, Chap. 25, "Phenotypic Characterization."

Blazevic, D. J., and E. M. Ederer, *Principles of Biochemical Tests in Diagnostic Microbiology*, Wiley, New York, 1975.

MATERIALS

4 tubes 1% tryptone broth
Kovacs' reagent
Nutrient-broth cultures of *Escherichia coli*, *Enterobacter aerogenes*, and *Proteus vulgaris*

PROCEDURE

1 Inoculate the tubes of tryptone broth with each of the three cultures. Use the uninoculated tube as the control.

2 Incubate the tubes at 35°C for 48 h.

NAME _____

27.B

Production of Indole

RESULTS Add approximately 1.0 ml of Kovacs' reagent to each of the tubes. Gently shake the tubes and allow them to stand to permit the reagent to rise to the top. The presence of indole is indicated by a bright red color, which develops in the reagent layer. Compare each inoculated tube with the uninoculated control. [See PLATE 18.] Record your observations in the chart provided.

PRODUCTION OF INDOLE		
Bacterial species	**Description of reagent layer**	**+ or –**
Escherichia coli		
Enterobacter aerogenes		
Proteus vulgaris		
+ = indole production; – = no indole production.		

191

QUESTIONS **1** If glucose is present at the same time as tryptophan in the culture medium, which compound does the bacterium generally prefer?

2 A positive indole test, as performed here, is indicated by the bright red color of the surface layer. Why is the red color not distributed throughout the culture medium?

3 Could any peptone serve satisfactorily to demonstrate production of indole? Explain.

EXERCISE 27.C

Additional Biochemical Characteristics
27.C Reduction of Nitrate

OBJECTIVE
To perform laboratory tests to detect nitrate reduction in the characterization of bacteria.

OVERVIEW
A biochemical characteristic of many bacteria is their ability to reduce certain compounds. Nitrate reduction is an example. The reduction of nitrates by some bacteria occurs in the absence of molecular oxygen and is therefore an anaerobic process. Such bacteria, for example, *Pseudomonas*, carry out anaerobic respiration in which inorganic substances such as nitrates or sulfates act as final electron acceptors during energy production. The biochemical reaction may be illustrated by the following equation:

$$NO_3^- + 2H^+ + 2e^- \xrightarrow{\text{nitrate reductase}} NO_2^- + H_2O$$
Nitrate → Nitrite + Water

This reaction can be detected by growing the organisms in a medium containing a nitrate and subsequently testing for the presence of its reduction product, nitrite.

Some bacteria have the enzymatic ability to react further on the nitrite to reduce it to ammonia NH_3 or molecular nitrogen N_2. These reactions are represented as follows:

$$NO_2^- \longrightarrow NH_3 \longrightarrow \tfrac{1}{2} N_2 \uparrow$$
Nitrite → Ammonia → Molecular nitrogen

The process by which bacteria convert nitrate to gaseous end products such as nitrogen is called **denitrification.**

REFERENCES
MICROBIOLOGY, Chap. 3, "Characterization of Microorganisms."

Blasevic, D. J., and G. M. Ederer, *Principles of Biochemical Tests in Diagnostic Microbiology*, Wiley, New York, 1975.
MM, Chap. 25, "Phenotypic Characterization."

MATERIALS
4 tubes trypticase-nitrate broth
Reagents for detection of nitrite: 5-ml sulfanilic acid solution and 5-ml dimethyl-α-naphthylamine solution
2 1-ml pipettes
Zinc dust
Nutrient-broth cultures of *Pseudomonas aeruginosa*, *Bacillus subtilis*, and *Serratia marcescens*

> **CAUTION: Although dimethyl-α-naphthylamine, unlike α-naphthylamine, has not been listed as a carcinogen by the Occupational Safety and Health Administration, Department of Labor, its structural similarity to α-naphthylamine would indicate that such safety precautions as the avoidance of aerosols, mouth pipetting, and contact with the skin should be followed.**

PROCEDURE
1 Inoculate tubes of trypticase-nitrate broth with each of the three cultures. Use the uninoculated tube as a control.
2 Incubate tubes at 35°C for 48 h.

Reduction of Nitrate

RESULTS

1 Add approximately 1.0 ml of sulfanilic acid solution and 1.0 ml of dimethyl-α-naphthylamine to each of the tubes, including the control. The development of a red, purple, or maroon color indicates the presence of nitrite. [See PLATE 19.] Record these results in the chart on page 196. The chemical reaction for this is shown below:

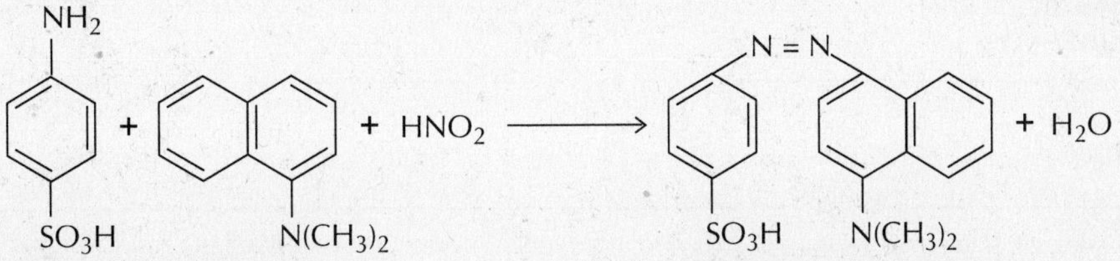

Sulfanilic acid (colorless) + Dimethyl-α-naphthylamine (colorless) + HNO_2 → Sulfobenzene azo-dimethyl-α-naphthylamine (red) + H_2O

2 A negative test (no development of color) is interpreted as follows:
  a Nitrate not reduced   OR
  b Nitrate reduction has occurred, but the reaction has gone beyond the nitrite stage to ammonia or gaseous nitrogen.

Therefore, it is desirable to test further those tubes giving a negative reaction for the presence of nitrite. This can be done by the addition of a very small amount (a trace) of powdered zinc. The zinc reduces nitrate to nitrite, and the color characteristic of a positive nitrate test develops. If the addition of zinc does not produce a color change, then the nitrates in the medium were reduced beyond nitrite to ammonia or gaseous nitrogen.

3 Compare the results obtained from inoculated tubes with those observed in the control. Record all results in the chart on page 196.

NITRATE REDUCTION TEST			
Bacterial species	Test for nitrite (color)	Zinc test, if nitrite negative	Interpretation of tests
Pseudomonas aeruginosa			
Bacillus subtilis			
Serratia marcescens			
Control			

QUESTIONS

1 Why is it preferable to perform this test by periodic testing of some of the culture over the duration of incubation?

2 Explain why an organism using nitrate or nitrite as the final electron acceptor can grow anaerobically.

3 Explain why a negative test (lack of color development in the presence of test reagents for nitrite) does not necessarily mean that the nitrate was not reduced.

EXERCISE 28

Morphological, Cultural, and Biochemical Characterization of an Unknown Culture

OBJECTIVE
To characterize an unknown bacterial culture using previously learned morphological, staining, cultural, and biochemical techniques.

OVERVIEW
The techniques performed in Exercises 9 through 27 are typical of the types of tests performed on bacteria to establish their identity in a conventional classification scheme. They do not, however, include all the tests required to identify bacteria of certain groups. For example, an organism found to be an aerobic Gram-negative diplococcus can be identified as to species on the basis of its ability or inability to ferment the carbohydrates glucose, levulose, maltose, and sucrose. On the other hand, identification of an unknown Gram-negative rod requires a much more elaborate scheme of testing. The purpose of this exercise is to characterize, though not necessarily identify, an unknown culture. TABLE 28.1 summarizes the biochemical tests presented in Part VI.

REFERENCES
MICROBIOLOGY, Chap. 3, "Characterization of Microorganisms."
MM, Chap. 25, "Phenotypic Characterization."

MATERIALS
Staining reagents for Gram stain, acid-fast stain, spore stain
1 tube each of motility-test agar, nutrient-agar slant, nutrient broth, nutrient-agar deep, nutrient gelatin, phenol-red fermentation broths (glucose, lactose, sucrose), tryptone broth, trypticase-nitrate broth, peptone-iron agar, and litmus milk
1 Petri-dish medium of nutrient agar, starch agar, and tributyrin agar
Reagents for the detection of starch hydrolysis, indole production, and nitrate reduction
Unknown culture identified by code number

PROCEDURE
1 Review the headings presented in the chart provided in the results section for this exercise, and familiarize yourself with the information to be obtained on the unknown culture. Review also TABLE 28.1 summarizing the biochemical tests in Part VI.
2 Prepare and examine Gram-, acid-fast-, and spore-stain smears of the unknown culture.
3 Inoculate and incubate all the media provided in the manner described in previous exercises. Specifically, inoculate a motility-test agar, a nutrient-agar slant, a nutrient broth, a nutrient-agar deep, and a nutrient gelatin.
4 Carry out biochemical tests on the unknown culture, specifically those for carbohydrate fermentation, indole production, nitrate reduction, hydrogen sulfide production, starch and lipid hydrolysis, and litmus milk reaction.

TABLE 28.1 Summary of Biochemical Tests

Test	Medium	Substrate	Positive test — End products	Positive test — Appearance	Type of reaction
Starch hydrolysis	Starch agar	Starch	Dextrins, maltose	Clear zone around growth in presence of iodine solution	Hydrolysis
Carbohydrate fermentation	Phenol red broth and carbohydrate, with Durham tube	Glucose, sucrose, or lactose, etc.	Acids, e.g., lactic, acetic, propionic; gases: carbon dioxide and hydrogen may or may not be produced	Indicator changes from red to yellow; gas collects in fermentation vial	Fermentation
Gelatin hydrolysis (liquefaction)	Nutrient gelatin	Gelatin	Peptones → peptides → amino acids	Failure of medium to gel at 20°C	Hydrolysis, proteolytic
Casein hydrolysis	Milk agar	Casein	(As above)	Clear zone around growth	Hydrolysis, proteolytic
Hydrogen sulfide production	Peptone-iron agar	Cysteine (sulfur-containing amino acid)	Hydrogen sulfide + pyruvic acid + ammonia	Blackening of media as result of iron sulfide formation	Amino acid degradation
Indole production	Tryptone broth	Tryptophan	Indole + pyruvic acid + ammonia	Red color in presence of Kovacs' reagent	Hydrolytic, amino acid degradation
Nitrate reduction	Trypticase nitrate broth	Potassium nitrate, KNO_3	Potassium nitrite, KNO_2	Red to maroon color develops in presence of sulfanilic acid and dimethyl-α-naphthylamine	Reduction
Lipid hydrolysis	Tributyrin agar	Tributyrin	Fatty acid and glycerol	Clear zone around growth	Hydrolysis, lipolytic

TABLE 28.1 Summary of Biochemical Tests *(Continued)*

Test	Medium	Substrate	Positive test		
			End products	**Appearance**	**Type of reaction**
Reactions in litmus milk	Litmus milk	Lactose	Acid(s); e.g., lactic, propionic, acetic, in various combinations; gases: carbon dioxide and hydrogen	Curd, coagulation of casein, litmus turns pink (acid curd); breaks or bubbles in curd	Fermentation
	Litmus milk	Casein	Calcium paracaseinate	Curd, coagulation of casein, neutral or alkaline reaction (litmus bluish), rennet curd	Coagulation
	Litmus milk	Casein	Peptones → peptides → amino acids	Clearing of milk (solubilization of casein) and development of blue color (alkaline reaction)	Proteolysis
	Litmus milk	Litmus	Leucolitmus (colorless compound)	Medium appears white except for ring at surface	Reduction

NAME _____

Morphological, Cultural, and Biological Characterization of an Unknown Culture

RESULTS 1 Record the results of the microscopic examinations, the cultural characteristics, and the biochemical tests in the chart provided.

CODE NO. OF UNKNOWN CULTURE ASSIGNED: _____	
Experimental procedure	**Observations**
Morphological characteristics	
Cell morphology and arrangement	
Staining reactions: Gram stain	
Spore stain	
Acid-fast stain	
Motility	
Cultural characteristics	
Colonial morphology	
Agar slant	
Nutrient broth	
Agar stab	
Gelatin stab (liquefaction)	

CODE NO. OF UNKNOWN CULTURE ASSIGNED: _____ (Continued)	
Experimental procedure	**Observations**
Biochemical characteristics (refer to TABLE 28.1)	
Fermentation of: 　Glucose	
Lactose	
Sucrose	
Indole production	
Nitrate reduction	
Hydrogen sulfide (H_2S) production	
Starch hydrolysis	
Lipid hydrolysis	
Reactions in litmus milk	

2 The name of the unknown culture (its characteristics are similar to a species used in the laboratory to date) is:

QUESTIONS 1 What is meant by *DNA hybridization?*

2 Briefly explain DNA probe technology.

CONTROL OF MICROBIAL POPULATIONS

PHYSICAL AGENTS

The physical environment has a profound influence on the growth and survival of microorganisms. As a group, microorganisms are capable of growing in a wide range of temperature, pH, osmotic pressure, and other physical conditions. However, members of a particular genus or species exhibit specific requirements.

Physical conditions or processes are commonly employed to kill, inhibit, or remove microorganisms from various materials or environments. The requirement may be complete destruction of all microbial life (***sterilization***) or a degree of destruction in the microbial flora where some survivors are permitted or can be tolerated (as in ***sanitization***).

The autoclave, the hot-air oven, and the bacteriological filter are representative of the type of apparatus available in the microbiological laboratory for sterilization of media, reagents, glass Petri dishes, test tubes, pipettes, and so on. Various items are preferentially sterilized by one of these methods, for example, bacteriological media are sterilized by autoclaving, whereas dry glassware is generally sterilized in the hot-air oven.

Microbial species differ in their susceptibility to destruction by physical agents, spore-forming bacteria being the most resistant group. Consequently, bacterial-spore suspensions are often used to test the efficiency of a sterilization process.

CHEMICAL AGENTS

An extensive array of chemical substances is available for control of microbial populations. Some, for example, are prescribed for use on inanimate surfaces (***disinfectants***), some are used on living tissue (***antiseptics***), and others are intended for use in treatment of infections (***chemotherapeutic agents***). The chemical characteristics of the substance, together with its intended application, are taken into consideration when tests are designed to determine antimicrobial activity. Laboratory in vitro testing of a chemical agent to determine its antimicrobial efficacy requires the use of designated test organisms. Specific strains of certain species have been selected for various assays, for example, *Staphylococcus aureus* ATCC 6538 for the phenol-coefficient technique.

Quantitative assay of chemotherapeutic antibiotics by microbiological methods represents a special application of microbial inhibition. These tests have been designed in such a way that a relationship exists between the degree of antimicrobial activity and the quantity of the antibiotic; that is, within certain limits of antibiotic concentration, proportionality exists between the amount of antibiotic and the degree of inhibition.

Three general methods for determining bacterial susceptibility to antibiotics are currently in use:

1. Serial broth dilution
2. Agar diffusion (impregnated disk)
3. Plate dilution

The first method is considered more reliable, provided constant conditions and care are maintained. The second is most widely used because of its simplicity. The third is much like the first and is chosen by some laboratories as a matter of individual preference.

FIGURE VII.1 shows three techniques for performing the agar-diffusion method: use of antibiotic-filled stainless steel cylinders (top left); use of antibiotic-impregnated paper disks (top right); and use of antibiotic-filled wells punched in the agar medium (bottom). Susceptibility of a bacterium to an antibiotic is expressed by a clear zone of growth inhibition on the lawn of bacterial growth.

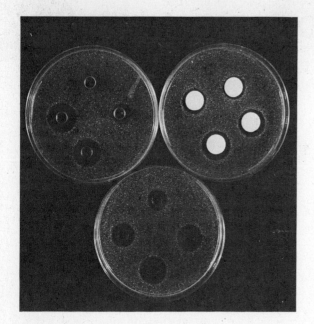

FIGURE VII.1
Three techniques for performing the agar-diffusion method.

The antimicrobial spectrum of any antibiotic is an important factor in its choice as a therapeutic agent. Unfortunately, great variations in susceptibility of microorganisms, not only from species to species but even from strain to strain, make it necessary to determine in vitro the effect of the available antibiotics upon the individual strain for optimum results. The correlation between the in vitro tests and the clinical responses, although not absolute, is usually accurate.

EXERCISE 29

The Effect of Temperature on Growth

OBJECTIVE
To study the effect of temperature on the growth of several microbial species.

OVERVIEW
Eubacteria, the bacteria commonly encountered, are capable of growing (increasing in cell numbers) over a wide range of temperature. At one extreme are bacteria that can grow at a temperature of 0°C, *psychrophiles*, and at the other are those requiring high temperatures, *thermophiles*, some of which grow at 75°C and even higher. Intermediate between these two groups are the *mesophiles*, which grow between the temperatures of 20 and 45°C. Each type of bacteria, psychrophile, mesophile, or thermophile, has a minimum, maximum, and optimum growth temperature.

It may be seen in FIGURE 29.1 that as the temperature of incubation rises, so does the **growth rate** (rate of doubling) of the microorganism until the optimum temperature is reached. This temperature gives the fastest growth rate, or the most rapid growth rate during a short period of time. As the temperature exceeds the optimum, the rate of growth falls rapidly until the maximum growth temperature is reached. Temperatures above the maximum usually result in cell death. Temperatures below the minimum slow or stop growth, but the cells do not necessarily die. On the contrary, most microorganisms survive for long periods of time at cold temperatures, and they will begin to grow again upon the return of higher temperatures. Thus, as learned in Exercise 23, it is even possible to store microorganisms in a mechanical freezer (−70°C) or in liquid nitrogen (−196°C).

REFERENCES
MICROBIOLOGY, Chap. 6, "Cultivation and Growth of Microorganisms."
MICROBIOLOGY, Chap. 7, "Control of Microorganisms: Principles and Physical Agents."

MATERIALS
16 nutrient-agar slants
8 Sabouraud's agar slants
Nutrient-broth cultures of *Staphylococcus aureus*, *Micrococcus roseus*, *Pseudomonas fluorescens*, and *Bacillus stearothermophilus*
Sabouraud's agar slant cultures of *Saccharomyces cerevisiae* and *Aspergillus niger*

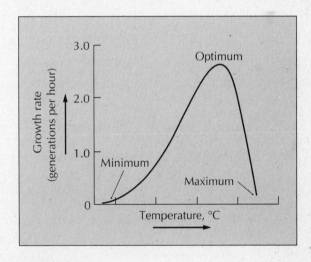

FIGURE 29.1
The effect of temperature on the growth rate of a microbial species.

Refrigerator
Incubators at 35 to 55°C

PROCEDURE

1 Inoculate four nutrient-agar slants with each of the bacterial cultures. Inoculate four Sabouraud's agar slants with *S. cerevisiae* and four others with *A. niger*.

2 Incubate one inoculated slant of each culture at the following temperatures: 4 to 6°C (refrigerator), 20 to 25°C (room), 35°C (incubator), and 55°C (incubator).

NAME _____

The Effect of Temperature on Growth

RESULTS Observe each of the tubes at the end of 24 to 48 h and again after 5 to 7 days' incubation. On the two charts that follow, record the amount of growth. Note any differences in appearance of growth, particularly pigmentation, at the different temperatures. [See PLATE 20.]

AMOUNT AND APPEARANCE OF GROWTH AFTER 24 TO 48 h				
Microbial species	4–6°C	20–25°C	35°C	55°C
Staphylococcus aureus				
Micrococcus roseus				
Pseudomonas fluorescens				
Bacillus stearothermophilus				
Saccharomyces cerevisiae				
Aspergillus niger				

0 = no growth; + = scant growth; ++ = moderate growth; +++ = profuse growth.

AMOUNT AND APPEARANCE OF GROWTH AFTER 5 TO 7 DAYS				
Microbial species	4–6°C	20–25°C	35°C	55°C
Staphylococcus aureus				
Micrococcus roseus				
Pseudomonas fluorescens				
Bacillus stearothermophilus				
Saccharomyces cerevisiae				
Aspergillus niger				

0 = no growth; + = scant growth; ++ = moderate growth; +++ = profuse growth.

QUESTIONS

1 What generalization can be made from the results of this experiment about temperature required for growth of bacteria versus temperature required for growth of yeasts and molds?

2 What effect does the temperature of incubation have on pigment production?

3 Explain how temperature would influence (a) the rate of growth and (b) the cell crop from a particular culture.

4 What is a *thermoduric bacterium?*

5 Give the range of temperatures for the growth of psychrophiles and thermophiles.

6 Compare the growth temperature requirements of the thermoacidophilic archaeobacteria with those of the thermophilic eubacteria.

EXERCISE 30

Resistance of Bacteria to Heat

OBJECTIVE
To determine the thermal death times of several bacterial species.

OVERVIEW
Vegetative cells of microorganisms exhibit differences in their tolerance to heat. The spores of bacteria are extremely resistant to heat; many can survive prolonged exposure to boiling water. The degree of heat tolerance of an organism can be assessed in the laboratory by exposing the cells to a fixed temperature for increasing periods of time, subcultures being made at the end of each time interval, or the time may be fixed and the temperature varied. The former method is called **thermal-death-time determination,** which refers to the shortest period of time required to kill a suspension of microorganisms at a given temperature under specific conditions. The latter procedure is referred to as the **thermal-death-point determination,** which refers to the lowest temperature at which microorganisms are killed in a given time. In this exercise we maintain the temperature constant and vary the time of exposure of cells to heat; that is, we shall make a thermal-death-time determination.

REFERENCE
MICROBIOLOGY, Chap. 7, "Control of Microorganisms: Principles and Physical Agents."

MATERIALS
21 tubes of nutrient broth
Water-bath at 80°C
Thermometer
Nutrient-broth cultures of *Escherichia coli, Staphylococcus aureus, Bacillus subtilis* (old culture, with spores), and *B. subtilis* (young culture, without spores)

Exercise courtesy of Dr. R. A. MacLeod, Macdonald College, McGill University.

PROCEDURE
1 Position 17 tubes of nutrient broth in a test-tube rack, and place the rack into a water-bath at 80°C. Introduce a thermometer into tube 17, and hold all the tubes in the bath until the broth has reached 80°C.

2 Inoculate each of the four unheated tubes with a different one of the four cultures, and set them aside as controls. Mark each tube as "Unheated control," and indicate the name of the organism introduced.

3 Have a beaker of cold water ready. Inoculate a tube with *E. coli* and place in the beaker of cold water. Remove to another rack and mark the tube "*E. coli.*" Repeat with each of the other three cultures and mark the tubes appropriately.

4 Inoculate a tube with *E. coli*, and hold it for 1 min in the 80°C bath. After 1 min, transfer the tube to the beaker of cold water and after cooling, to the other rack. Mark the tube "*E. coli*, 1 min." Repeat with each of the other organisms.

5 Inoculate each of four tubes at 80°C with a different organism at 1-min intervals using a loop. Five minutes after the first inoculation, transfer the tube to the beaker of cold water. At the end of 6 min, when the second tube has been exposed for 5 min, remove the second tube and cool in the beaker. Repeat at 1-min intervals with each of the other tubes. This gives you four tubes each inoculated with a different culture, each of which has been exposed to 80°C for 5 min. Label all tubes "Name of organism, 5 min."

6 Repeat the procedure but expose each culture to 80°C for 15 min. Incubate all cultures at 35°C for 48 h.

NAME _____

Resistance of Bacteria to Heat

RESULTS Examine all tubes for growth. Compare the inoculated tubes exposed to 80°C for various times with the unheated control tubes. Record evidence of survival (growth) as + and kill (no growth) as 0 in the chart provided.

THERMAL-DEATH-TIME DETERMINATION AT 80°C					
Bacterial species	0 min	1 min	5 min	15 min	Unheated, control
Escherichia coli					
Staphylococcus aureus					
Bacillus subtilis, no spores					
Bacillus subtilis, spores					

QUESTIONS **1** Are any bacterial spores killed when a spore suspension is heated at a high temperature, for example, at 80°C for 10 min?

2 Compare the heat resistance of mold and yeast spores to the heat resistance of bacterial spores.

3 How could the isolation of spore-forming bacteria from a soil sample be facilitated?

EXERCISE 31

Bactericidal Effect of Ultraviolet Radiations

OBJECTIVE
To study the killing effect of ultraviolet rays on bacteria.

OVERVIEW
Radiations in the ultraviolet region (generally between 100 to 400 nm) of the light spectrum are lethal to microorganisms. The most lethal wavelength is 265 nm, which corresponds to the optimum absorption wavelength of deoxyribonucleic acid (DNA). However, effective germicidal activity occurs from about 240 to 300 nm. This is illustrated by the diagram in FIGURE 31.1.

The germicidal effectiveness of ultraviolet light is influenced by several factors, including the specific wavelength employed and the time and intensity of the exposure. Ultraviolet (UV) radiations have very little penetrating power; unless the microorganisms are directly exposed, they are likely to escape destruction. These radiations induce the formation of pyrimidine dimers in the microbial nucleic acid, resulting in a mutation in the following manner. When a UV ray hits a pair of thymine molecules, which are next to each other on the same strand of DNA, they become dissociated from the complementary bases of the other strand and join together to form the structure called **thymine dimer**. This covalent bonding distorts the shape of one DNA strand. When this strand is later being replicated, incorrect bases may be inserted into a new strand of DNA as it forms along and complements the old distorted DNA strand. The result is a mutation in the genetic code. This is why a UV lamp is often used to increase the frequency of mutation in microorganisms in the laboratory. However, all microorganisms have enzymes that will repair the damage to the DNA strand providing that the damage caused by UV radiation was not too great. When thymine dimers are exposed to visible light, the enzyme pyrimidine dimerase is activated and splits the thymine dimer. This is called **photoreactivation** and results in a reversal of the killing action of UV light.

The sun gives off UV radiation of many wavelengths, but only the longer wavelengths of 300 to 400 nm are able to penetrate the earth's atmosphere. This shielding is attributed to the ozone layer, which is being dissipated by human-made gases like Freon and is a concern of environmentalists. Most of the wavelengths that finally reach the earth's surface do not affect living cells adversely, although some have a germicidal effect. A germicidal lamp that produces UV light of around 260 nm is used to decontaminate hospital operating rooms and food processing areas. Its sterilization activity is confined to surfaces only since UV light is not very penetrating.

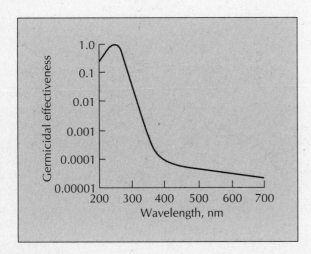

FIGURE 31.1
The efficiency of germicidal activity at various wavelengths of light.

REFERENCES

MICROBIOLOGY, Chap. 7, "Control of Microorganisms: Principles and Physical Agents."
MM, Chap. 14, "Gene Mutation."

MATERIALS

5 Petri plates of nutrient agar
1 sterile cotton swab
Ultraviolet (germicidal) lamp
Cardboard mask with design to cover bottom half of Petri dish
Nutrient-broth culture of *Serratia marcescens*

PROCEDURE

1 Inoculate the agar surface evenly and completely in each plate, using a cotton swab soaked with the broth culture of *S. marcescens*.
2 Label one plate as the control. It will not receive further treatment.
3 Expose the remaining four plates to the ultraviolet light as follows: Replace each cover from three of the dishes with a square piece of cardboard in which a narrow letter has been cut. The letter should be within the circumference of the dish [FIGURE 31.2]. Expose the dishes covered in this manner to the ultraviolet light source, one for 30 s, one for 1 min, and the other for 3 min (or for other time periods as directed by the laboratory instructor). Remove the dishes after the exposure period and replace their covers. Expose one Petri dish with the cover in place for 3 min.

> **CAUTION: Precautions should be observed during the performance of this exercise to shield the eyes from direct exposure to the ultraviolet radiations. This can be done by using a light source housed in a box with one side open, through which the Petri dishes can be inserted. Some commercial ultraviolet-light sources have this safety feature built in.**

4 Incubate all the Petri dishes at 25°C or room temperature for 48 h.

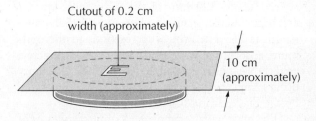

FIGURE 31.2
Cardboard cover with letter cutout for entry of ultraviolet light.

NAME _____

Bactericidal Effect of Ultraviolet Radiations

RESULTS

1 Sketch the pattern of growth on the Petri dishes that were exposed to ultraviolet light with the cardboard-mask cover.

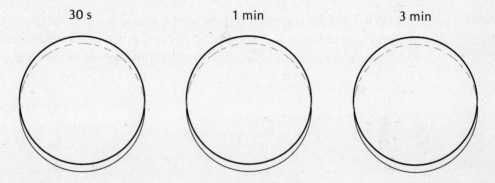

30 s   1 min   3 min

2 Compare the growth on the control plate with that on the plate exposed with the glass cover intact.

QUESTIONS 1 Are all wavelengths in the ultraviolet region equally germicidal?

2 What did the results of this exercise reveal about the penetrating ability of ultraviolet radiations? How do climatic and other conditions of our atmosphere influence the amount of ultraviolet radiations that reach the earth's surface?

3 Give an example of an industrial application of ultraviolet radiation to accomplish a reduction of the microbial flora.

4 What other types of radiations are lethal to microorganisms?

EXERCISE 32

Effect of Osmotic Pressure on Microbial Growth

OBJECTIVE

To appreciate the relationships between solute concentration and osmotic pressure and to learn the differential responses of bacteria and fungi to solute concentrations.

OVERVIEW

Osmotic pressure is the force developed when two solutions of different concentrations are separated by a semipermeable membrane that is only permeable to the solvent but not to the solute. In a solution, the *solute* is the substance dissolved in the liquid, or *solvent*. Because microorganisms are separated from their aqueous environment by a semipermeable membrane, they are affected by changes in osmotic pressure as a result of changes in solute concentrations.

When high concentrations of substances, such as salts and sugars, are dissolved in a liquid, the medium is said to be *hypertonic* with respect to the cells in the medium. Such a medium has a marked osmotic effect on the cells suspended in it. Many microorganisms in a medium with abnormally large amounts of dissolved materials (osmotic concentration of the medium exceeds that of the cells) will be restricted in their growth since there will be a tendency for establishment of an equilibrium between the solution enclosed within the cellular membrane and the solution exterior to the membrane. Water will pass from the cell interior into the solution. In effect, the cells become dehydrated, or *plasmolysed*; this phenomenon is characterized by the cytoplasmic membrane shrinking away from the rigid cell wall. This is illustrated in FIGURE 32.1.

When the osmotic concentration of the medium is considerably lower than that of the cell (a *hypotonic* medium), diffusion of water into the cell occurs in excess, causing increased outward pressure [FIGURE 32.1]. Cells that are not enclosed by a rigid

FIGURE 32.1
Effect of osmotic pressure on a bacterial cell.

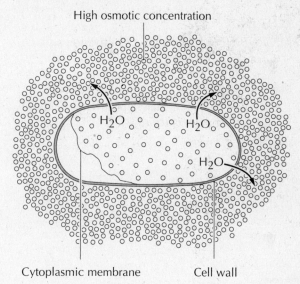

[A] Hypertonic solution

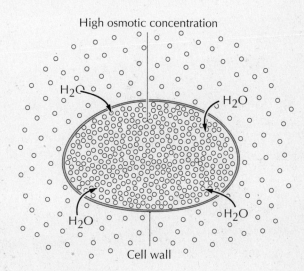

[B] Hypotonic solution

cell wall, such as erythrocytes, may undergo lysis or *plasmoptysis*, that is, the exudation of protoplasm from a cell as a result of cell rupture.

Many dissolved substances are inside a microbial cell. In general, the solute concentrations are much higher inside the cell than outside. A typical microorganism grows best in a medium with a slightly lower osmotic concentration than the cell itself. This enables water to flow into the cell—a condition necessary for the diffusion of substances into the cell and for the maintenance of *turgor*, or outward pressure. However, some microorganisms are capable of tolerating very high concentrations of salt or sugar; others grow only in a medium that contains a relatively large amount of salt.

REFERENCE

MICROBIOLOGY, Chap. 6, "Cultivation and Growth of Microorganisms."

MATERIALS

1 Petri dish each of nutrient agar with 0, 5, 10, 15, and 20% sodium chloride

2 tubes each of malt-extract broth with 0, 10, 25, and 50% sucrose (5 ml per tube)

Nutrient-broth cultures of *Staphylococcus aureus*, *Streptococcus faecalis* subsp. *liquefaciens*, *Brevibacterium linens*

Halobacterium medium liquid culture of *Halobacterium salinarium*

Sabouraud's agar slant culture of *Saccharomyces cerevisiae*

Spore suspension of *Aspergillus niger*

PROCEDURE

1 Mark the bottom of each plate into quadrants, using a glass-marking pen. Inoculate each of the bacterial cultures (streak inoculation) in one quadrant of a plate; repeat this process with all the plates.

2 Incubate the plates at 35°C for 48 h.

3 Inoculate one set of the malt-extract-broth media (0, 10, 25, and 50% sucrose) with *S. cerevisiae* and another set with the spores of *A. niger*. Incubate at room temperature. Observe after 48 h and 5 to 7 days' incubation.

Effect of Osmotic Pressure on Microbial Growth

RESULTS   1 Record the growth on the control plate (nutrient agar without sodium chloride) as +++. Compare the amount of growth of each organism on the sodium chloride plates with that of the control, and record the results as +++, ++, +, or 0 in the chart provided.

EFFECT OF SODIUM CHLORIDE CONCENTRATION ON THE GROWTH OF BACTERIA					
Bacterial species	Sodium chloride content of medium, percent				
	5	10	15	20	0, control
Staphylococcus aureus					
Halobacterium salinarium					
Streptococcus faecalis					
Brevibacterium linens					

2 Observe the tubes of malt-extract broth inoculated with the yeast and mold, and compare the growth in each tube containing sucrose with that of the control (no sucrose). Record the results in the chart provided, as for sodium chloride concentration effect.

EFFECT OF SUCROSE CONCENTRATION ON THE GROWTH OF FUNGI				
Species, incubation time	Sucrose content of medium, percent			
	10	25	50	0, control
Saccharomyces cerevisiae, 48 h				
S. cerevisiae, 5–7 days				
Aspergillus niger, 48 h				
A. niger, 5–7 days				

QUESTIONS

1 Distinguish between the terms *plasmolysis* and *plasmoptysis*.

2 What are *halophilic bacteria*? Where do they occur in nature?

3 A sucrose solution, approximately 10%, is usually used in the preparation of protoplasts. What is the function of this solution?

4 Is the absence of growth, as observed under some conditions of this experiment, due to a microbicidal or microbistatic effect? How could this be proven experimentally?

EXERCISE 33

Comparative Evaluation of Antimicrobial Chemical Agents

OBJECTIVE
To determine the rate of killing of bacterial species by chemical agents.

OVERVIEW
One of the best techniques employed for assessing the antimicrobial capacity of a chemical agent is to add a "test" organism to a solution of the chemical agent and make periodic transfers from this mixture into a suitable medium to determine whether the test organism has been killed. Many refinements of this general procedure have been developed in order to provide tests suitable for the evaluation of the antimicrobial efficacy of a great variety of chemical substances.

REFERENCES
MICROBIOLOGY, Chap. 8, "Control of Microorganisms: Chemical Agents."
Block, S. S., ed., *Disinfection, Sterilization and Preservation*, 4th ed., Lea & Febiger, Philadelphia, 1991, 1162 pp.

MATERIALS
18 tubes of nutrient broth
Chemical agents: group A—0.5% phenol, 70% alcohol, 3% hydrogen peroxide, tincture of iodine; group B—several proprietary disinfectants and antiseptics, such as Lysol and Listerine
4 1-ml and 4 10-ml sterile pipettes
4 sterile stoppered test tubes
Nutrient-broth cultures of *Staphylococcus aureus* and *Bacillus subtilis*

PROCEDURE
1 Select one of the solutions from group A, and add 5 ml of it to a sterile test tube. To this add 0.5 ml of the *S. aureus* culture, and gently agitate the tube to distribute the cells uniformly. Note the time; at intervals of 2½, 5, 10, and 15 min, transfer one loopful of the mixture to a tube of nutrient broth [FIGURE 33.1].

2 Repeat step 1, using one of the products from group B.

3 Repeat steps 1 and 2, using *B. subtilis* as the test organism.

NOTE: By staggering the times of addition, for example, every 2½ min, of the test organism to the chemical agent, it is possible to conduct the tests simultaneously.

4 Inoculate a tube of nutrient broth with a loopful of *S. aureus* and another with a loopful of *B. subtilis* to serve as controls.

5 Incubate all tubes at 35°C for 48 h.

FIGURE 33.1
Procedure to determine the rate of killing of bacteria by chemical agents.

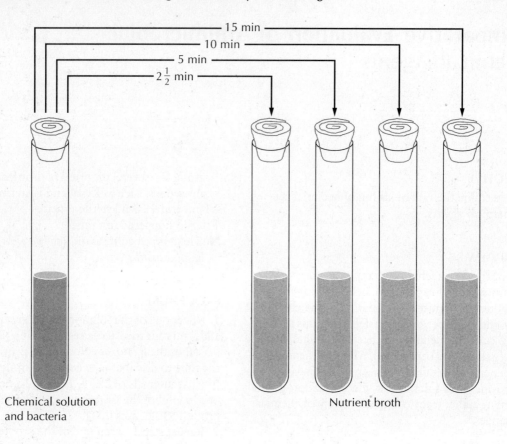

Comparative Evaluation of Antimicrobial Chemical Agents

RESULTS Observe all the inoculated tubes for evidence of growth. Denote presence of growth as + and absence of growth as 0. Record results in the chart provided. (Include results obtained by other members of the class who used different chemical agents.)

	KILLING RATE OF BACTERIA BY ANTIMICROBIAL CHEMICAL AGENTS						
Test organism	Agent	Control	Minute intervals				
			2.5	5	10	15	

QUESTIONS 1 List three reasons why the results of a test-tube evaluation of a disinfectant may not accurately reflect its performance in practical application.

2 List three criteria contributing to an ideal disinfectant.

EXERCISE 34

Antibiotics: Agar-Diffusion Method

OBJECTIVE
To perform an antibiotic susceptibility test using a filter paper disk agar-diffusion method adapted from the standard method of Kirby-Bauer or the National Committee for Clinical Laboratory Standards (NCCLS).

OVERVIEW
The susceptibility of microorganisms to many different antibiotics can be expediently determined by the agar-diffusion technique. This procedure entails inoculating the organism of interest on an agar plate, placing disks impregnated with antibiotics on this inoculated surface, and, after incubation, measuring zones of inhibition. The size of the zone of inhibition is influenced by a complex of factors, such as the rate of diffusion of the drug through agar, the size of the inoculum, the rate of growth of the bacterium, and the bacterium's susceptibility to the antibiotic. For this reason, carefully standardized techniques are needed for acceptable intra- and interlaboratory reproducibility. For example, the amount of antimicrobial agent impregnated in the disk must be standardized. In the United States, only one disk potency is recommended for each antimicrobial agent in a standardized test.

The standardized disk-diffusion test currently recommended by the U.S. Food and Drug Administration and by the National Committee for Clinical Laboratory Standards is a slight modification of the *Kirby-Bauer* test (Bauer, A. W., W. M. M. Kirby, J. C. Sherris, and M. Turch, "Antibiotic Susceptibility Testing by a Standardized Single Disk Method," *Am. J. Clin. Pathol.* 45:493–496, 1966) and is described in the following publication: *Performance Standards for Antimicrobial Disk Susceptibility Tests*, 4th ed., Approved Standard, National Committee for Clinical Laboratory Standards, 771 E. Lancaster Avenue, Villanova, PA 19085, 1988. The material in this manual is constantly being updated, and the latest information should be obtained from NCCLS before carrying out a standardized test.

NOTE: Since meticulous details of the test are impractical to perform in a large class, a modified method embodying the principles of the standard test will be carried out in this exercise.

The test is standardized in order to avoid variation. It is expected that every laboratory that correctly performs the antibiotic disk-diffusion test according to the NCCLS procedure will have the same accurate and comparative results. The test is designed to evaluate only one variable: the susceptibility of a pathogen to an array of antibiotics. All other variables are held constant. Although automated and computerized susceptibility test modules and instruments are available, especially in big city hospitals, the NCCLS procedure is still used worldwide.

TABLE 34.1 presents standardized zones of inhibition, in millimeters, for various antibiotics used in this experiment. The degree of susceptibility of a test organism to these antibiotics, denoted as resistant, intermediate, or susceptible, can be determined through comparison with figures in this table.

REFERENCE
MICROBIOLOGY, Chap. 21, "Antibiotics and Other Chemotherapeutic Agents."

MATERIALS
2 Petri dishes with Mueller-Hinton agar
Commercial antibiotic-impregnated disks as follows: ampicillin, 10 μg; chloramphenicol, 30 μg; penicillin G, 10 units; streptomycin, 10 μg; tetracycline, 30 μg
Small forceps in beaker of 70% ethanol
2 cotton sterile swabs on wooden applicator sticks
8-h nutrient-broth cultures of *Escherichia coli* and *Staphylococcus aureus*

TABLE 34.1 Zone Diameter Interpretative Standards

Antibiotic	Disk concentration	Zone diameters, mm		
		Resistant	Intermediate	Susceptible
Chloramphenicol	30 μg	≤ 12	13–17	≥ 18
Penicillin G	10 units	≤ 20	21–28	≥ 29
Streptomycin	10 μg	≤ 11	12–14	≥ 15
Tetracycline	30 μg	≤ 14	15–18	≥ 19
Ampicillin (Gm. − bact.)	10 μg	≤ 11	12–13	≥ 14
Ampicillin (Gm. + bact.)	10 μg	≤ 20	21–28	≥ 29

PROCEDURE

1 To inoculate the agar medium, dip a sterile cotton swab into a bacterial suspension and remove excess broth by pressing and rotating the swab firmly against the tube above the fluid level.

2 Inoculate the surface of each plate evenly and entirely with the swab, using a different organism for each plate. Allow the plate to dry for 5 min.

3 Remove the pair of forceps from the alcohol and burn off the excess alcohol. Pick up an antibiotic disk and place it on the surface of the agar medium, gently pressing down on it to ensure complete contact of the disk with the agar surface [FIGURE 34.1]. In this manner, apply five different disks to each plate, making sure that they are evenly spaced and not too close to the edge of the plate.

4 Incubate the plates at 35°C for 16 to 18 h before examination.

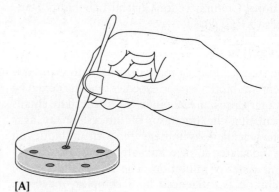

[A]

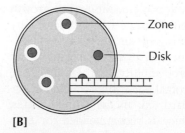

[B]

FIGURE 34.1
[A] Placement of antibiotic disks on agar surface and [B] metric measurement of resultant antibiotic inhibitory zones.

Antibiotics: Agar-Diffusion Method

RESULTS 1 With a millimeter ruler, measure the diameter of all zones of inhibition to the nearest whole millimeter as shown in FIGURE 34.1 and record the results in the chart provided. [See PLATE 21.]

ZONES OF INHIBITION IN MILLIMETERS		
Antibiotic (concentration)	*Staphylococcus aureus*	*Escherichia coli*
Chloramphenicol, 30 µg		
Penicillin, 10 units		
Ampicillin, 10 µg		
Streptomycin, 10 µg		
Tetracycline, 30 µg		

2 Using TABLE 34.1 on standard zone diameters, interpret your results as to degrees of susceptibility, for example, resistant, intermediate, or susceptible.

Staphylococcus aureus to:	Susceptibility
Chloramphenicol	
Penicillin G	
Streptomycin	
Tetracycline	
Ampicillin	

Escherichia coli to:	Susceptibility
Chloramphenicol	
Penicillin G	
Streptomycin	
Tetracycline	
Ampicillin	

QUESTIONS

1 Name the microorganisms used to produce five commonly employed antibiotics.

2 What is an *antimicrobial spectrum?*

3 Assume that you have performed a susceptibility test using antibiotic A and antibiotic B against *S. aureus*. The zone of inhibition produced by A is 15 mm; that produced by B is 30 mm. Does this mean that B is twice as effective as A? Explain.

4 Describe one way that a microbe can acquire antibiotic resistance.

EXERCISE 35

Effect of Enzyme on Bacteria: Spheroplast Formation by *Pseudomonas putida*

OBJECTIVE
To use an enzyme to degrade the peptidoglycan layer in the cell wall of a Gram-negative bacterium in order to form spheroplasts.

OVERVIEW
Microbial cells completely devoid of the cell-wall structure but otherwise intact and metabolically active are called *protoplasts*. Bacterial protoplasts can be produced experimentally by dissolving the cell wall with the enzyme lysozyme. Protoplasts are most easily produced from Gram-positive bacteria. In Gram-negative bacteria, the action of lysozyme, while acting on the peptidoglycan layer, does not remove other cell-wall layers of lipopolysaccharides and lipoproteins. Thus, even though osmotically fragile round bodies are produced, the remnants of the cell wall still cling to them. Such bodies are called *spheroplasts*, even though they superficially resemble protoplasts. Only the electron microscope can reveal the difference between the two kinds of spherical bodies from thin sections.

REFERENCE
MICROBIOLOGY, Chap. 4, "Procaryotic and Eucaryotic Cell Structures."

MATERIALS
24-h nutrient-broth culture of *Pseudomonas putida*, 50 ml
Centrifuge bottle for refrigerated centrifuge
Refrigerated centrifuge
$0.5\ M$ sucrose, 100 ml
EDTA solution (disodium ethylenediaminetetraacetic acid), 32.7 mg/ml distilled water and adjusted to pH 7 with $N/1$ NaOH
Lysozyme, 1 mg/ml distilled water
Tris buffer, 15.1 mg/ml distilled water

PROCEDURE
1 You are provided with a 24-h culture of *P. putida* in 50 ml of nutrient broth.
2 Centrifuge the cells at $13,200 \times g$ in the refrigerated centrifuge for 15 min.
3 Wash the cells three times with $0.5\ M$ sucrose using 20 ml for each wash.
4 Resuspend the organism in 20 ml sucrose solution.
5 Add into 8.5 ml of the suspension 1 ml of each of the following solutions: EDTA, lysozyme, and Tris buffer.
6 Make wet mounts of the suspension at intervals of 0, 5, 10, and 15 min, and examine them for spheroplast formation using phase-contrast illumination and the oil-immersion objective [FIGURE 35.1].

Exercise courtesy of Dr. R. A. MacLeod, Macdonald College, McGill University.

FIGURE 35.1
[A] *Pseudomonas putida* cells and [B] spheroplasts as seen by phase-contrast microscopy.

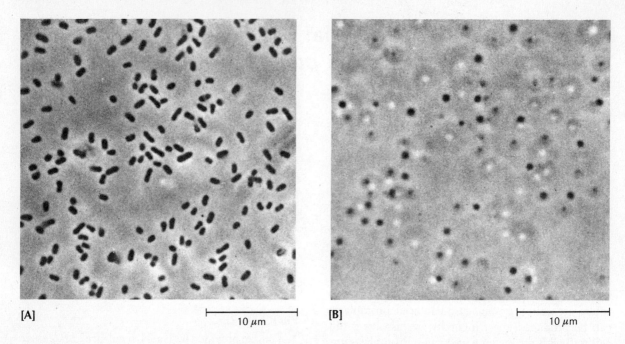

NAME _____

Effect of Enzyme on Bacteria: Spheroplast Formation by *Pseudomonas putida*

RESULTS 1 Draw the morphology of the cells in the circles below:

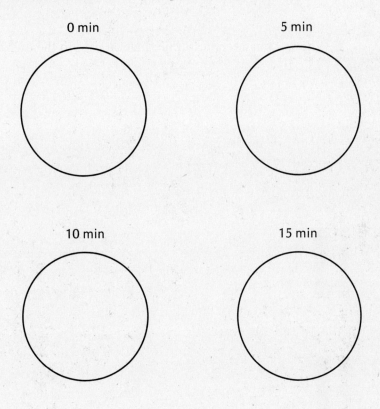

231

QUESTIONS 1 What is the specific mode of action of lysozyme?

2 Why was phase-contrast illumination used in the exercise you performed?

BACTERIAL GENETICS

Like biochemical principles, genetic principles are universal. These principles were first explored in the early 1860s by the Austrian monk, Gregor Mendel, in his studies on the heredity of peas. It was only in the 1940s that genetic processes in microorganisms were discovered; specifically, George Beadle and Edward Tatum showed a correlation between chromosomal mutations and protein deficiencies in the bread mold *Neurospora*. Since then the use of microorganisms as subjects for investigating many fundamental problems in genetics has contributed greatly to our basic knowledge of genetics at the molecular level.

There are many distinct advantages in the use of microorganisms, especially bacteria, for genetic experiments. These may be outlined as follows:

1 A bacterial culture contains millions of individual cells. By appropriate selective and/or differential techniques, rare genetic events can be discovered. Such events can be spontaneous or induced.
2 A procaryotic cell contains a single chromosome. Thus a change in the genetic material of a procaryote results in an immediate, observable change in characteristics. That is, there is no masking effect due to the presence of an unaffected member of a paired chromosome.
3 The rapid growth rate of bacteria allows observation of transmission of a trait through many generations.
4 Selected bacteria are relatively inexpensive to maintain and easy to propagate.
5 A constant environment can be maintained for the bacteria.
6 A wide spectrum of metabolic types among bacteria is available.

Genetics is the study of the inheritance and the variability of the characteristics of an organism. **Inheritance** concerns the exact transmission of genetic information from parents to their progeny. **Variability** is associated with two fundamental properties of an organism: its genotype and its phenotype. **Genotype** refers to the entire genetic capability of an organism as found in its DNA and represents the *inheritable total potential* of a cell. The **phenotype** represents the portion of the genetic potential that is actually expressed by the cells under a given set of environmental conditions. Thus variability of the inherited characteristics can be accounted for by a change either in the genetic makeup (genotype) of a cell or in environmental conditions.

EXERCISE 36

Bacterial Variation Due to Environmental Change

OBJECTIVE
To learn that phenotypic changes may be due to an alteration of the environment.

OVERVIEW
Variation of the inherited characteristics of a bacterium may be due to either *a change in the environmental conditions* or *a change in genotype*. Such changes are usually accompanied by a change in **phenotype**. Phenotypic changes brought about by altering the environmental conditions are not of a permanent nature, for example, formation of capsular material by a bacterium only in the presence of the sugar sucrose. Such changes affect virtually all the cells in the culture.

REFERENCE
MICROBIOLOGY, Chap. 13, "Inheritance and Variability."

MATERIALS
Nutrient-agar slants of *Serratia marcescens* and *Proteus vulgaris*
Gram-stain reagents
1 nutrient-agar plate
Nutrient-agar plate with 0.1% phenol incorporated
2 nutrient-agar slants

PROCEDURE
1 You are provided with an agar slant inoculated with *S. marcescens* and incubated at 37°C. Subculture this organism on two nutrient-agar slants; incubate one at room temperature and the other at 37°C.
2 Examine the tubes after 24 h for pigment production.
3 Make a Gram stain for the growth obtained at each temperature to be certain that the same bacteria are present on both slants.
4 You are provided with an agar-slant culture of *P. vulgaris* and two Petri plates, one containing nutrient agar and the other nutrient agar plus 0.1% phenol [FIGURE 36.1].
5 Make a hanging-drop preparation to check the motility of the culture.
6 Streak both plates for isolated colonies with *P. vulgaris*, and incubate them at 37°C for 24 h.
7 Examine the type of growth obtained. Look for surface growth with a background light.
8 Make a hanging-drop preparation for motility examination from each plate.

FIGURE 36.1

[A] Flagella stain of cells of *Proteus vulgaris* from nutrient-agar medium and **[B]** from nutrient-agar + 0.1% phenol medium. Note the absence of flagella when the cells have been grown in the presence of phenol. *(Courtesy of Liliane Therrien and E. C. S. Chan, McGill University.)*

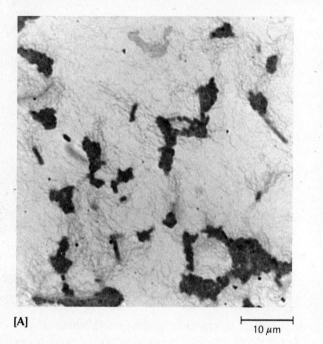
[A] 10 μm

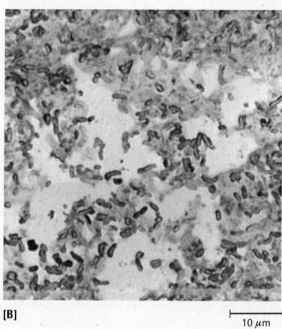
[B] 10 μm

NAME _____

Bacterial Variation Due to Environmental Change

RESULTS 1 Describe the pigmentation of *S. marcescens* after incubation at:

37°C	
25°C (room temperature)	

2 Describe the Gram-stain reaction and morphology of *S. marcescens* incubated at:

37°C	
25°C	

3 Describe the type of growth obtained with *P. vulgaris* on:

Nutrient-agar plate	
Nutrient-agar with 0.1% phenol plate	

4 Describe the motility of the *P. vulgaris* cultures examined from:

Nutrient-agar plate	
Nutrient-agar with 0.1% phenol plate	

QUESTIONS 1 What was the phenotypic change in the cells of *P. vulgaris* grown in the presence of phenol?

2 What two other activities of the cells of *P. vulgaris* were correlated with the phenotypic change observed?

3 Did the phenotypic changes observed in your exercise affect virtually all the cells in each culture?

EXERCISE 37

Bacterial Variation Due to Genotypic Change

OBJECTIVE
To learn that phenotypic changes may be due to an alteration in genotype.

OVERVIEW
Genotypic changes in bacteria affect only small numbers of cells in a population, and such changes are permanent. Changes in genotype may occur by **mutation** of the genetic material, that is, the deoxyribonucleic acid (DNA), or by **recombination**. Recombination is any alteration of genotype resulting from the interaction of one microorganism with another or with genetic material derived from another microorganism; the phenomenon leads to new combinations of genes on a chromosome. More specifically, recombination may occur by the following processes:

1 *Conjugation:* A mating process between sexually compatible bacterial strains; it allows for one-way transfer of genetic material from one cell to another. Cell-to-cell contact is required. In some types of conjugation, only a **plasmid** (a circular small DNA molecule that can replicate independently in the cytoplasm of a bacterial cell) may be transferred from the donor bacterium to the recipient bacterium.
2 *Transformation:* Introduction of cell-free DNA from one cell to another, resulting in a genetic alteration in the recipient cell.
3 *Transduction:* A bacterial virus-mediated transfer of DNA from one cell to another.

One type of bacterial mutant is the fermentation mutant. In this exercise you will work with a mutant of *Escherichia coli*. The wild-type *E. coli* ferments the sugar, lactose. The mutant strain has lost the capacity to ferment lactose; it is lactose-negative. However, it is not a very stable mutant and frequently reverts back to its original state, that is, to the wild-type strain. The process is a back mutation to the lactose-positive state. Such a change from the lactose-negative state to the lactose-positive state can be detected very easily on a differential medium. On MacConkey agar medium, lactose-positive colonies appear red and lactose-negative colonies appear colorless [FIGURE 37.1].

REFERENCE
MICROBIOLOGY, Chap. 13, "Inheritance and Variability."

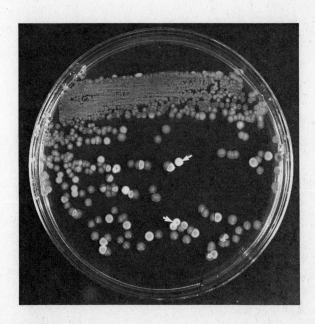

FIGURE 37.1
An unstable lactose fermentation mutant of *Escherichia coli* on MacConkey agar medium. Lactose-negative colonies are colorless; lactose-positive colonies are red in color (arrows). *(Courtesy of Liliane Therrien and E. C. S. Chan, McGill University.)*

MATERIALS

Escherichia coli ATCC 15939 on nutrient agar
3 MacConkey agar plates

PROCEDURE

1 You are provided with a mutant strain of *E. coli* grown on a nutrient-agar slope.
2 Streak the organism on MacConkey agar for isolated colonies, and incubate at 37°C overnight.
3 Observe the plate and examine for both lactose-positive and lactose-negative colonies. (On MacConkey agar, lactose-positive colonies appear red, and lactose-negative, colorless.)
4 Pick one lactose-positive and one lactose-negative colony, and streak them on MacConkey agar plates.
5 Incubate at 37°C, and examine them the next day.

NAME _____

Bacterial Variation Due to Genotypic Change

RESULTS

1 Describe the appearance of the colonies on the initial MacConkey agar medium plate.

2 What kind of results did you obtain upon subsequent transfer of the lactose-positive and lactose-negative colonies?

QUESTIONS 1 Explain the chemical basis for the differential property of MacConkey agar medium with respect to fermentation mutants of lactose.

2 What other medium, which you have used previously, can accomplish the same purpose as the MacConkey agar medium used in this exercise?

EXERCISE 38

Nutritional Mutants

OBJECTIVE
To show that a number of bacterial mutants are blocked at different steps in the synthesis of tryptophan.

OVERVIEW
In bacteria, the genes that code for the enzymes of a metabolic pathway usually are arranged in a consecutive manner to form a functional unit called an *operon*. A mutation in any one of these genes can lead to an inactive enzyme, so that an important intermediate is not synthesized, and there results a nonfunctional metabolic pathway. The end result may be either death or nongrowth of the cell.

Nutritional mutants are bacteria that have lost their ability to synthesize a substance essential to their growth because of a mutation in their genes. Such mutants can therefore grow only when that substance is provided in the medium. They have one or more enzyme deficiencies preventing the synthesis of the essential substance, for example, an amino acid.

The following experiment has been designed to show that a number of mutants, all derived from the same wild-type able to synthesize tryptophan and now requiring its addition to the minimal medium for growth to occur, are blocked at different steps in the synthesis of this amino acid. A portion of the pathway of tryptophan biosynthesis is as follows:

$$\xrightarrow{A} \text{Anthranilic acid} \xrightarrow{B} \text{indole} \xrightarrow{C} \text{tryptophan}$$

If the mutant is blocked in the conversion of indole to tryptophan, that is, at point C, it will grow only when tryptophan is added to the medium. If, however, the mutant is blocked at point B, it cannot convert anthranilic acid to indole and will grow when either indole or tryptophan is added to the medium but not when anthranilic acid is added. The reason for this, of course, is that the enzyme which converts anthranilic acid to indole has been lost as a result of the mutation. If the mutation has caused a block at point A, the mutant will grow when any one of the three substances is added to the medium. By the method outlined below, it is possible to establish at which point a mutation has occurred in the biosynthesis of tryptophan.

REFERENCES
MICROBIOLOGY, Chap. 13, "Inheritance and Variability."
Freifelder, D., *Microbial Genetics*, Jones and Bartlett Publishers, Boston, 1987, p. 601.

MATERIALS
Nutrient-agar slants of *E. coli* strains ATCC 23716, 23717, 23718, 23719, and 23720 (12 to 18 h)
20 minimal-agar-medium plates
10 ml of sterile 1% acid-hydrolyzed casein
Sterile glass spreaders
10 ml of sterile tryptophan, 2 mg/ml
10 ml of sterile anthranilic acid, 2 mg/ml
10 ml of sterile indole, 2 mg/ml
5 tubes of liquid minimal medium of 5 ml each

PROCEDURE

1 You are provided with four mutants blocked at various points in the tryptophan pathway and also with the wild-type parent organism, all grown for 12 to 18 h on nutrient-agar slants. Inoculate heavily into liquid minimal medium and incubate overnight before use.

2 Prepare the minimal-medium plates in the following manner. Add 3 to 4 drops of 1% casein hydrolysate (acid-hydrolyzed) to each of 20 plates. Spread these drops over the surface with a clean sterile glass spreader. The small amount of casein hydrolysate, which contains no tryptophan, helps to get the organism started on simple glucose-salt medium.

3 To 5 of the above plates, spread in a similar manner anthranilic acid; to another 5, indole; and to another 5, tryptophan. The remaining 5 serve as controls and will detect the wild-type.

4 Streak a loopful of each organism on one of each of the 4 different plates, in the following manner:

a Minimal medium plus casein hydrolysate
b Minimal medium plus casein hydrolysate plus anthranilic acid
c Minimal medium plus casein hydrolysate plus indole
d Minimal medium plus casein hydrolysate plus tryptophan

5 Incubate at 37°C for 24 to 48 h.

NAME _____

Nutritional Mutants

RESULTS   Locate the block in each mutant, and designate which strain is the wild-type:

E. coli ATCC 23716	
E. coli ATCC 23717	
E. coli ATCC 23718	
E. coli ATCC 23719	
E. coli ATCC 23720	

QUESTIONS   **1** Define the following terms:
Auxotroph:

Prototroph:

2 What is the difference between an inducible enzyme and a constitutive enzyme?

EXERCISE 39

Isolation of Streptomycin-Resistant Mutants of *Escherichia coli*

OBJECTIVE
Attempt to isolate streptomycin-resistant strains (mutants) of *Escherichia coli* from a parent (wild-type) strain that is streptomycin-susceptible by means of the gradient-plate technique.

OVERVIEW
During normal growth of a bacterial culture, a few cells undergo **mutation**, which is a change in the nucleotide sequence of a single gene. Such cells that carry a mutated gene are called **spontaneous mutants** because no obvious mutagenic agent was imposed on the culture. If they are able to grow in the environment provided, the altered DNA is reproduced in successive generations; consequently, these organisms exhibit some characteristics different from those of the parent strain. In normal cell populations, the incidence of spontaneous mutants is very low, for example, 1 in 10,000 (10^{-4}) or 1 in 1,000,000 (10^{-6}). Nevertheless, since bacterial cultures ordinarily reach populations in the hundreds of millions or billions, an appreciable number of mutants is produced. The fate of the mutant is determined by its ability to survive in the environment in which it occurs; in many instances, the environmental conditions are less favorable for the mutant than for the parent strain, and the mutant is crowded out.

However, spontaneous mutants that are resistant to an antibiotic are easily detected because they grow in concentrations of antibiotics that are inhibitory to the growth of other microorganisms. The selective action of antibiotics added to the medium allows such mutants to overgrow the susceptible microorganisms in the culture.

REFERENCES
MICROBIOLOGY, Chap. 13, "Inheritance and Variability."

Freifelder, D., *Microbial Genetics*, Jones and Bartlett Publishers, Boston, 1987, p. 601.

MATERIALS
18 nutrient-agar deeps, 15 ml per tube
6 nutrient-agar slants
6 sterile Petri dishes
Sterile streptomycin stock solution, 10 mg/ml
18- to 24-h broth culture of *Escherichia coli*
1 glass spreader immersed in beaker with 70% ethanol
Reagents for Gram's stain

PROCEDURE
1 Prepare three streptomycin gradient plates as follows. Pour one tube of melted, cooled (45°C) nutrient agar into a Petri dish; elevate one side of the plate so that the medium is in a thin layer on one side and a thick layer on the other [FIGURE 39.1A].

2 When this layer of medium solidifies, place the plate on a horizontal surface and pour in a tube of nutrient agar to which you previously added 1 ml sterile streptomycin solution. The plate should now appear as shown in FIGURE 39.1B. The streptomycin will establish a concentration gradient across the plate; the high concentration of streptomycin will be where the streptomycin agar is thick, and the low concentration on the opposite side. Mark your plate accordingly.

3 Place the agar plates in the 35°C incubator for approximately 1 h to reduce surface moisture.

4 Inoculate each plate by placing a single drop of *E. coli* culture on the agar surface and then spreading it out uniformly over the entire surface with a sterile glass spreader. It is important that the surface be uniformly and completely inoculated. The spreader may be sterilized by leaving it immersed in 70% ethanol; the alcohol is then flamed off and

FIGURE 39.1
Preparation of a streptomycin-gradient plate.

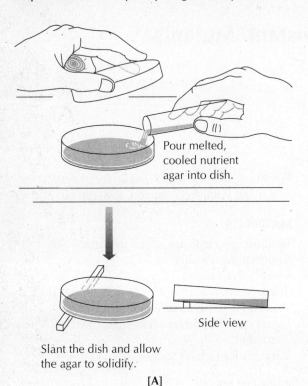

[A]

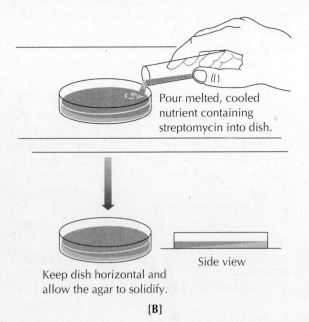

[B]

the spreader cooled by contact on the sterile surface of the gradient plate before spreading.

5 Incubate the plates from 2 to 4 days; then observe them for the presence of isolated colonies in the region of the high streptomycin concentration. Transfer one or more of these colonies (maximum of five) to a nutrient-agar slant, label them adequately, and incubate, along with a transfer from the parent strain, for 24 to 48 h.

6 Compare the streptomycin resistance of the parent strain and mutant strains. Prepare one gradient plate for each mutant as well as the parent strain as before. Make a single streak of each culture across each plate from the side with low streptomycin concentration to the side with high streptomycin concentration. Be sure that the plate is labeled so that you can identify the culture used for each streak. You will, in effect, have each culture inoculated on media ranging from a low level to a high level of streptomycin. Incubate plates at 35°C and observe them for growth after 24 to 48 h.

7 At the same time that step 6 is performed, characterize each of the strains morphologically using the Gram-stain technique.

Isolation of Streptomycin-Resistant Mutants of *Escherichia coli*

RESULTS

1 Draw the appearance of the three plates after the initial incubation.

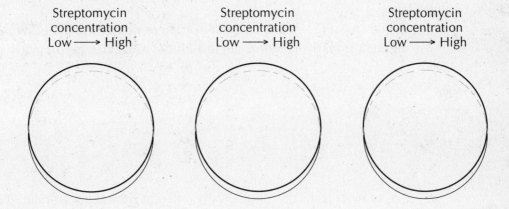

2 Draw the appearance of the control plate and the isolated mutant plates after the second incubation.

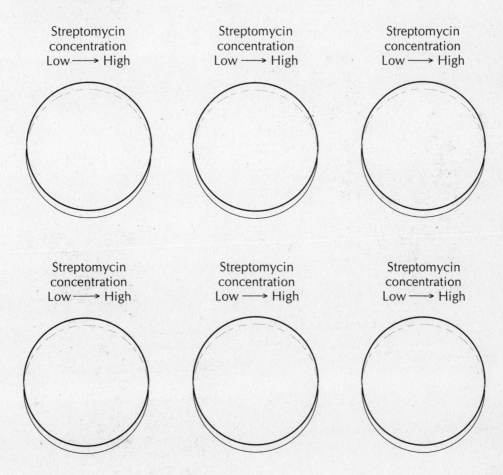

QUESTIONS 1 From your drawings, state whether you have isolates that are more resistant than the parent or wild-type strain. Provide reasons to support your statement.

2 Did any of the strains that you isolated exhibit morphological characteristics different from the parent strain in addition to greater streptomycin resistance?

3 Can mutants be morphologically different from the wild-type strain?

EXERCISE 40

Bacterial Conjugation

OBJECTIVE
To carry out conjugation between two compatible types of bacteria for the purpose of demonstrating recombination in procaryotic cells.

OVERVIEW
Conjugation is one of three recombination processes in haploid procaryotic microorganisms. It is a mating process that involves the one-way transfer of DNA through cell-to-cell contact between two sexually different cell types. The differentiation between cell types is based on the presence of a *fertility factor*, or *F factor*, within the cell. Cells possessing the F factor have the ability to act as genetic donors (males) during mating; they are designated as either **F$^+$** or **Hfr**. The F factor in F$^+$ cells is extrachromosomal; that is, it exists in the cell as a *plasmid*. The F factor in Hfr cells is integrated into the bacterial chromosome, and a high rate of recombinants occurs upon transfer of DNA from Hfr donors; that is why the abbreviation stands for *high-frequency recombinants*. Recipient cells (females) with no F factor are designated **F$^-$**. Thus, conjugation involves the transfer of genetic material between a donor F$^+$ or Hfr cell and a recipient F$^-$ cell.

In an F$^+$ × F$^-$ cross, it is usually only the donor's F plasmid that moves to the F$^-$ cell. Chromosomal genes can be transferred along with the F plasmid, but this is a rare event, perhaps only 1 in 10 million matings. However, in a Hfr × F$^-$ mating, a donor Hfr strain donates its genes to the recipient strain in a linear fashion from a starting point that includes a *small* piece of the F plasmid on the bacterial chromosome. Genes located near the starting point are transferred earlier during mating than those located farther away. Because mating is often interrupted, it is rare that a whole set of genes (including the remainder of the F plasmid) or the whole chromosome is transferred to the recipient.

Therefore, those genes closest to the starting point in each strain are most often represented in the recombinants. The entire integrated F plasmid itself is rarely transferred during conjugation. Therefore, most recipient cells remain F$^-$ after conjugation with Hfr cells.

In this exercise, two *auxotrophic* mutants that cannot grow on a minimal medium will be used. If, after the cultures have been mixed and incubated for a short time, they are then plated on minimal medium and growth occurs, conjugation must have taken place.

REFERENCES
MICROBIOLOGY, Chap. 13, "Inheritance and Variability."
Freifelder, D., *Microbial Genetics*, Jones and Bartlett Publishers, Boston, 1987, p. 601.

MATERIALS
Escherichia coli strains ATCC 23741 (Hfr, *met*$^-$) and 23724 (F$^-$, *thr*$^-$, *leu*$^-$, *thi*$^-$) grown in 10 ml of heart-infusion broth for 18 h
10 ml of sterile saline
2 centrifuge tubes
Bench centrifuge
10 minimal-medium-agar plates, supplemented with 20 µg/ml threonine and 2 µg/ml thiamine
4 heart-infusion-agar plates
Sterile 1-ml pipettes

PROCEDURE
1 You are provided with two strains of *E. coli* grown overnight in 10 ml of heart-infusion broth at 37°C.
2 Centrifuge each culture and wash twice with sterile saline. After the final centrifugation, resuspend the cells in 3 ml of sterile saline.

3 Combine 2 ml from each culture, retaining the other 1 ml for control plates.
4 Incubate at 37°C for 15 to 20 min.
5 Plate samples from each culture and from the mixture by spread-plate technique on minimal-medium-agar plates as follows:

Mixture:
0.05 ml on two minimal-medium-agar plates
0.10 ml on two minimal-medium-agar plates
0.20 ml on two minimal-medium-agar plates

Pure cultures (separately):
0.2 ml on two minimal-medium-agar plates
0.2 ml on two heart-infusion-agar plates

6 Incubate at 37°C for 24 to 72 h and examine periodically for growth of prototrophs from the mixed culture plated on minimal medium agar. There should be no growth on the minimal medium agar with the original pure (unmixed) cultures.

Bacterial Conjugation

RESULTS Report any growth obtained as + or no growth obtained as − in the chart provided.

| GROWTH OF TWO BACTERIAL STRAINS BEFORE AND AFTER CONJUGATION ||||||
Culture	Medium	Inoculum volume, ml	Growth, +	No growth, −
Mixed	Minimal	0.05		
Mixed	Minimal	0.05		
Mixed	Minimal	0.1		
Mixed	Minimal	0.1		
Mixed	Minimal	0.2		
Mixed	Minimal	0.2		
Hfr	Minimal	0.2		
Hfr	Minimal	0.2		
Hfr	Infusion	0.2		
Hfr	Infusion	0.2		
F⁻	Minimal	0.2		
F⁻	Minimal	0.2		
F⁻	Infusion	0.2		
F⁻	Infusion	0.2		

QUESTIONS 1 Explain the differences between F^-, F^+, and Hfr strains of bacteria.

2 Briefly describe the morphology and the function of a filamentous appendage that is involved in bacterial conjugation.

3 Explain why there was no growth of the pure cultures on the minimal medium agar plates.

4 Explain why there was growth of the mixed cultures on the minimal medium agar plates.

EXERCISE 41

Regulation of Enzyme Synthesis: Enzyme Induction and Catabolite Repression

OBJECTIVE

To demonstrate experimentally the mechanisms of enzyme induction and catabolite repression in *Escherichia coli*.

OVERVIEW
ENZYME INDUCTION

Many bacteria can use a wide range of organic compounds as carbon and energy sources. However, in many cases the structural genes specifying the enzymes required to degrade the organic compounds are not transcribed unless the substrates, that is, the organic compounds, are actually present in the cell's environment. (In reality, a low level of synthesis is always taking place, but it may be so small as to be undetectable.) In the presence of the substrate, the enzyme is **induced**; that is, the appropriate structural gene is transcribed by the RNA polymerase, and the resulting *m*RNA is translated to yield the active enzyme. In this way the cell does not waste energy and amino acids for the synthesis of enzymes which are not needed because the substrates for these enzymes are not present. Of course, there are some enzymes, such as those of the Embden-Meyerhof pathway, which are so important to the growth of cells that they must always be present. These enzymes are not induced but are always present. Such enzymes are said to be **constitutive**.

The mechanism of enzyme induction is as follows. In the absence of the *inducer* (the substrate), *repressor proteins* coded by a regulator gene (and continuously synthesized) bind to a specific nucleotide sequence next to the structural gene for the enzyme. This specific nucleotide sequence where the repressor binds is called the *operator*. As long as a repressor protein is bound to the operator, the RNA polymerase cannot transcribe the structural gene for the enzyme. If the inducer is present, however, it or some derivative of it binds to the repressor protein; this inactivates the repressor so that it can no longer bind to the operator. The repressor protein is an *allosteric* (that is, it has two reactive sites) protein that changes its conformation in response to the attachment of a smaller molecule. With an inducer attached to the repressor, the repressor is changed into an inactive conformation. With the repressor no longer bound to the operator, the RNA polymerase can bind to the DNA at a site called the *promoter* and then proceed to transcribe the structural gene coding for the enzyme.

After the enzyme has been synthesized, it can catalyze the chemical reaction involved in the degradation of the substrate. Thus the substrate for the enzyme will eventually be removed from the environment, and there will no longer be a need for continued synthesis of the enzyme. Fortunately, the inducer (substrate) is bound to the repressor protein by very weak secondary bonds (hydrogen bonds, salt linkages), so that the link between the two can be rapidly made or broken. As the concentration of the inducer decreases, less and less of the repressor protein will have inducer attached to it, and the repressor will again be able to attach to the operator and prevent transcription of the structural gene for the enzyme.

FIGURE 41.1 illustrates the mechanism for induction of the enzyme β-galactosidase in *Escherichia coli*. This enzyme catalyzes the hydrolysis of lactose into glucose and galactose, and this hydrolysis is an essential first step in the metabolism of lactose by *E. coli*. The control mechanism for the synthesis of β-galactosidase was one of the earliest studied control mechanisms, and it is still the best understood.

CATABOLITE REPRESSION

Although an inducer may be present and thus a bacterial cell may theoretically be capable of synthesizing an inducible enzyme, it often happens that

FIGURE 41.1
The Jacob-Monod model of gene control in enzyme induction.

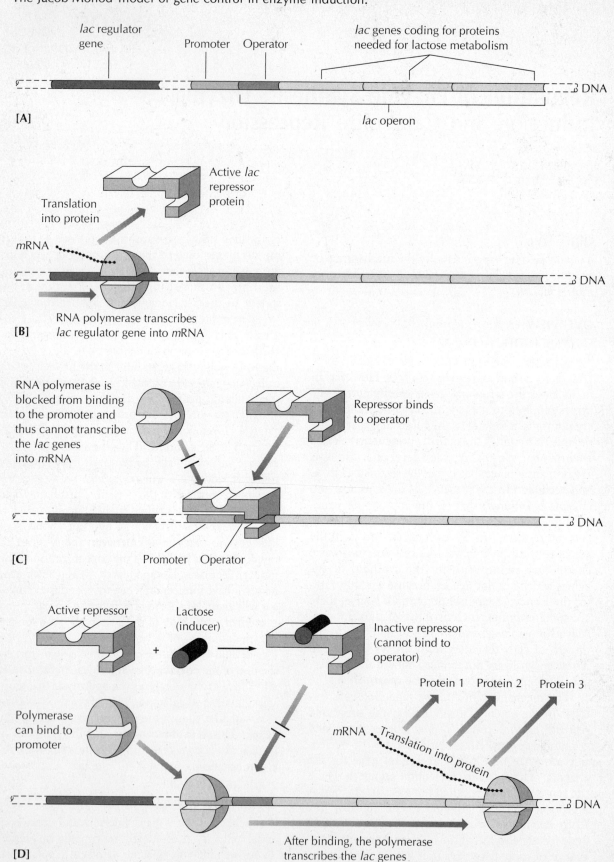

the enzyme may still not be made if a rapidly utilizable carbon source is present. For example, in *E. coli*, the enzyme β-galactosidase may not be made even though the lactose is present if the sugar glucose is also present. When this phenomenon was first observed, it was called the "glucose effect," but it is now called **catabolite repression**. The action of glucose in inhibiting transcription of the structural gene for β-galactosidase is not direct. However, before we can discuss *how* glucose prevents enzyme induction, we must first look at the overall sequence of events.

When enzyme induction takes place, RNA polymerase binds to the promoter and then transcribes the structural genes. The resulting *m*RNA is then translated to yield protein (enzyme). However, the binding of RNA polymerase to the promoter is not a straightforward process. In order for binding to occur, a protein called CAP must already be bound to the promoter at a site approximately 25 base pairs long. This protein has cyclic AMP (adenosine-3', 5'-monophosphate), or cAMP, attached to it. The CAP protein is a dimer of two identical subunits, each containing 210 amino acids, and each molecule of CAP binds two molecules of cAMP. (The complete nucleotide sequence of CAP has been deduced from the nucleotide sequence of the cloned CAP gene.) Once the CAP-cAMP complex is bound to the promoter, RNA polymerase can now easily bind to the promoter. Thus as long as there is enough CAP-cAMP in the cell, RNA polymerase will bind to the promoter and transcribe the structural gene for β-galactosidase. However, when *E. coli* metabolizes certain sugars such as glucose, specifically sugars that are transported into the cell by a *phosphoenolpyruvate phosphotransferase system (PEP-PTS)*, the level of cAMP in the cell decreases to a very low level. This results in too little cAMP being available to allow formation of the CAP-cAMP complex that is required to ensure good binding of RNA polymerase to the promoter. Thus even if lactose, maltose, arabinose, or another sugar whose metabolism requires the synthesis of inducible enzymes is present, induction of those enzymes will not occur as long as glucose is also present.

A reasonable question to ask at this point is, "How does glucose metabolism decrease the level of cAMP in cells?" The answer is not entirely clear, but there is evidence that supports the model which is now described. The PEP-PTS functions to transport certain sugars such as glucose into the cell. However, one of the enzymes of the PEP-PTS, namely, Enzyme III, also seems to play an important role in controlling the level of cAMP in the cell. Available evidence indicates that Enzyme III, when phosphorylated to form Enzyme III-phosphate, can activate adenylate cyclase. Adenylate cyclase is an enzyme that catalyzes the conversion of ATP to cAMP + pyrophosphate. This results in the production of a high level of cAMP in the cell. However, when the PEP-PTS is being used to transport glucose into the cell, Enzyme III is only briefly phosphorylated. This is because it very quickly passes its phosphate group along to Enzyme II, which in turn transports glucose into the cell and converts the glucose to glucose-6-phosphate. Without a phosphate group, Enzyme III does not activate adenylate cyclase, and thus cAMP is not formed from ATP. Consequently, the CAP-cAMP complex is not formed, RNA polymerase is unable to bind well to the promoter for inducible enzymes, and inducible enzymes are not synthesized despite the presence of their inducers.

REFERENCES

MICROBIOLOGY, Chap. 13, "Inheritance and Variability."

Feucht, B. U., and M. H. Saier, Jr., "Fine Control of Adenylate Cyclase by the Phosphoenolpyruvate: Sugar Phosphotransferase Systems in *Escherichia coli* and *Salmonella typhimurium*," *Journal of Bacteriology* 141:603–610, 1980.

Osumi, T., and M. H. Saier, Jr., "Regulation of Lactose Permease Activity by the Phosphoenolpyruvate Sugar Phosphotransferase System: Evidence for Direct Binding of the Glucose-Specific Enzyme III to the Lactose Permease," *Proceedings of the National Academy of Sciences U.S.A.* 79:1457–1461, 1982.

GENERAL PLAN OF EXERCISE

A culture of *E. coli* strain K12 will be grown to mid-exponential phase in a minimal medium containing glycerol as the sole carbon source. The culture will be divided into five portions:

1 No addition
2 Glucose addition
3 Lactose addition
4 Lactose + glucose addition
5 Lactose + glucose + cAMP addition

The production of β-galactosidase in the cells in each of these five portions will be monitored for 60 min. Lactose will be used as the inducer of β-galactosidase. The activity of β-galactosidase will be assayed by using an artificial substrate, an analog of lactose called o-nitrophenyl-β-D-galactoside (ONPG). This compound is colorless. β-galactosidase catalyzes the hydrolysis of ONPG to galactose and o-nitrophenol. The free o-nitrophenol has a yellow color whose intensity can be measured spectrophotometrically at 420 nm.

MATERIALS

Culture of *Escherichia coli* strain K12: Grow the culture in glycerol minimal medium for 2 or 3 days prior to the experiment to adapt it to glycerol. From this culture, inoculate a flask containing 60 ml of glycerol minimal medium on the day of the experiment. Incubate the flask for about 4 h, or until the optical density is about 0.05 at 420 nm. Shortly before class, dispense portions of this culture from the flask into each of five sterile culture tubes so that each tube contains 5 ml of culture. Keep the tubes in a 37°C water-bath.

Water-bath at 37°C (to hold test-tube rack)

Test-tube rack

Colorimeter or spectrophotometer [to read color intensity (optical density or equivalent) at 420 nm]

2 optically matched cuvettes for the colorimeter or spectrophotometer (to measure absorbance of samples having a volume of 3.7 ml)

1 tube sterile distilled water

25 culture tubes, each containing 1 drop of toluene and 0.5 ml of Z buffer. Keep these tubes chilled until the beginning of the experiment.

1 empty culture tube

Z buffer, composition per liter of distilled water:

$Na_2HPO_4 \cdot 7H_2O$	16.1 g
$NaH_2PO_4 \cdot H_2O$	3.5 g
KCl	0.7 g
$MgSO_4 \cdot 7H_2O$	0.2 g
β-mercaptoethanol	2.7 ml

1 tube containing 7 ml (o-nitrophenyl-β-galactoside) solution, 4 mg/ml. Store in refrigerator. (ONPG can be obtained from various biochemical supply companies; for example, cat. no. N1127 from Sigma Chemical Co., St. Louis, MO, is suitable.)

100 ml of 0.5 M Na_2CO_3 solution

2-ml portions of sterile carbohydrate solution (if necessary, dissolve by gentle heating and sterilize by filtration):
Glucose 200 mg/ml
Glucose 400 mg/ml
Lactose 200 mg/ml
Lactose 400 mg/ml

0.5 ml of sterile cAMP (sodium salt) solution, 80 mg/ml. (cAMP can be purchased from various chemical supply companies; for example, cat. no. A 6885 Sigma Chemical Co. is suitable.) Sterilize by filtration.

35 sterile 1-ml serological pipettes with 0.1-ml or 0.01-ml graduations

2 nonsterile 5-ml pipettes with 0.1-ml graduations

Ice bucket with crushed ice

PROCEDURE

1 Label the five tubes of *E. coli* culture A, B, C, D, and E. Keep them in the 37°C water-bath.

2 Label each of the 25 tubes containing 1 drop toluene and 0.5 ml of Z buffer, numbering them 1 to 25. Label an empty 26th tube "Blank." Keep tubes 1 through 25 on ice so that the toluene does not evaporate.

3 At the indicated times, make the following additions to tubes A through E:

Tube	Time, min*	Addition
A	0	0.4 ml sterile distilled water
B	0 + 2	0.4 ml sterile glucose solution (200 mg/ml)
C	0 + 4	0.4 ml sterile lactose solution (200 mg/ml)
D	0 + 6	0.2 ml sterile glucose solution (400 mg/ml) + 0.2 ml sterile lactose solution (400 mg/ml)
E	0 + 8	0.2 ml sterile glucose solution (400 mg/ml) + 0.2 ml sterile lactose solution (400 mg/ml) + 0.1 ml sterile cAMP solution (80 mg/ml)

*Additions will be made on a staggered schedule. Tubes will be prepared at intervals of 2 min. Thus, though time of incubation remains constant, samples will be removed at intervals of 2 min.

Immediately after making the additions to each tube, mix the contents of the tube by gentle agitation and withdraw 0.5 ml. Place the 0.5 ml into a toluene-Z buffer tube as follows:

0.5-ml sample from tube A ⟶ tube 1
0.5-ml sample from tube B ⟶ tube 2
0.5-ml sample from tube C ⟶ tube 3
0.5-ml sample from tube D ⟶ tube 4
0.5-ml sample from tube E ⟶ tube 5

Keep tubes 1 through 5 on ice. The samples taken at this time will give the basal, or "0 time," uninduced β-galactosidase activity for tubes A through E. No more enzyme is synthesized after the cells are placed in the toluene.

4 Follow the protocol below to remove a 0.5-ml sample at 15 min from tubes A through E to a toluene-Z buffer tube. Keep the tubes in ice.

Tube	Time of removal of 0.5-ml sample	Sample removed to tube no.	Period during which cells were incubated with substrate(s)
A	0 + 15 min	6	15 min
B	0 + 17 min	7	15 min
C	0 + 19 min	8	15 min
D	0 + 21 min	9	15 min
E	0 + 23 min	10	15 min

5 Follow the protocol below to remove a 0.5-ml sample at 30 min from tubes A through E to a toluene-Z buffer tube. Keep the tubes in ice.

Tube	Time of removal of 0.5-ml sample	Sample removed to tube no.	Period during which cells were incubated with substrate(s)
A	0 + 30 min	11	30 min
B	0 + 32 min	12	30 min
C	0 + 34 min	13	30 min
D	0 + 36 min	14	30 min
E	0 + 38 min	15	30 min

6 Follow the protocol below to remove a 0.5-ml sample at 45 min from tubes A through E to a toluene-Z buffer tube. Keep the tubes in ice.

Tube	Time of removal of 0.5-ml sample	Sample removed to tube no.	Period during which cells were incubated with substrate(s)
A	0 + 45 min	16	45 min
B	0 + 47 min	17	45 min
C	0 + 49 min	18	45 min
D	0 + 51 min	19	45 min
E	0 + 53 min	20	45 min

7 Follow the protocol below to remove a 0.5-ml sample at 60 min from tubes A through E to a toluene-Z buffer tube. Keep the tubes in ice.

Tube	Time of removal of 0.5-ml sample	Sample removed to tube no.	Period during which cells were incubated with substrate(s)
A	0 + 60 min	21	60 min
B	0 + 62 min	22	60 min
C	0 + 64 min	23	60 min
D	0 + 66 min	24	60 min
E	0 + 68 min	25	60 min

8 You now have 25 enzyme samples. Remove the toluene from them by placing all 25 tubes into the 37°C water-bath to warm them. **Do this in a chemical hood so that the toluene vapors will not enter the room.** If necessary, blow air across the tubes, by means of a fan, to hasten the evaporation of toluene. The toluene damaged the cytoplasmic membrane of the cells and thus made the cells permeable to the artificial substrate ONPG; that is, ONPG can now get into the cells and be hydrolyzed by β-galactosidase, if present.

9 Now add 0.2 ml of ONPG solution to tubes 1 through 25 at 30-s intervals (i.e., add 0.2 ml of ONPG to tube 1, then 30 s later add 0.2 ml of ONPG to tube 2, then 30 s later add 0.2 ml of ONPG to tube 3, and so on). Continue in this manner until all 25 tubes have received the ONPG.

10 Prepare a blank by adding 0.5 ml of uninoculated glycerol minimal medium, 0.5 ml of Z buffer, and 0.2 ml of ONPG solution to the tube marked "Blank" (tube 26).

11 Exactly 30 min after ONPG was added to tube 1, add 3.0 ml of 0.5 M Na_2CO_3 to the tube to stop the reaction. Thirty seconds later add 0.5 M Na_2CO_3 to tube 2, etc. Continue in this fashion until all 25 tubes have received the Na_2CO_3 solution. Then add Na_2CO_3 to tube 26.

NOTE: In this experiment we did not measure bacterial growth during the 60-min incubation period for tubes A through E. We did not do this because we know from experience that there is very little difference in growth between tubes A through E during the 60 min of the experiment. However, if a longer incubation time is used, or if the growth pattern of the microorganism being tested is not known, some measurement of cell growth (absorbance, dry weight, cell number, or total protein) should be made. This is because the cells may grow much more with some substrates than with others, and, even if an enzyme is not induced, there may appear to be an increase in enzyme activity merely because more cells are present. Thus, if you found five times more cells during incubation with one substrate compared with another substrate, there would appear to be five times more enzyme present, even though the amount of enzyme per cell or per milligram of dry weight might be identical.

Regulation of Enzyme Synthesis: Enzyme Induction and Catabolite Repression

RESULTS

1 Pour the contents of tube 26 into a cuvette, and set the colorimeter or spectrophotometer to read 0 absorbance at 420 nm. Then, using a second optically matched cuvette, measure and record the absorbance of the contents of each tube successively from 1 through 25. (After measuring the absorbance of each sample, rinse out the cuvette with distilled water and drain before adding the subsequent sample.)

2 The absorbance of each sample is proportional to the amount of β-galactosidase activity present in the cells before enzyme synthesis was stopped by toluene. Let an absorbance value of 0.001 equal 1 unit of enzyme activity (i.e., divide each absorbance value by 0.001).

3 Complete the following chart by recording the units of enzyme activity below each tube number.

	UNITS OF ACTIVITY OF β-GALACTOSIDASE IN TUBES				
Incubation time of cells with substrates	A	B	C	D	E
0 min	Tube 1	Tube 2	Tube 3	Tube 4	Tube 5
15 min	Tube 6	Tube 7	Tube 8	Tube 9	Tube 10
30 min	Tube 11	Tube 12	Tube 13	Tube 14	Tube 15
45 min	Tube 16	Tube 17	Tube 28	Tube 19	Tube 20
60 min	Tube 21	Tube 22	Tube 23	Tube 24	Tube 25

4 Prepare a plot on linear graph paper of the units of enzyme activity (vertical axis) versus time in minutes (horizontal axis) for each of the five enzyme systems (tubes A through E).

QUESTIONS 1 Is β-galactosidase in *E. coli* inducible by lactose? By glucose?

2 What is the effect of exposing *E. coli* to both lactose and glucose at the same time? Describe the mechanism involved.

3 What is the effect of exposing *E. coli* to a combination of lactose, glucose, and cAMP at the same time? Describe the mechanism involved.

4 What would be the result of a mutation in the operator that no longer allowed the repressor protein to bind?

5 What would be the result of a mutation in the gene for CAP such that the CAP would no longer bind to the promoter?

EXERCISE 42

The Ames Test: Using Bacteria to Detect Potential Carcinogens

OBJECTIVE
To learn the principles involved in performing the Ames test to screen for potential chemical carcinogens.

OVERVIEW
It is generally accepted that some forms of cancer in humans are caused by certain types of chemicals. Therefore, it is important for us to know what these chemicals are, and, knowing them, we can prevent them from being released into the environment or contaminating our foods. One way to detect potential *carcinogenic*, or cancer-causing, chemicals is to use laboratory animals. However, a more humane way is to use bacteria since very few known animal carcinogens have failed detection by the use of bacteria. In any case, since there are so many chemical substances that have to be tested, it is far easier, faster, and more cost-effective to use microbial systems than rats, mice, rabbits, or other laboratory animals in screening programs.

Research has shown that more than 90 percent of the chemical carcinogens are also mutagens. From a genetic standpoint, strong evidence links carcinogenicity with mutagenicity. This suggests that human cancers result from mutational alterations in cellular genetic material, or DNA. This observation spurred Dr. Bruce Ames and his colleagues at the University of California to develop a simple and inexpensive test that uses a bacterial mutant to screen for mutagens. The *Ames test* detects mutagenic agents, and hence probable carcinogens, by using special histidine-dependent strains of *Salmonella typhimurium*. (See BOX 42.1 on the special attributes of the mutant strains.) When these histidine-requiring (his^-) auxotrophs (nutritionally deficient) are inoculated on a chemically defined agar medium containing a trace of histidine, only a few cells revert to histidine-independence (his^+) and are able to form colonies. If a mutagenic chemical is added to this histidine-poor medium, the reversion rate to histidine independence is greatly increased. The trace of histidine is used in the test medium to enable all cells to undergo a few cell divisions, which are usually necessary for mutagenesis. This is because some mutagenic agents act only on replicating DNA. This trace concentration cannot lead to colony formation. The Ames test takes only about 3 days and costs several hundred dollars per chemical tested compared to 2 to 3 years and more than $100,000 per chemical when using laboratory animals for testing.

If you were working professionally and carrying out the Ames test, you would be required to follow a standard set of procedures. This ensures reliable and reproducible results from independent laboratories. The actual test involves many controls, various mutant strains, and various concentrations of the chemical to be tested. The exercise you will do is an abbreviated and modified test that can be performed expediently in a large class to enable you to learn the principles underlying the actual Ames

---

BOX 42.1 About the Ames Mutants

There are many mutant strains of *Salmonella typhimurium* used in the Ames test. Each of them is auxotrophic because of a different type of mutation in the histidine operon. All appear to be either frame-shift or base-substitution mutants, and each is designed to test for a different type of reversion or back mutation. These mutants do not commonly revert. They also lack a DNA excision-repair mechanism, which would normally correct such deletion errors. In addition, these mutants have a defective lipopolysaccharide layer on their cell surface that enables mutagens to penetrate more easily into the cell. Certain strains used in this test also carry R plasmids that render them more susceptible to some weak mutagens.

test. For example, in the standard Ames test, every chemical to be tested is treated with an induced liver-enzyme preparation in order to *activate* it. This is because many chemical substances are not carcinogenic or mutagenic unless they are metabolized to active substances. In humans and other animals these metabolic conversions to active carcinogens occur primarily in the liver. This step will not be performed in this exercise because of the kind of carcinogens used, namely, the nitrocarcinogens which are converted to their active forms by bacterial nitroreductase of the test organism.

REFERENCES

MICROBIOLOGY, Chap. 13, "Inheritance and Variability."
MM, Chap. 14, "Gene Mutation."

MATERIALS

Nutrient-broth culture of *Salmonella typhimurium* Ames strain TA 1538 (ATCC 29631), 24-h culture
3 Petri plates with Ames minimal agar medium
3 2-ml Ames soft-agar tubes containing traces of histidine [two of these contain a different chemical each; they are labeled "Chemical 1" for 4-nitro-ortho-phenyldiamine (NPD) and "Chemical 2" for 4-nitroquinoline-N-oxide (NQNO)]
0.1-ml sterile pipettes
Water-bath set at 45°C

> **CAUTION:** The instructor will prepare these tubes from stock solutions in a chemical fume hood. Dissolve enough of the NPD or NQNO in distilled water to obtain 10 μg/ml. Students will use 0.1 ml per overlay. Also, students performing this exercise must wear protective latex gloves. They should discard waste materials in special containers provided.

PROCEDURE

1 Label the three Petri plates containing Ames minimal agar medium in the following manner: "Control," "Chemical 1" for NPD, and "Chemical 2" for NQNO.

2 Obtain three 2-ml Ames soft-agar tubes from the water-bath set at 45°C: one contains no chemical, one has chemical 1, and another has chemical 2. Leave them in the water-bath and work with them one at a time.

3 Aseptically add 0.1 ml *S. typhimurium* culture into the control tube containing no chemical. Mix the suspension by rolling the tube between the palms of your hands or gently shaking it. Pour it into the Petri plate marked "Control" and gently tilt the Petri plate around to distribute the molten agar evenly over the surface of the minimal agar base layer.

4 Repeat step 3 with the other two soft-agar tubes containing chemicals.

The procedure for carrying out the exercise is shown in FIGURE 42.1.

5 Incubate all plates at 30°C for 48 to 72 h.

42 THE AMES TEST: USING BACTERIA TO DETECT POTENTIAL CARCINOGENS

FIGURE 42.1
Procedure for performing the modified Ames test.

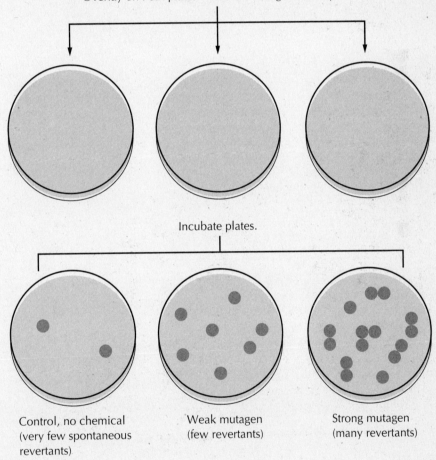

NAME _____

The Ames Test: Using Bacteria to Detect Potential Carcinogens

RESULTS Count the number of large colonies on each plate and record your results in the chart below.

Control plate (no chemical)	
Chemical 1 plate	
Chemical 2 plate	

NOTE: The control plate gives you the count of *spontaneous revertants*.

QUESTIONS 1 From your results, state, with reasons, whether the chemicals tested are mutagenic or potentially carcinogenic.

2 Give the advantages and disadvantages of using the Ames test for screening potential carcinogens.

3 What medium could you have added to your exercise to show that a histidine-deficient, chemically defined medium is necessary to carry out the Ames test?

VIRUSES

The laboratory propagation of viruses requires stable viable host cells in which the viral agent multiplies. The host cells are bacterial for bacterial viruses (***bacteriophages*** or ***phages***); molds for mold viruses (***mycophages***); animal-cell cultures, embryonated chicken eggs, or animals for animal viruses; and plant-cell cultures and whole plants for plant viruses.

The bacterium-bacteriophage system is the most convenient since it is relatively much easier to work with bacteria in the laboratory than with other virus-host systems and therefore serves as a model for investigating phenomena in host-cell–virus interaction. For instance, bacteria can be produced in abundance economically, rapidly, and easily. Action of the bacteriophage can be assayed in broth cultures or upon surface growth on nutrient-agar plates. This system is amenable to many manipulations for experimental purposes.

Propagation of animal or plant viruses in tissue cultures requires the maintenance and cultivation of appropriate host-cell lines in nutrient solutions dispensed in shallow layers in bottles or special flasks or tubes. The embryonated chicken egg is also a suitable and convenient means for the propagation of many animal viruses.

Direct viewing of a virus is possible by means of electron microscopy [FIGURE IX.1]. Some of the characteristic virus shapes are helical, polyhedral, and spermlike.

Measurement of the concentration of virus preparations, particularly animal viruses, may be performed through the use of a variety of serological tests in which the virus functions as the antigen. Alternatively, ***plaque,*** that is, an area of cell lysis, assays may be performed on monolayers of animal-cell cultures.

FIGURE IX.1
Human immunodeficiency virus, type 1 (HIV-1), budding from the surface of an infected human T cell. *(Courtesy of Advanced Biotechnologies, Inc., Columbia, Maryland.)*

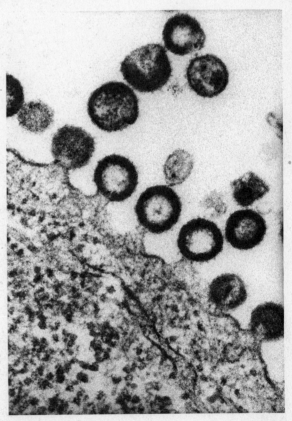

0.1 μm

EXERCISE 43

Bacterial Lysis by Bacteriophage: Phage Titer and the Plaque Assay

OBJECTIVE
To determine the concentration (titer) of a bacteriophage suspension by means of a modified plaque assay.

OVERVIEW
Viral agents that attack bacteria, namely, bacteriophages (phages), may be considered as representative of the general nature of viruses. Thus they are used as models for understanding viral replication in host-cell–virus interaction. They are very specific in action: a bacteriophage for one strain of bacteria within a species may not attack another strain of the same species. The *virulent* bacteriophages lyse bacterial cells. This can be demonstrated in broth cultures, as clearing of broth, or on agar surface cultures, as clear zones or plaques. Such viruses may be used in epidemiological studies to identify specific strains of bacteria. There are other phages which do not carry out a lytic cycle within the host cell; instead, they have a *lysogenic cycle* within the host cell—the phage nucleic acid becomes incorporated into the host cell's chromosome (now called a *prophage*) and is replicated right along with the host chromosome through many bacterial generations with no lysis occurring. Such *temperate phages* are also very important because they have been shown to be responsible for the production of diphtheria toxin, botulism toxin, and other toxins by their bacterial hosts; such phages have also been implicated in the development of antibiotic resistance in bacteria. The two "lifestyles" of bacteriophages are depicted in FIGURE 43.1.

The host bacterium most commonly used for bacteriophage work is *Escherichia coli*. A well-studied group of phages belong to the T series, numbered 1 through 7, which attack the nonmotile strain B of *E. coli*. These are designated T1, T2, T3, . . . , T7 (T for type). The T phages are tadpole or sperm-like in form, with polyhedral heads (about 50 μm in diameter) and a narrow tail. The head has a single linear molecule of nucleic acid arranged in one continuous loop, which may be 50 μm long, and consists of double-stranded DNA. This DNA is tightly condensed in the phage head and is covered with a layer of protein. The phage tail is literally a hollow tube of protein enclosed within a contractile sheath. A typical bacteriophage is shown in FIGURE 43.2.

What happens in the lytic cycle may be described briefly as follows [see FIGURE 43.1]. The phage attaches to the cell wall of the bacterium by the tip of the viral tail; the sheath contracts, and the molecule of phage DNA is injected into the host cell. The protein coat of the phage remains outside the cell [FIGURE 43.3A]. There follows a redirecting of the host cell's synthetic capability by the phage genetic material so that the bacterium makes phage material instead of cell material. Many new phages are replicated inside the bacterium; the cell wall ruptures with the burst release of new phage particles. These new virions reinfect and subsequently lyse more cells; this continuing cycle of lysis/reinfection/lysis produces a visible plaque on a lawn of cells [FIGURE 43.3B].

REFERENCES
MICROBIOLOGY, Chap. 15, "Viruses: Morphology, Classification, Replication."
MICROBIOLOGY, Chap. 16, "Viruses: Cultivation Methods, Pathogenicity."

MATERIALS
2 Petri plates with trypticase agar
10 tubes with 9 ml in each of trypticase broth
Sterile glass spreader
9 sterile 1-ml pipettes (mouths plugged with cotton)
Specimen of *Escherichia coli* bacteriophage adjusted to have a titer of approximately 10^5 plaque-forming units (pfu)/ml
Trypticase-broth culture of 6- to 12-h bacteriophage-sensitive strain of *E. coli*

FIGURE 43.1
The two cycles of virus replication within a bacterium.

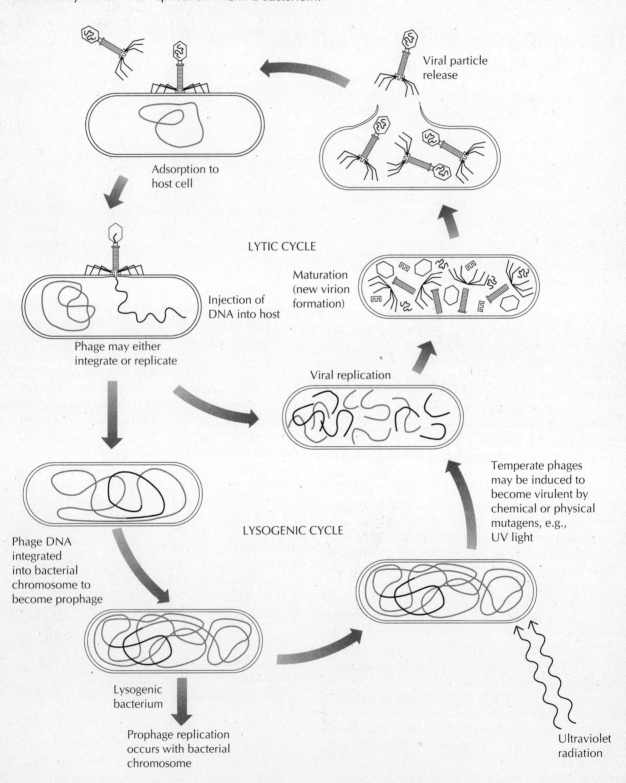

PROCEDURE

1 Arrange 10 tubes of trypticase broth in a row in a test-tube rack; label them 1 to 10.

2 Using aseptic precaution, transfer 1 ml of the phage specimen into tube 1. **Discard pipette.** With a fresh pipette mix the phage and broth by drawing up about 0.5 ml of the mixture, then releasing the volume back into the tube. Repeat this about five times; then draw up 1 ml and place in tube 2. **Discard pipette.** With a fresh pipette draw up and release back the mixture in tube 2 several times; then draw up 1 ml and place in tube 3. Continue this procedure until tube 9 is reached. Discard the 1 ml removed from tube 9. Tube 10 does not receive any of the diluted phage. FIGURE 43.4 illustrates the preparation of the phage dilutions.

3 Place one drop of *E. coli* on each of the two plates of trypticase agar provided. Using a sterile glass spreader, spread the inoculum over the entire surface of the plate. Extreme care must be exercised to ensure complete and uniform inoculation of the agar surface. Allow the inoculated plates to stand for several minutes to permit the agar surface to dry.

4 Invert the plates, draw five lines across the bottom of each, and label the lines numerically as shown below:

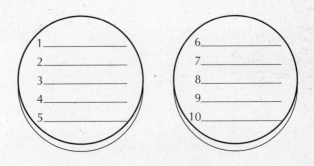

5 Now remove one loopful of material from tube 1 and streak this on the surface of the agar plate following line 1. Do the same with each of the remaining tubes; one loopful from each tube is streaked along the line corresponding to the tube number. Incubate plates inverted at 35°C.

6 Add 1 drop of the broth culture *E. coli* to each of the 10 tubes prepared in step 2 of this procedure. Incubate these tubes at 35°C.

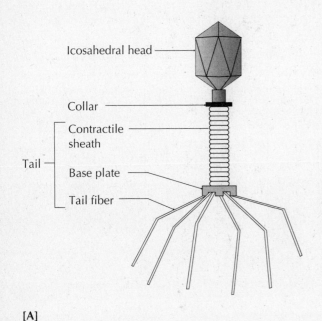

[A]

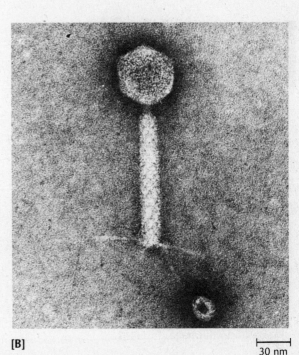

[B]

FIGURE 43.2
[A] Drawing of a typical bacteriophage. [B] Electron micrograph of a bacteriophage. *(Courtesy of H. W. Ackermann, Laval University.)*

FIGURE 43.3
[A] A phage lambda-infected *Escherichia coli* cell negatively stained. Because the phages have injected the DNA into the cell, the heads are empty and stain dark. *(Courtesy of Shuangyong Xu and Michael Feiss, University of Iowa.)* [B] Plaques of phage lambda. This phage forms very tiny plaques.

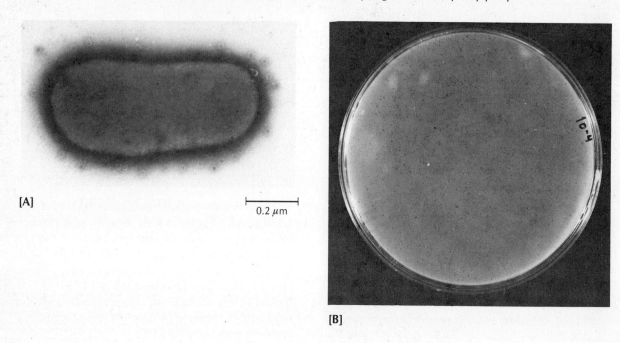

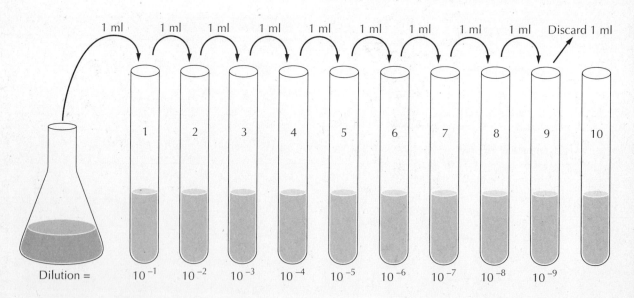

FIGURE 43.4
Preparation of phage dilutions.

NAME _____

Bacterial Lysis by Bacteriophage: Phage Titer and the Plaque Assay

RESULTS

1 Observe the plates and tubes after 6 to 8 h incubation and again after 8 to 12 h. **This time of observation *must* be adhered to.** Compare the growth in each tube to that in tube 10. Tube 10 is a positive control since it contains no phage. Record growth in tube 10 as +++, lesser amounts of growth as ++ or +, and no growth as 0.

Tube	GROWTH OF HOST BACTERIUM, *ESCHERICHIA COLI*	
	6–8 h	8–12 h
1		
2		
3		
4		
5		
6		
7		
8		
9		
10 (control)		

2 Observe carefully the growth along each of the marked lines on the plate. Is there any clearing along any of these lines? Sketch one of the plates that exhibits clearing along one or more of the streaks. What is the *estimated* phage titer in your exercise?

Estimated phage titer: _____

QUESTIONS **1** Briefly describe how an alternate method, the seeded-agar-layer technique differs from the one you used to titrate a phage suspension.

2 How does one determine that a bacterial culture has a lysogenic phage?

EXERCISE 44

Phage-Typing

OBJECTIVE
To learn the principle of phage-typing by identification of unknown bacterial cultures.

OVERVIEW
Phage-typing is a procedure for identifying bacterial strains by exposing them to known strains of bacteriophages. This method of identification is possible because phages are very specific for a given strain or species of bacteria. For example, a bacteriophage for one strain of bacteria within a species may not attack another bacterial strain of the same species. Epidemiological studies can be performed with phage-typing methods.

For instance, strains of *Staphylococcus aureus* involved in outbreaks of food poisoning or other infections can be traced by comparing their susceptibility to lysis by known phage strains. More than 20 *S. aureus* strains have been classified into five lytic groups on the basis of staphylophage susceptibility. The staphylophages are coded by number, and the different strains of staphylococci are identified by the number of the phage responsible for its lysis.

NOTE: This exercise will not use coagulase-positive *S. aureus* strains, that is, strains that coagulate blood plasma, since they are potentially pathogenic. Less virulent bacteria will be used.

REFERENCE
MICROBIOLOGY, Chap. 16, "Viruses: Cultivation Methods, Pathogenicity."

MATERIALS
Trypticase-broth cultures of 18- to 24-h *Escherichia coli* B ATCC 11303 (sensitive to phage T2)
Trypticase-broth culture of 18- to 24-h *Bacillus subtilis* ATCC 15563 (sensitive to phage 15563B1)

NOTE: Only the instructor will know the identity of the bacterial cultures. The tubes will not bear the names of the cultures, only coded numbers.

Separate suspensions of phage T2 and phage 15563B1
2 1-ml pipette (sterile)
2 tubes of 3-ml melted nutrient-tryptone soft agar kept in a water-bath at 45 to 50°C
2 nutrient-agar plates

PROCEDURE
1 For each unknown bacterial culture, label one tube of melted nutrient-tryptone soft agar and one nutrient-agar plate using the coded numbers.
2 Pipette aseptically 0.2 ml of coded bacterial culture and transfer into a correspondingly marked tube of melted nutrient-tryptone soft agar. Roll the tube in the palms of the hands to mix the cells evenly in the agar. Pour the contents on top of a nutrient-agar baseplate with the same coded number. Rotate the plate to distribute the soft-agar layer evenly.
3 After the top layer has solidified, draw two squares at the bottom of the plate. Label each square with the specific phage strain being used. Repeat the procedure with the other bacterial culture.
4 With aseptic technique, remove a loopful of each phage strain and place it on the appropriate square on each plate.
5 Allow the inoculated areas to dry at room temperature. Incubate the plates in an inverted position at 37°C for 18 h.

NAME _____

Phage-Typing

RESULTS Examine the plates after incubation for 18 h, and record your results in the following chart:

Phage T2 has plaque on plate _____ .

Plate _____ bacterium is _____ .

Phage 15563B1 has plaque on plate _____ .

Plate _____ bacterium is _____ .

QUESTIONS 1 Why are some bacterial strains susceptible and others resistant to certain phage strains? That is, what determines such specificity?

2 What other techniques are there to trace the source of bacterial contamination in a disease outbreak?

EXERCISE 45

Propagation of Viruses by Tissue Culture (Demonstration)

OBJECTIVE
To become familiar with some materials associated with the cultivation of viruses by use of tissue cultures.

OVERVIEW
The replication of viruses takes place only within susceptible living host cells. Animal viruses may be propagated in cells in the tissues of a living animal or in cells in vitro (tissue culture). The tissue-culture–virus system is both convenient and economical for the maintenance and study of viruses. The cost of keeping cell cultures is much less than the expense of using laboratory animals. The ability to confine viruses in closed containers (in tissue-culture vessels) is much safer than handling and observing animals inoculated with viruses. For humane and ethical considerations, the use of tissue cultures for virus research is preferred although it is recognized that use of animals is necessitated at some point or in some kinds of experiments.

For animal cells to grow successfully in vitro, all the necessary nutrients that are supplied to the cells in the animal body must be present in the medium in appropriate concentrations. There is no individual cell-culture medium that will support the continued growth of all types of cells. Individual cell types require particular formulations of culture media. Culture media for tissue cells are very complex and are laborious to make. Fortunately, many biological supply houses sell prepared tissue-culture media; this makes the propagation of animal cell cultures much less tedious and labor-intensive. In addition to supplying media and biochemicals for the cultivation of animal cells, many of these companies (e.g., Sigma Chemical Company, St. Louis, MO 63178-9916) also have supplies for the cultivation of insect cell cultures for the propagation of *arboviruses,* that is, viruses transmitted by arthropod vectors.

As a result of virus replication in tissue culture, cytopathic effects (CPEs) may be observed. These may be macroscopic or microscopic degenerative changes or abnormalities in the cells of the tissue culture. They provide a visible manifestation of the effect of virus infection.

REFERENCES
MICROBIOLOGY, Chap. 16, "Viruses: Cultivation Methods, Pathogenicity."
Goldstein, G., *Introductory Experiments in Virology*, Wm. C. Brown Publishers, Dubuque, Iowa, 1992.

MATERIALS
Samples of commercially prepared tissue-culture media and supplements, for example, Dulbecco's modified Eagle medium, Hank's balanced salt solution, animal sera, and so on
Tissue-culture vessels, for example, Leighton tubes, tissue-culture flasks, roller bottles, multiple-well tissue-culture plates, and others
Normal uninfected tissue culture (control)
If available, virus growing in tissue culture in which the CPE is apparent
Stained virally infected and uninfected monolayer tissue cultures

PROCEDURE
1 Examine the tissue-culture media and the nutritional supplements. Note the composition of the various mixtures.
2 Examine the tissue-culture vessels on display.
3 Observe the inoculated tissue-culture specimen using low-power (100×) and high-power (400×)

magnification, and look particularly for areas showing CPEs. Compare this with an uninoculated tissue culture. Cytopathic effects are characteristic for each virus and may appear as follows: the cells may exhibit rounding and clumping, granularity, thinning, and tearing of the cell sheet. An illustration of a Giemsa-stained monolayer, comparing a normal monkey-kidney cell tissue culture with one that was infected with poliovirus, is shown in FIGURE 45.1.

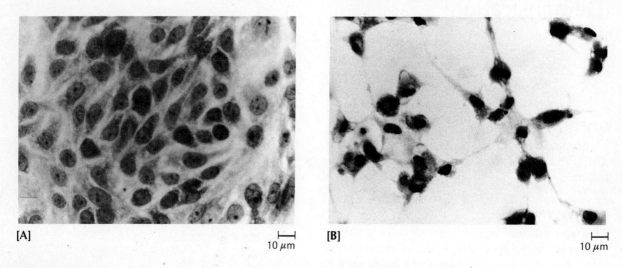

FIGURE 45.1
[A] Normal Giemsa-stained monkey-kidney cell culture. [B] Cells infected with poliovirus. Note tearing of the monolayer cell sheet.

NAME _____

Propagation of Viruses by Tissue Culture (Demonstration)

RESULTS

1 Make a sketch of the cells in the infected tissue culture and in the uninoculated tissue culture.

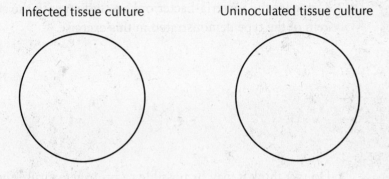

Infected tissue culture   Uninoculated tissue culture

2 Describe the appearance of the cells in the stained monolayer tissue-culture slides.

Uninfected cells	
Infected cells	

283

QUESTIONS   **1** Compare the media used for cultivation of tissue-culture cells with those used for the cultivation of bacteria.

2 Compare the bacteria-bacteriophage system with the animal-virus-propagation system of the type demonstrated in this exercise.

3 Do you think it may be possible to propagate viruses in a system free of living host cells? Explain why.

RECOMBINANT DNA TECHNIQUES

Since genetic material is composed of deoxyribonucleic acid (DNA), **recombinant DNA technology** is also known as **gene technology** or **genetic engineering**. Use of this technology in recent years has led to the very exciting advances of *molecular biology*—the study of the biological molecules DNA, RNA (ribonucleic acid), and protein. Whether *holistic microbiologists* (those who study whole microbial organisms) like it or not, molecular biology has permeated every discipline, and students of microbiology should know something about this science.

The era of genetic engineering was ushered in in the late 1960s when enzymes were discovered in *Escherichia coli* that recognized specific DNA sequences and cleaved the DNA at highly specific sites within the recognized sequence. These enzymes were called *restriction endonucleases* because they functioned to restrict the entry of foreign DNA into the cell by cutting it up. Soon, some restriction enzymes were found that were not random in activity but were able to recognize and make double-stranded cuts at specific sequences in DNA. Now, over 250 restriction enzymes from different bacteria have been discovered so that there is a pool of enzymes that permit an investigator to cleave DNA at virtually any site within a genome.

It was determined that some restriction enzymes leave short, single-stranded tails called *sticky ends* that develop from the staggered cuts made in the two strands of DNA. The single-stranded overhangs could base-pair with complementary overhangs from other pieces of similarly cleaved DNA. Thus any two fragments of DNA, regardless of their origin, can be combined into a single DNA molecule by means of their sticky ends, giving rise to a recombinant DNA molecule. For example, a piece of DNA containing a gene of interest can be inserted into a plasmid carrier, or vector, by cutting the plasmid and DNA of interest with the same restriction enzyme, generating fragments with sticky ends. Such fragments are reannealed in the presence of another enzyme called *DNA ligase,* creating a recombinant DNA molecule. This recombinant plasmid can then be inserted, or transformed, across the cell membrane into a suitable bacterial cell host. The gene is then said to be *cloned* in the bacterial cell. Bacterial cells containing a recombinant plasmid can produce large quantities of an economically useful gene product.

The microbiology student, in shifting to work in molecular biology from working with whole organisms, has to change gears with respect to working habit. He or she has to deal with microvolumes, for instance, μl are used instead of ml or liters! In the same way that the student in microbiology acquires "aseptic technique" as a second nature, in working with recombinant DNA technology, the student has to think and perform "small."

The mushrooming growth of gene technology has given rise to a proliferation of biotechnology companies that sell specialized supplies required in molecular biology. This happy circumstance not only saves time but also the tedium in preparation of materials. Indeed some commercial companies even market *kits* for molecular biology procedures that represent a new height of convenience for the consumer in the discipline. The instructor should explore this avenue in planning exercises in recombinant DNA technology, especially when the expertise of the technical support staff of each company is just a phone (or fax) call away. In this part of the lab manual, an introduction is provided for the basic skills of molecular biology, such as extraction and purification of DNA from bacterial cells, restriction digests of DNA, and agarose gel electrophoresis. Attention of the instructor will also be directed to the commercial availability of some materials for teaching recombinant DNA technology in the marketplace.

# EXERCISE 46

Isolation of DNA from Bacterial Cells

OBJECTIVE

To carry out (A) the conventional method and/or (B) a commercially available ion-exchange method (Qiagen, Inc.) for isolation of DNA from bacterial cells.

OVERVIEW

Purified DNA extracted from bacterial cells can be used for several purposes, such as:

1. Determination of the overall nucleotide base composition of the DNA (moles percent guanine + cytosine, or mol% G + C, of the DNA)
2. DNA homology experiments to determine the relatedness among various bacteria for purposes of bacterial classification or identification (*gene probing*)
3. Bacterial transformation
4. A source of bacterial genes for genetic engineering

The initial step in DNA extraction involves the breakage of the bacterial cells without extensive damage to the DNA. In general, Gram-negative cells are easier to disrupt than Gram-positive cells, and a detergent such as sodium dodecyl sulfate can readily destroy the integrity of bacterial membranes. For disruption of Gram-positive bacteria, the thick cell wall is usually destroyed by an enzyme such as lysozyme before a detergent is applied. Once the cells are lysed, the subsequent steps involve purification of the DNA by destroying or eliminating other cellular constituents, particularly protein and RNA.

The method described in Section A of the exercise can be applied to a wide variety of Gram-negative bacteria and yields DNA fragments of high molecular weight and high purity. It is the conventional and time-tested method of isolating DNA from bacterial cells introduced by Marmur and has been used for decades by microbiologists.

However, recent years have seen the introduction of more rapid and simpler procedures for the extraction of chromosomal DNA from cells. They are based on the principle of *ion-exchange chromatography*. Nucleic acids carry a large negative charge, which enables them to be separated from other cellular components. The separation is a two-step procedure of *adsorption* and *desorption*.

Adsorption

Initially, counter ions in the environment bind to the positively charged matrix (anion-exchange resins) to maintain charge neutrality at the surface of the matrix. When lysed cells are added, the environmental counter ions are exchanged for the negatively charged nucleic acid molecules. Because of their greater affinity, the nucleic acids will adsorb selectively to the matrix and remain bound there. Other cellular components do not bind to the column and are washed out.

Desorption

A buffer of high ionic strength is applied to the column (a pH gradient can also be used). The change in ionic strength (or pH) changes the electrostatic interactions between the nucleic acid molecules and the ion-exchange matrix. Desorption takes place when the DNA/RNA is exchanged for negatively charged counter ions in the eluting buffer. After the nucleic acids have desorbed, they are eluted from the column and collected. The DNA in the eluate can be concentrated by ethanol precipitation after the RNA has been removed by RNase treatment.

Many biotechnology companies sell kits that facilitate the extraction and purification of DNA from cells using ion-exchange chromatography. We mention only two here simply to introduce the instructor to explore this convenient method of teaching molecular biology. The instructor should contact these companies for details of cost and procedure. The

Extractor System is from Molecular Biosystems, Inc., 10030 Barnes Canyon Road, San Diego, California 92121. The other one is the Qiagen method for bacterial genomic DNA preparation from Qiagen, Inc., 9259 Eton Avenue, Chatsworth, California 91311.

REFERENCES

MICROBIOLOGY, Chap. 14, "Microbes and Genetic Engineering,"
MICROBIOLOGY, Chap. 31, "Biotechnology: The Industrial Applications of Microbiology."
MM, Chap. 26, "DNA Sequence Relatedness."
Marmur, J., "A Procedure for the Isolation of DNA from Microorganisms," *Journal of Molecular Biology* 3:208–218, 1961.

MATERIALS

A. Conventional Method

2 liters *Escherichia coli* culture in trypticase-soy broth grown at 37°C (Cells should be in the log phase of growth for best disruption.)

Refrigerated high-speed centrifuge

50-ml polypropylene screw-cap centrifuge tubes

250-ml polypropylene or polycarbonate screw-cap centrifuge bottles

Wrist-action shaker

2 glass stirring rods (End of each rod should be etched about an inch with a file so that the surface is frosted in appearance.)

Saline-EDTA buffer: $0.15M$ NaCl; $0.01M$ sodium ethylenediamine tetraacetate (EDTA); pH adjusted to 8.0

Sodium dodecyl sulfate (SDA), 20% (w/v) solution: Warm to dissolve. Store at room temperature. The SDS may precipitate if the ambient temperature becomes too cool; if this occurs, merely place the solution in a bath of warm water until the SDS redissolves.

Sodium perchlorate, $5M$ solution

Chloroform-isopentanol mixture: 24 parts of chloroform to 1 part isopentanol, v/v

95% ethanol, stored in a freezer

Lysozyme, egg white, crystalline

Pronase, crystalline

Stock solution of SSC, 20X: $3.0M$ NaCl; $0.3M$ trisodium citrate; pH 7.0 [NOTE: Other concentrations of SSC are indicated by a number such as 1X (1:20 dilution) or 0.1X (1:200 dilution). After such dilutions are made, check that the pH is still 7.0 unless otherwise indicated.]

0.1X SSC-MES buffer: To 0.1X SSC add 2-(N-morphololino) ethanesulfonic acid (MES buffer) to a concentration of $0.01M$

Ribonuclease A, bovine pancreatic (Sigma Chemical Co., St. Louis, MO): Dissolve 10 mg in 10 ml of 0.1X SSC (adjusted to pH 5) and heat in a water-bath at 80°C for 20 min to inactivate any traces of deoxyribonuclease (DNase). Add 10 ml of cold 0.1X SSC, and chill the mixture on ice. Dispense 5 ml amounts into screw-cap tubes and store in a freezer.

Ribonuclease T_1, from *Aspergillus oryzae* (Sigma Chemical Co.): Dissolve 5000 units in 10 ml of 0.1X SSC-MES buffer. Heat as described for RNase above. Add 10 ml of 0.1X SSC-MES and chill on ice. Dispense 5-ml amounts into screw-cap tubes and store in a freezer.

Ultraviolet spectrophotometer with quartz cuvettes (optional item)

B. QIAGEN > BacterialGenomic < Protocol from Qiagen, Inc.

QIAGEN>Genomic DNA<Kits. The starter kit consists of the following: 5 QIAGEN-Tip 20 (*Tip* is the term used by the company for column); 5 QIAGEN-Tip 100; 1 QIAGEN-Tip 500; Proteinase K; RNase A; and Buffers G1, QBT, QC, and QF. Obtain a company brochure for other kits. All kits contain QIAGEN columns, all reagents and buffers ready to use, and protocols.

PROCEDURE

A. Conventional Method

1 Harvest the bacteria by centrifugation in 250-ml centrifuge bottles and discard the supernatant medium into a container of disinfectant. Pool the centrifuged cells.

2 Suspend the centrifuged cells in a flask containing 100 ml of saline-EDTA buffer (i.e., suspend to a final volume of one-twentieth that of the original culture). The EDTA in this buffer helps to inactivate any DNase activity by binding magnesium ions, which are required for DNase activity.

3 Add 0.5 ml of ribonuclease A and 0.5 ml of ribonuclease T_1.

4 Add 5 ml of SDS.

5 Place flask in a water-bath at 60°C and swirl. Lysis should occur within 10 min and is indicated by a marked increase in viscosity and clearing of the mixture.

46 ISOLATION OF DNA FROM BACTERIAL CELLS

6 After lysis occurs, add a pinch of lysozyme and a pinch of Pronase and incubate for 1 h at 50°C.

7 Add 25 ml of sodium perchlorate. This reagent helps to separate protein from DNA so that protein can later be precipitated by chloroform-isopentanol treatment.

8 Add 35 ml of chloroform-isopentanol.

9 Place the mixture in a 500-ml flask with the top covered by aluminum foil, and shake the flask on a wrist-action shaker at room temperature for 20 min, using a shaking speed that is just sufficient to produce an emulsion. Very vigorous shaking is not desirable.

10 Centrifuge the emulsion in polypropylene tubes at 17,000 × g for 15 min in a refrigerated centrifuge at 0 to 4°C.

11 Using an inverted 10-ml serological pipette with a propipette, slowly and carefully remove and save the upper aqueous phase from each tube [FIGURE 46.1]. Be careful to avoid pulling up any portion of the white layer of precipitated denatured protein that forms at the interface between the two phases. Discard the lower chloroform-isopentanol phase.

12 Using a fresh centrifuge tube, repeat steps 8, 9, 10, and 11 with the supernatant material that you have collected.

13 If the amount of white precipitate between the chloroform-isopentanol and aqueous phases is still abundant, repeat steps 8 through 11 until the amount is small or absent.

14 Place the aqueous phase into two 250-ml beakers (about 50 ml in each) and chill in an ice bath. Then slowly pour ice-cold ethanol down the inside wall of each beaker until you have overlaid the aqueous phase with alcohol. The total amount of alcohol should be about twice the volume of the aqueous phase.

15 The DNA will begin to precipitate at the interface between the alcohol overlay and the aqueous phase. For each beaker, use a glass stirring rod to gently stir the two phases while rotating the rod. The DNA will "wind up," or "spool," onto the rod [FIGURE 46.2]. Continue to collect as much DNA as you can on the rod.

16 Wash the spooled DNA by placing each rod in a tube of ice-cold 80% ethanol for a few minutes. Repeat the washing using fresh tubes of 80% ethanol.

17 Press each rod against the inside wall of a beaker to remove excess ethanol. Then allow the rod to drain vertically (with DNA end up) for a few minutes to remove most of the remaining ethanol; however, do not let the DNA become completely dry.

18 Place both rods, DNA-coated end down, into a tube containing 30 ml of 0.1X SSC. Add 3 to 5 drops of chloroform and incubate overnight in a refrigerator. Gently remove the rods, making sure that the DNA on the end has dissolved in the 0.1X SSC. Allow to stand longer if necessary until the DNA is completely dissolved.

19 After the DNA has dissolved completely, adjust the SSC concentration to 1X by adding 20X SSC in a ratio of 1 part of 20X SSC to 21 parts

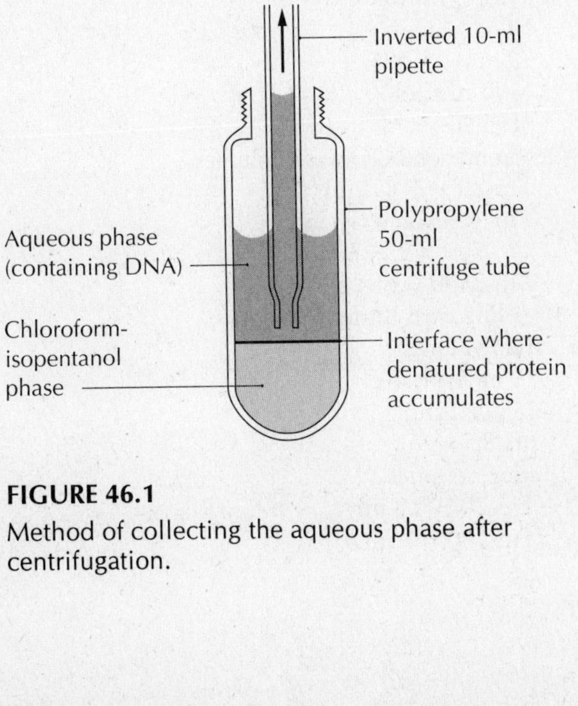

FIGURE 46.1
Method of collecting the aqueous phase after centrifugation.

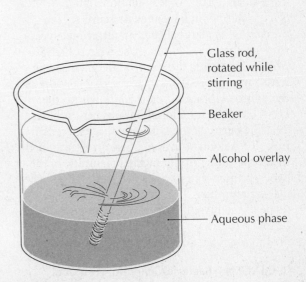

FIGURE 46.2
Method of collecting DNA on a glass rod.

of 0.1X SSC-DNA solution. This is necessary in order to allow alcohol precipitation of the DNA in subsequent steps.

20 Transfer to a flask and add 0.5 ml of ribonuclease A and ribonuclease T_1 to the DNA solution. Incubate for 30 min at 37°C to destroy residual RNA.

21 Add 5 to 10 ml of chloroform-isopentanol and shake on wrist-action shaker for 20 min, using a shaking speed just sufficient to produce an emulsion.

22 Centrifuge the emulsion in polypropylene tubes at 17,000 × g for 15 min in a refrigerated centrifuge at 0 to 4°C.

23 Carefully remove and save the upper (aqueous) layer from each tube as in step 11.

24 Repeat steps 21 through 23 until little or no denatured protein can be seen at the interface between the chloroform-isopentanol and aqueous phases.

25 Precipitate the DNA with ethanol and dissolve in 0.1X SSC as you did previously in steps 14 through 19 (use only one beaker and one glass rod this time). Repeat steps 14 through 19 two or three times to remove any residual RNA. Be sure to adjust the SSC concentration to 1X each time before each subsequent alcohol precipitation—DNA will not precipitate from a 0.1X SSC solution!

26 After the final alcohol precipitation, dissolve the DNA on a glass rod in 10 ml of 0.1X SSC as in step 18.

27 Store the DNA in a freezer.

28 If desired, you can quantify the DNA by diluting a small portion of the solution 1:20 in 0.1X SSC, placing it in a 1-cm quartz cuvette, and measuring its absorbance in an ultraviolet spectrophotometer at 260 nm. (Use plain 0.1X SSC as a blank for setting the spectrophotometer to an initial absorbance of 0.) The absorbance value of the diluted DNA solution is equivalent to the DNA concentration in milligrams per milliliter.

29 The purity of the DNA, that is, its freedom from RNA contamination, is usually determined spectrophotometrically by the magnitude of the hyperchromic shift of the DNA during the thermal denaturation. This method is beyond the scope of the present exercise; however, details can be found in the article by J. L. Johnson (1993) in *MM* in the reference section of this exercise.

B. QIAGEN > BacterialGenomic < Protocol

1 Pellet bacteria by centrifugation.

2 Resuspend bacterial pellet in 1/50th volume of:
  10 mM Tris-Cl
  100 mM NaCl
  5 mM EDTA, pH 7.0

3 Add lysozyme to a final concentration of 2 mg/ml. Incubate at 37°C for 15 to 20 min.

4 Add 9 volumes of the following:
  10 mM Tris-Cl
  250 mM NaCl
  1.2% Triton X-100
  100 μg/ml RNase A
  12 mM EDTA
  0.5M Guanidine-HCl
  pH 8.0
Place on ice for 20 min.

5 Add Proteinase K to a final concentration of 2 mg/ml. Incubate at 50°C for 2 h with gentle agitation.

6 Centrifuge at 15,000 to 20,000 × g for 10 min in order to pellet insoluble debris.

7 Decant supernatant into a clean tube. Particularly viscous solutions should be diluted with 0.5 to 1 volume of column equilibration buffer, QBT.

8 Load material onto the appropriate QIAGEN column, preequilibrated with Buffer QBT:
  750 mM NaCl
  50 mM MOPS
  15% ethanol
  0.15% Triton X-100
  pH 7.0
Equilibration volumes:
  Tip-20 = 1 ml
  Tip-100 = 3 ml
  Tip-500 = 10 ml
  Tip-2500 = 50 ml

9 Wash with Buffer QC:
  1.0M NaCl
  50 mM MOPS
  15% ethanol
  pH 7.0
Recommended QC wash volumes:
  Tip-20 = 4 × 1 ml
  Tip-100 = 2 × 10 ml
  Tip-500 = 2 × 30 ml
  Tip-2500 = 6 × 50 ml

10 Elute with Buffer QF:
  1.25M NaCl
  50 mM Tris-Cl
  15% ethanol
  pH 8.5
Elution volumes:
  Tip-20 = 0.8 ml
  Tip-100 = 5 ml

Tip-500 = 15 ml
Tip-2500 = 50 ml

11 Add 0.7 volumes of room-temperature isopropanol and invert gently for a few times. A visible precipitate should form within the tube.

Pellet DNA by centrifuging at 15,000 × g for 30 min at 4°C.

12 Wash the pellet 1X with ice-cold 70% ethanol. Air-dry for 10 min and resuspend in an appropriate volume of TE.

NAME _____

Isolation of DNA from Bacterial Cells

RESULTS

(A) Conventional Method

1 Describe the appearance of the DNA wound around the glass rods.

(A) Conventional or (B) QIAGEN Method

2 What was the absorbance reading of the diluted DNA solution?

3 Calculate the concentration of DNA that you have isolated using the absorbance value obtained.

QUESTIONS 1 How might you use the DNA you have extracted to demonstrate bacterial transformation?

2 How might you use the DNA you have extracted to determine the relatedness of *E. coli* to *Salmonella typhi*?

3 How might you use the DNA you have extracted as donor DNA for genetic engineering?

EXERCISE 47

Restriction Enzyme Analysis of DNA

OBJECTIVE
To carry out the genotypic analysis of DNA using restriction enzymes and gel electrophoresis.

OVERVIEW
Recombinant DNA methods involve the in vitro enzymatic manipulation of DNA. Specific fragments of DNA may be created by digesting DNA with enzymes called **restriction endonucleases**. These enzymes cleave DNA at specific sequences of four to six base pairs called **recognition sequences**. The ends of such fragments may be staggered ("sticky ends") or blunt, depending on the point of cleavage within the recognition sequence. DNA generated in this manner may be joined with other species of DNA to form recombinant molecules.

In this exercise, phage lambda (λ) DNA will be digested separately with two restriction endonucleases (*Eco*R I and *Bam*H I), and the fragments will be separated by gel electrophoresis. Purified DNA from phage λ is very suitable for demonstrating the principle of DNA restriction. It is inexpensive, readily available, and is not large like a bacterial genome. Each enzyme has five or more restriction sites in λ DNA and therefore produces six or more restriction fragments of varying length [FIGURE 47.1]. The gel matrix acts as a sieve through which smaller DNA molecules migrate faster than larger ones; restriction fragments of differing sizes separate into distinct bands during electrophoresis.

The characteristic pattern of bands produced by each restriction enzyme is made visible by staining with a dye that binds to the DNA molecule. Two dyes usually used are ethidium bromide and methylene blue. Ethidium bromide staining is the more rapid and sensitive method. Ethidium bromide is a fluorescent dye. The molecule intercalates between nucleotides, and the nucleic acid–ethidium bromide complex fluoresces when exposed to ultraviolet light.

The recognition sequences (sites) of the two endonucleases are as follows:

*Eco*R I	*Bam*H I
–G↓AATT C–	–G↓GATC C–
–C TTAA↑G–	–C CTAG↑G–

λ DNA can exist both as a circular molecule and as a linear molecule. FIGURE 47.1 shows λ DNA as a linear molecule and is 48,502 base pairs (bp) in length. Also shown are the cutting sites of *Bam*H I and *Eco*R I on the molecule (→← marks the site of cleavage); the length of each cut fragment is also shown.

REFERENCES
MICROBIOLOGY, Chap. 14, "Microbes and Genetic Engineering."

Davis, L. G., M. D. Dibner, and J. F. Battey, *Basic Methods in Molecular Biology*, Elsevier, New York, 1986.

Micklos, D. A., and G. A. Freyer, *DNA Science: A First Course in Recombinant DNA Technology*, Carolina Biological Supply Company, Burlington, N.C., 1990.

MATERIALS
Phage λ DNA (obtainable from Sigma Chemical Co., St. Louis, Mo.)

Microcentrifuge tubes, 1.5 ml (obtainable from Fisher Scientific Co., Silver Spring, Md.)

Micropipettes and tips (obtainable from Fisher Scientific Co.)

After an exercise courtesy of Dr. R. C. Bates, Virginia Polytechnic Institute and State University.

FIGURE 47.1

Restriction maps of the linear lambda genome. Cleavage sites and fragment lengths reported in base pairs are also shown.

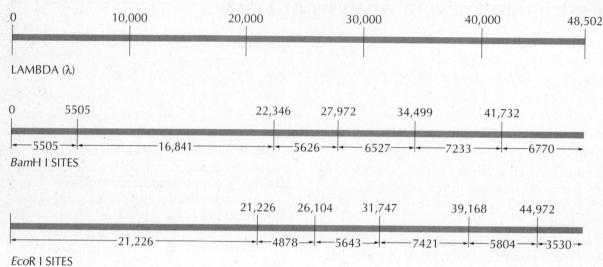

- Restriction enzymes *Bam*H I and *Eco*R I, and 10X enzyme buffer (obtainable from International Biotechnologies, Inc., New Haven, Conn.)
- Endo R. Stop solution (obtainable from Bethesda Research Laboratories, Gaithersburg, Md.)
- 10X TAE electrophoresis buffer: 0.4 M Tris-HCl buffer (pH 7.8) containing 0.2 M sodium acetate and 0.02 M disodium ethylenediamine tetraacetic acid (EDTA)
- Agarose (obtainable from BioRad Laboratories, Inc., Richmond, Calif.) (Prepare a 1% gel in 1X electrophoresis buffer for gel electrophoresis.)
- Ethidium bromide (obtainable from Sigma Chemical Co.) (Prepare in distilled water to a concentration of 1 μg/ml.)
- Horizontal-slab gel-electrophoresis apparatus with comb (e.g., model H5, Bethesda Research Laboratories)
- Hand-held Polaroid camera (obtainable from Fotodyne, Inc.) and films
- Ultraviolet (UV) source lamps, such as used for detecting fluorescence of mineral specimens (obtainable from Fisher Scientific Co.) (UV lamps with wavelengths from 254 to 302 nm can be used.)
- UV-absorbing safety goggles (obtainable from Fisher Scientific Co.)

NOTE: Other commercial companies can also supply many of the materials listed. Two others that gear themselves to provide for teaching classes are the following: Carolina Biological Supply Company, 2700 York Road, Burlington, NC 27215, and Fotodyne, Incorporated, 16700 West Victor Road, New Berlin, WI 53151-4131.

PROCEDURE

1 Label three microcentrifuge tubes as follows and place the tubes on ice:
"No enzyme"
"*Bam*H I"
"*Eco*R I"

2 In the following step, add reagents directly to the bottom of the three tubes by means of a micropipette, changing the pipette tip each time to avoid contamination. Make additions as in the table that follows to give a total volume of 20 μl in each microtube.

3 Place the tubes in a 37°C water-bath and incubate for 30 min.

4 Stop the enzyme reaction by adding 2.0 μl of Endo R. Stop solution to each tube.

5 Store the tubes at 4°C until ready to perform gel electrophoresis.

6 For use of your horizontal-slab gel electrophoresis, consult the directions provided by the manufacturer. Prepare a 1% agarose gel in 1X electrophoresis buffer for use in the apparatus, and after it has solidified, overlay it with 1X electrophoresis buffer.

	Tube		
Addition	No enzyme	BamH I	EcoR I
10X restriction enzyme buffer	2.0 μl	2.0 μl	2.0 μl
Phage λ DNA (1 μg)*	2.0 μl	2.0 μl	2.0 μl
BamH I (4 units)**	—	2.0 μl	—
EcoR I (4 units)**	—	—	2.0 μl
Distilled water	16.0 μl	14.0 μl	14.0 μl

*If the stock solution is too concentrated, dilute in sterile water to give 1 μg in 2.0 μl.
**If the stock solution is too concentrated, dilute in 1X restriction enzyme buffer to give 4 units in 2.0 μl.
NOTE: The unit is the standard measure of restriction enzyme activity and is defined as the amount of enzyme needed to digest to completion 1 μg of λ DNA in a 50-μl reaction in 1 h.

7 Load samples by layering into wells using a micropipette. Perform electrophoresis at 100 V for 2 h or until the tracking dye has migrated two-thirds to three-fourths the length of the gel.

8 Remove gel from the apparatus and transfer to a glass dish containing ethidium bromide solution.

CAUTION: Wear plastic or rubber gloves, as this chemical is a carcinogen!

9 After 20 min remove the gel, rinse it in distilled water, and lay it on a piece of plastic. Using UV-safe goggles, examine the gel with a UV lamp to visualize the stained DNA bands.

CAUTION: Never look at an unshielded UV light source with naked eyes. View only through a filter or safety glasses.

10 Mark the position of the bands with toothpicks and leave in place for use in determining the relative migrations of the DNA fragments produced by the restriction endonucleases after turning the UV lamp off.

Alternatively, the gel may be photographed directly on the UV light source using a special handheld Polaroid camera. Follow the simple directions of the manufacturer. Use Polaroid high-speed film Type 667 (ASA 3000) for photographing ethidium bromide-stained gels in UV light.

Restriction Enzyme Analysis of DNA

RESULTS

1 Make a sketch that shows the number of fragments and the relative positions of the bands as they appear in the gel for each DNA sample.

2 Determine from the number of fragments observed how many sites the DNA had for each restriction enzyme used in the experiment.

QUESTIONS 1 Why do restriction endonucleases cleave DNA into fragments of *specific sizes?*

2 Why do some restriction endonucleases cleave DNA into many pieces while others cleave the same DNA into a few pieces or not at all?

3 Why do DNA fragments of different sizes separate into bands during electrophoresis through an agarose gel?

4 How does ethidium bromide stain DNA?

EXERCISE 48

Transformation of *Escherichia coli* with Plasmid DNA Containing Genes for Antibiotic Resistance

OBJECTIVE
To learn how bacteria may be transformed using plasmid DNA.

OVERVIEW
Genetic recombination and transfer can occur by three different processes in bacteria: *conjugation*, *transduction*, and *transformation*. **Transformation** is the uptake and incorporation of cell-free DNA (naked DNA) by bacterial cells. Strains of some species of bacteria, for instance, *Streptococcus pneumoniae*, *Haemophilus influenzae*, *Neisseria gonorrhoeae*, and *Bacillus subtilis*, have demonstrated the capability of being transformed by picking up pieces of DNA from their environment. Other species have not been shown to have this ability—unless they are made **competent.**

At low temperatures and in the presence of calcium ions, *Escherichia coli* is made competent and will take up DNA molecules and become transformed. The calcium chloride used destabilizes the cell membrane, making it permeable and therefore competent. A calcium-phosphate complex is formed and adheres to the cell surface. The DNA is then taken up during a heat-shock step when the cells are exposed briefly to temperatures of 37°C or higher. Selection for cells containing transformed DNA is greatly enhanced if markers carried by the DNA are first expressed in the absence of selective conditions by growing the cells in a rich medium for a given time (from 30 min to 2 h; 1 h in this exercise). Recombinant DNA molecules prepared from bacterial plasmids or viral chromosomes are often used to transform bacteria, either to confer new properties on the bacteria or as a means for synthesis of the product of a cloned gene. The plasmid pBR322 carries a gene for resistance to the antibiotic ampicillin and also a gene for resistance to the antibiotic tetracycline. Bacteria that are transformed with this plasmid become resistant to these two antibiotics. This antibiotic resistance can be used as a selective marker to isolate colonies of bacteria containing the plasmid. Plasmids such as pBR322 can also be modified by recombinant DNA techniques so that they carry useful foreign genes.

In this exercise we shall make *E. coli* cells competent using $CaCl_2$ and will transform the cells using the plasmid pBR322, which has several antibiotic-resistance properties that the *E. coli* cells do not have. The plasmid map of pBR322 is shown in FIGURE 48.1. Antibiotic resistance will be used to select for the cells transformed by this plasmid.

REFERENCES
MICROBIOLOGY, Chap. 14, "Microbes and Genetic Engineering."
MICROBIOLOGY, Chap. 31, "Biotechnology: The Industrial Applications of Microbiology."
Miklos, D. A., and G. A. Freyer, *DNA Science: A First Course in Recombinant DNA Technology*, Carolina Biological Supply Company, Burlington, N. C., 1990.

MATERIALS
Escherichia coli strain Le392 (ATCC 33572) or strain HB101 (ATCC 33694) (obtainable from the American Type Culture Collection, Rockville, Md.)
Plasmid pBR322 (obtainable from Bethesda Research Laboratories, Gaithersburg, Md.)
NaCl solution, 10 mM, sterile
Tris buffer, 25 mM, pH 7.5 (Sterilize by filtration.)
Tris buffer, 25 mM, containing 50 mM $CaCl_2$ (Sterilize by filtration.)
Refrigerated centrifuge
Centrifuge tubes, sterile, screw-capped

After an exercise courtesy of Dr. R. C. Bates, Virginia Polytechnic Institute and State University.

FIGURE 48.1

Plasmid map of pBR322. This plasmid is a circular, double-stranded DNA and carries genes that confer tetracycline and ampicillin resistance. It was constructed from portions of several naturally occurring plasmids.

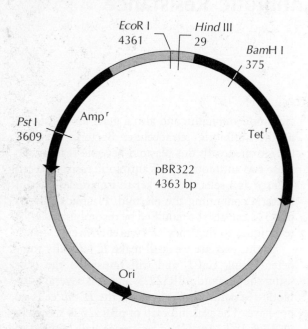

Antibiotics (Stock solutions for addition to media are sterilized by filtration. Ampicillin and tetracycline are obtainable from various biochemical supply companies, for example, Sigma Chemical Co., St. Louis, Mo.)
YT (yeast-extract-tryptone) broth
YT agar plates
  (a) No antibiotics
  (b) With ampicillin, 25 µg/ml
  (c) With tetracycline, 10 µg/ml
  (d) Ampicillin plus tetracycline
Antibiotic stock solutions sterilized by filtration are added to molten agar medium at 45 to 50°C.
5 small sterile glass or polypropylene tubes
Spectrophotometer with cuvettes for measuring cell density

PROCEDURE

1 Inoculate *E. coli* into a tube of YT broth and incubate overnight at 37°C. Inoculate 50 ml of YT broth with 1 to 2 ml of the overnight culture and incubate with vigorous shaking on a mechanical shaker at 37°C until the optical density of the culture is 0.3 as measured at a wavelength of 600 nm.

2 Chill the 50-ml culture on ice until cold. Transfer to a sterile centrifuge tube and centrifuge at 4000 r/min for 5 min at 4°C.
3 Decant the supernatant fluid and suspend the sedimented cells in 10 mM NaCl to the original volume (50 ml).
4 Centrifuge at 4000 r/min at 4°C.
5 Decant the supernatant fluid and suspend the sedimented cells in 25 ml Tris-CaCl$_2$ buffer. Place on ice for 10 min.
6 Centrifuge at 4000 r/min.
7 Decant the supernatant fluid and suspend the cells in 0.5 ml of Tris-CaCl$_2$ buffer.
8 Obtain 5 sterile, chilled small glass or polypropylene tubes. Dilute the plasmid pBR322 in 25 mM Tris buffer (no CaCl$_2$) so that each tube will contain 10 µl (0.01 ml) of fluid and various amounts of plasmid DNA according to the following table:

Tube number	Amount DNA contained in 10 µl volume
1	0.0001 µg
2	0.001 µg
3	0.01 µg
4	0.1 µg
5	No DNA

9 Add 50 µl (0.05 ml) of the cells from step 7 to each of the tubes, flicking each tube with a finger as the cells are added to mix the cells with the DNA.
10 Incubate the tubes on ice for 10 min to precipitate the DNA onto the cells.
11 Transfer all tubes to a 37°C water-bath for 2 min to heat-shock the cells.
12 Add 1 ml of YT broth to each tube and incubate for 1 h at 37°C.
13 Plate a 0.1-ml volume from each culture (using the spread-plate method and a glass rod) onto four plates:
  YT agar containing no antibiotics
  YT agar containing ampicillin only
  YT agar containing tetracycline only
  YT agar containing ampicillin plus tetracycline
This makes a total of 20 plates for the experiment.
14 Incubate the plates at 37°C for 24 h.

Transformation of *Escherichia coli* with Plasmid DNA Containing Genes for Antibiotic Resistance

RESULTS

1 Count the number of transformants obtained with each level of plasmid DNA:

From tube 1: _____

From tube 2: _____

From tube 3: _____

From tube 4: _____

From tube 5: _____

2 Determine the transformation efficiency, expressed as the number of colonies per microgram of plasmid DNA used in the transformation assay.

3 Record your observations for the control plates (no antibiotics), and give an explanation for any differences between these plates and the other plates.

QUESTIONS

1 Why do the bacteria containing the plasmid DNA form colonies on agar plates containing an antibiotic?

2 Why is it necessary to allow the bacteria to recover following transformation with DNA before plating onto antibiotic-containing media?

3 Do both ampicillin and tetracycline need to be present in the medium to select bacteria that have been transformed with plasmid pBR322? Explain.

4 What is the role of calcium ions in transformation of *E. coli* with plasmid DNA?

5 What properties do plasmids have that make them useful in the cloning of foreign genes?

EXERCISE 49

Plasmid-Mediated Transformation of the Ampicillin-Resistance and the *lac* Genes

OBJECTIVE

To demonstrate phenotypic changes in bacteria that have been transformed with an antibiotic-resistance gene and the *lac* gene using a plasmid.

OVERVIEW

With this exercise we wish to remind the instructor of the availability of instructional recombinant DNA technology kits from commercial sources. One that we have laboratory tested with our students with good success is from Carolina Biological Supply Company, The Biotechnology Company, 2700 York Road, Burlington, N.C. 27215, in cooperation with the DNA Learning Center of Cold Spring Harbor Laboratory and Dr. Greg Freyer who developed the *p*BLU™ plasmid used in the kit.

The kit is called the *p*BLU™ Colony Transformation Kit (Cat. No. 21-1146G). The kit uses a specially developed *p*BLU™ plasmid, which carries genes for two identifiable phenotypes:

1 The amp^r gene codes for resistance to the antibiotic ampicillin
2 The lac^z gene codes for β-galactosidase, which breaks down the galactose analog X-gal (5-Bromo-4-chloro-indoyl β-D-galactopyranoside) to form a visible blue product

Students observe the phenotypic effect of inserting new genes into bacteria. A mutant strain of *Escherichia coli* strain JM101, incapable of producing β-galactosidase necessary for degradation of lactose and sensitive to ampicillin, is induced to take up *p*BLU plasmid DNA. The transformed cells are plated on medium containing ampicillin and medium containing ampicillin and X-gal. Transformed cells appear as white colonies on ampicillin medium and as blue colonies on ampicillin/X-gal medium; that is, their phenotype is lac+ and amp^r. The kit comes with an instructor's manual and offers minimal instructor preparation; it contains all necessary materials sufficient for six teams of students.

In this exercise, *E. coli* cells are picked from two large colonies growing on a Luria broth (LB) agar plate and suspended in two tubes containing a solution of calcium chloride. *p*BLU plasmid is added to one cell suspension, and both tubes are incubated at 0°C for 15 min. Following a brief heat shock at 42°C, cooling, and addition of Luria broth, samples of the cell suspensions are plated on three types of media: plain LB agar, LB agar with ampicillin (LB/Amp), and LB with ampicillin plus X-gal (LB/Amp/X-gal).

The plates are incubated for 12 to 24 h at 37°C and then checked for bacterial growth. Only cells that have been transformed by taking up the plasmid DNA with the ampicillin-resistant gene will grow on the LB/Amp or LB/Amp/X-gal plates. The growth of blue colonies on LB/Amp/X-gal plates demonstrates the expression of β-galactosidase by transformed cells. Each colony growing on these selective media represents a single transformation event. The results obtainable with this kit are illustrated in FIGURE 49.1 and PLATE 22.

REFERENCES

MICROBIOLOGY, Chap. 14, "Microbes and Genetic Engineering."
MICROBIOLOGY, Chap. 31, "Biotechnology: The Industrial Applications of Microbiology."
Micklos, D. A., and G. A. Freyer, *DNA Science: A First Course in Recombinant DNA Technology*, Carolina Biological Supply Company, Burlington, N. C., and Cold Spring Harbor Laboratory Press, Cold Spring Harbor, N.Y. 1990.

NOTE TO INSTRUCTOR: The usual exercise sections on materials, procedure, results, and questions will be furnished in the kit from Carolina Biological Supply Company, The Biotechnology Company.

FIGURE 49.1

Phenotypic appearance of *Escherichia coli* cells transformed by the *p*BLU™ plasmid. Top row: Cells not transformed with plasmid plated on plain LB agar, LB/Amp agar, and LB/Amp/X-gal agar. Note the complete absence of growth (no colonies) on the selective media (with ampicillin). Bottom row: Plasmid-mediated transformed cells plated on the same kind of media. Note the appearance of white colonies on LB/Amp medium and blue colonies on LB/Amp/X-gal medium. Satellite colonies are seen around the large colonies; they are composed of ampicillin-susceptible nontransformed cells growing in the zone of "antibiotic shadow" where ampicillin has been broken down by the large resistant colony.

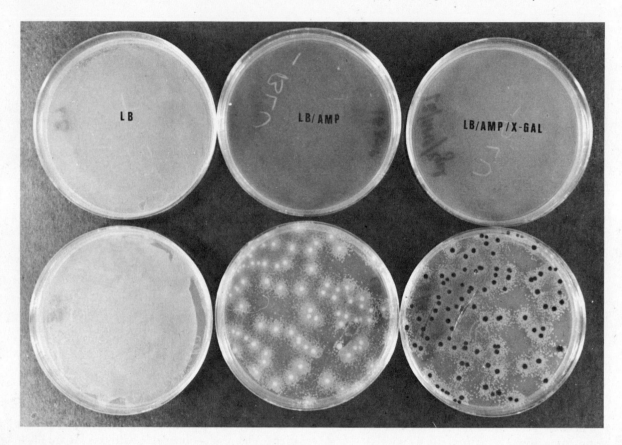

EUCARYOTIC MICROORGANISMS

Eucaryotic microorganisms include protozoa, algae, and fungi. The laboratory study of these eucaryotic microbes is performed in essentially the same manner as that already described for the bacteria. In general, their cells are larger, and they exhibit more anatomical details and structures. The DNA of a eucaryotic cell is enclosed within a membrane-bound nucleus. In addition, eucaryotic cells possess membrane-bound organelles, which have specialized structures and perform specific functions. Microscopic examinations are frequently made from microcultures or wet preparations; fixed, stained smears, commonly used for the observation of bacteria, are used less often here. Morphological characteristics are the most important single contribution to the identification and classification of most of these organisms. Consequently, the routine biochemical tests that are used for the characterization of bacteria have less application to most of the higher protists. However, studies of the metabolic processes of yeasts and molds (fungi), protozoa, and algae are conducted by the same general procedures that are performed for bacteria.

Laboratory cultivation of the eucaryotic microbes is likewise similar to that described for the bacteria. Broth and agar media are conventionally employed. As a group, the higher protists are nutritionally less versatile than the bacteria.

EXERCISE 50

Characteristics of Protozoa

OBJECTIVE
To become familiar with the distinguishing characteristics of protozoa.

OVERVIEW
Protozoa are a diverse group of unicellular eucaryotic microorganisms. Most are free-living, but some are parasites. They occur generally as single cells, although some may occur as colonies. Protozoa are distinguished from other eucaryotic protists by their ability to move at some stage of their life cycle and by their lack of cell walls. (Some members may possess a flexible layer, a pellicle, or a rigid shell of inorganic material exterior to the cytoplasmic membrane.) The majority of protozoa are microscopic in size. They have a heterotrophic nutrition in which the free-living forms generally ingest particulate foods such as bacteria, yeasts, and algae and are found widely distributed in ponds and streams. The parasitic forms generally derive nutrients from the body fluids of their hosts.

The many forms of protozoa have been grouped together taxonomically much like bacteria; that is, protozoan classification is based more on convenience and practicality than on criteria of evolution. Means of locomotion and fine structure from electron microscopy have both played a significant role in the classification of these protists.

REFERENCES
MICROBIOLOGY, Chap. 4, "Procaryotic and Eucaryotic Cell Structures."
MICROBIOLOGY, Chap. 10, "The Major Groups of Eucaryotic Microorganisms: Fungi, Algae, and Protozoa."

MATERIALS
Living culture of *Paramecium caudatum*
5 ml of 10% methyl cellulose
Thread
Carmine powder and toothpick
Yeast stained with congo red (Preparation: Mix a cake of yeast in water and let stand for 24 h. Boil the mixture. Add 0.1 g of congo red powder and boil 8 min longer. Allow to cool before using.)
Sample of stagnant pond water
Cylinder of sterile Pasteur pipettes
Rubber bulb (for Pasteur pipettes)
Prepared slide of *Amoeba* sp.
Stained smear of human blood showing *Plasmodium vivax*
Chart showing various stages of development of *Plasmodium vivax* in red blood cells

PROCEDURE
MORPHOLOGY AND MOTILITY OF *PARAMECIUM*

1 Prepare a wet mount of the living culture of *Paramecium* provided. (Use Pasteur pipettes with a rubber bulb to handle the protozoan culture.) Before placing a cover slip on the preparation, add a drop of methyl cellulose. The addition of a drop of methyl cellulose slows down the movement of the organisms and facilitates observation.

2 Observe the wet mount using the low-power and high-power objectives of the microscope. After scanning several fields, select a field with a single organism in the center. Use FIGURE 50.1 to assist in the identification of cellular structures. Note the characteristic shape of *Paramecium* and whether both ends of the organism are the same.

3 Prepare another wet mount of *Paramecium*; add a drop of methyl cellulose as well as a few grains of carmine powder with a toothpick and a few pieces of thread before placing the cover slip over the preparation. Observe the movement of the carmine particles along the surface of the organism. Note the structures on the surface of the cell that are

FIGURE 50.1
Major cellular structures of a paramecium.

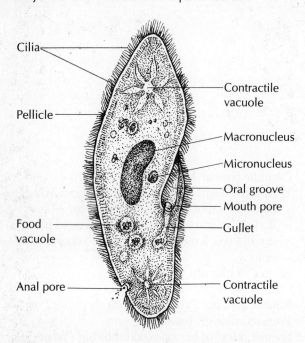

responsible for this. Describe the movement of the organisms after they hit the thread.

NUTRITION AND EXCRETION BY PARAMECIUM

1 Prepare a wet mount of *Paramecium* with a drop of methyl cellulose, but before adding a cover slip, add a drop of yeast stained with congo red. Examine as quickly as possible under the high-power objective.
2 Observe the paramecia as they ingest the stained yeast. Note how the yeasts are directed into the organism, where they enter it, and where they are collected. Watch for food vacuoles to appear, and note where they go once they are formed. Determine how undigested particles are eliminated from the cell.
3 Observe and account for color changes of the yeast in the vacuoles. (Congo red is an acid-base indicator. It is red at pH 5 and blue-violet at pH 3.)
4 Prepare another wet mount of the *Paramecium* culture with methyl cellulose. Observe under the microscope using the high-power objective. Locate the contractile vacuoles within the organisms. Determine the number of contractile vacuoles, and note whether they move throughout the cytoplasm or are fixed in place.

EXAMINATION OF STAGNANT POND WATER

1 With a Pasteur pipette obtain a drop of pond water from the bottom of the sample of stagnant pond water. Place it in the center of a clean glass slide, add a drop of methyl cellulose to slow down the movement of the protozoa, and make a wet mount as before.
2 Examine under the low-power and high-power objectives. Make a few drawings of the different protozoa present.

EXAMINATION OF A PREPARED SLIDE OF AMOEBA SP.

1 Examine the prepared slide of *Amoeba* sp. using the low-power and high-power objectives.
2 Use FIGURE 50.2 to assist in the identification of cellular structures.

EXAMINATION OF SMEAR OF HUMAN BLOOD SHOWING PLASMODIUM VIVAX

1 Review the chart provided by the instructor showing the various stages of development of *Plasmodium vivax* in red blood cells.
2 Examine the prepared smear and locate infected red blood cells. Make a sketch of several such cells in the results section that follows.

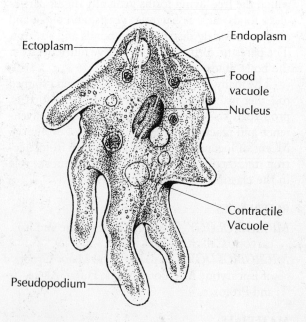

FIGURE 50.2
Major cellular structures of an amoeba.

NAME _____

Characteristics of Protozoa

RESULTS

1 Identify the cellular structures of the paramecium in the wet mount with the aid of FIGURE 50.1. Tick off the structures identified.

- _____ Cilia
- _____ Food vacuole
- _____ Contractile vacuole
- _____ Macronucleus
- _____ Micronucleus
- _____ Oral groove
- _____ Pellicle
- _____ Mouth pore
- _____ Gullet
- _____ Anal pore

2 Describe the movement of the paramecium when it hits a thread.

3 Describe the movement of the carmine particles over the surface of the paramecium.

4a Observe the ingestion of yeast cells by paramecia. Where do the yeast cells enter the paramecium?

4b Where are the yeast cells collected in the paramecium?

4c Describe the movement of the food vacuoles.

4d How does the paramecium eliminate undigested particles?

5 Observe and account for the color changes of yeast cells in the vacuoles.

6 Determine the number of contractile vacuoles in the paramecium, and note whether they move throughout the cytoplasm or are fixed in place.

7 Describe your overall impression of living forms in the sample of pond water.

8 Make drawings of several protozoa present in stagnant pond water.

NAME _____ 50 CHARACTERISTICS OF PROTOZOA *(Continued)*

9 With the aid of FIGURE 50.2, identify the cellular structures of the amoeba in the prepared slide provided. Tick off the structures identified.

_____ Ectoplasm   _____ Nucleus

_____ Endoplasm   _____ Contractile vacuole

_____ Food vacuole   _____ Pseudopodium

10 Sketch infected and uninfected red blood cells from the smear of human blood showing *Plasmodium vivax* infection.

QUESTIONS 1 Name a function of each of the structures of the paramecium listed below.

FUNCTIONS OF THE STRUCTURES OF THE PARAMECIUM	
Structure	Function
Pellicle	
Cilia	
Oral groove	
Mouth pore	
Gullet	
Food vacuole	
Anal pore	
Contractile vacuole	
Cell membrane	
Macronucleus	
Micronucleus	

2 Compare the mode of locomotion in the amoeba and in the paramecium.

EXERCISE 51

Morphology and Cultivation of Algae and Cyanobacteria

OBJECTIVE
To gain a familiarity with a group of photosynthetic microorganisms.

OVERVIEW
Algae and cyanobacteria contain the green pigment chlorophyll, which is also a characteristic property of plants. This pigment is used for photosynthesis, a process that employs light energy as a power source for fixing carbon dioxide and transforming it into organic compounds. The algae are eucaryotic, while the cyanobacteria are procaryotic. Algae and cyanobacteria range in size from microscopic single cells to extremely large forms such as the seaweeds. Algae and cyanobacteria are found in a variety of environments, including both fresh water and salt water, soil and moist surfaces. Because of their ability to carry out photosynthesis and their simple nutritional requirements, they are **primary producers** in a food chain. They have been labeled as "grasses of the oceans," and they are indispensable in a food chain supporting other marine life as food and oxygen producers. They are also found in association with certain species of fungi, protozoa, and animals. For example, the lichens often have cyanobacteria as one of the partners living in mutualism with fungi. The study of algae is called **phycology**.

REFERENCES
MICROBIOLOGY, Chap. 4, "Procaryotic and Eucaryotic Cell Structures."

MICROBIOLOGY, Chap. 10, "The Major Groups of Eucaryotic Microorganisms: Fungi, Algae, and Protozoa."

MATERIALS
6 Petri plates containing calcium nitrate-salts agar
6 Petri plates containing nutrient agar with 0.5% glucose
Nutrient-agar slant cultures of the algae *Chlorella* sp. and *Chlamydomonas* sp. and of the cyanobacterium *Nostoc* sp.

PROCEDURE
1 Streak two plates of each medium with each culture provided. Incubate one set of plates, that is, each of the organisms on both media, in the *dark* (such as in a desk drawer) and the other set of plates in the *light* (near a window). Incubate the plates at room temperature (25°C) for 5 to 7 days.
2 Prepare a wet mount of each culture and examine the cells under low-power magnification and high-power magnification. Observe the cells for size, shape, and evidence of external and internal structures. Use FIGURE 51.1 as a guide for your drawings. Note particularly any similarities and/or differences with bacteria, yeasts, molds, and protozoa.

FIGURE 51.1
Morphology of some photosynthetic microorganisms.

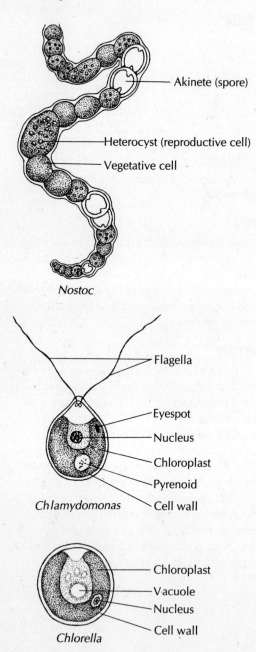

NAME _____

Morphology and Cultivation of Algae and Cyanobacteria

RESULTS   1 Make a drawing of several cells of each species as seen by high-power magnification, and label all structures.

Chlorella sp.

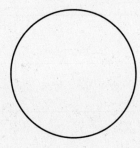

Chlamydomonas sp.

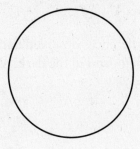

Nostoc sp.

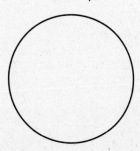

2 At the end of the incubation time, examine the Petri plates with cultures. Record the presence or absence of growth, and in the following chart give a description of any growth obtained.

\multicolumn{5}{c}{GROWTH DESCRIPTION OF PHOTOSYNTHETIC MICROORGANISMS}
Medium	Conditions of incubation	*Chlorella* sp.	*Chlamydomonas* sp.	*Nostoc* sp.
CSA	Light			
CSA	Dark			
NAG	Light			
NAG	Dark			

CSA = calcium nitrate-salts agar; NAG = nutrient agar with glucose.

QUESTIONS 1 Were any differences observed in the growth of the algae on the two different media? In the light and in the dark? What accounts for the differences?

2 What is another name for *oxygenic* photosynthetic bacteria?

EXERCISE 52

Morphological and Cultural Characteristics of Molds

OBJECTIVE
To be introduced to a group of eucaryotic, multicellular, filamentous microorganisms called *molds*.

OVERVIEW
The *molds* (filamentous fungi) are much larger and morphologically more complex than bacteria. They are characterized as *eucaryotic* microbes, in contrast to bacteria, which are *procaryotic* microbes. Morphological details can be differentiated in molds by microscopic examination of unstained specimens. In fact, identification of these fungi is dependent to a large extent on morphological descriptions. They are cultivated in the laboratory in the same general manner as bacteria. Unlike bacteria, which are aerobic, facultative, or anaerobic, molds are usually aerobic.

Molds are normally grown on solid media and incubated at room temperature. A mold colony usually consists of a web of branched, interwoven hyphae, called a *mycelium*, part of which grows into the agar, forming a tough adhesive mat. The portion of the mycelium growing on or in the nutrient medium is called the *vegetative mycelium*, as it extracts necessary nutrients. The portion of the mycelium that grows upward from the surface is called the *aerial mycelium* and has specialized *reproductive hyphae* responsible for producing sexual or asexual spores. [See PLATE 23.]

Sexual or asexual spores may be produced on the aerial hyphae, but some species remain nonfertile. The hyphal strand of most molds is composed of individual cells separated by crosswalls, or *septa*; the hypha is said to be *septate*. However, some molds have a hyphal strand devoid of crosswalls; the hypha is said to be *nonseptate* or *coenocytic*.

Because of the particular morphology of molds, the bacteriological inoculating loop is not a suitable tool for transferring fungal cultures, and a heavy-gauge, stiff wire needle with the tip flattened into a cutting edge is used.

When subculturing a mold, a small block approximately 1 to 2 mm square should be cut from the edge of the colony and placed in the center of a fresh agar-medium slant or in an agar medium in a Petri dish. When possible, sporulating areas should be avoided for the following reasons:

1 Contamination of the laboratory with airborne mold spores
2 Seeding of multiple colonies on the new medium which may distort typical morphological characteristics, for example, rate of growth and edge of colony
3 Increasing frequency of mutant colonies (Some molds have a pronounced tendency to mutation in laboratory culture, and cultivation of young vegetative hyphae gives the most constant characteristics.)

There are few chemical tests that can be applied to molds. Identification rests primarily on morphological and microscopic characteristics such as growth rate, type of mycelium, pigmentation, type of sporulating structures, and sexual reproduction (when present). It is therefore essential to master the technique of culturing molds in the laboratory.

REFERENCES
MICROBIOLOGY, Chap. 4, "Procaryotic and Eucaryotic Cell Structures."
MICROBIOLOGY, Chap. 10, "The Major Groups of Eucaryotic Microorganisms: Fungi, Algae, and Protozoa."

MATERIALS
4 slants Sabouraud's agar (large tubes)
Stiff wire needle

Beaker of 70% ethanol
4 slides and 4 cover slips
5-ml syringe containing paraffin-petrolatum mixture
1 pair forceps
4 tubes Sabouraud's agar (approximately 2 ml each)
4 sterile Pasteur pipettes with 1 rubber bulb
4 Petri dishes containing a sterile glass support and a small ball of cotton soaked with water
Sabouraud's agar slant cultures of *Penicillium notatum*, *Aspergillus niger*, *Rhizopus stolonifer*, and *Alternaria* sp.

PROCEDURE

1 You are supplied with four slants of Sabouraud's agar.
2 Make a spot inoculation on each slant from each mold culture provided. Use inoculum from the edge of growth of each mold. To transfer the inoculum, use a stiff wire needle (discussed previously). Flame the needle to sterilize it; then dip it in alcohol to cool it. The needle is finally passed through the flame to burn off the alcohol. The sterile needle is now ready for use. Incubate the inoculated slants at 25°C for 3 to 5 days.
3 Morphological characteristics of molds can best be observed from slide cultures. The following technique [see FIGURE 52.1] will be used to prepare slide cultures:

a Select a clean slide and pass it through the flame of the burner. Allow it to cool.
b Make two ridges on the slide with paraffin-petrolatum (approximately 1 in apart). The thin column of paraffin-petrolatum can be forced out of the syringe more effectively if the loaded syringe is warmed beforehand (kept in the 35°C incubator).

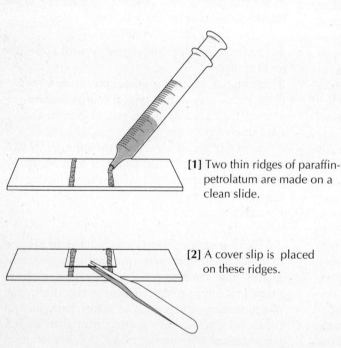

[1] Two thin ridges of paraffin-petrolatum are made on a clean slide.

[2] A cover slip is placed on these ridges.

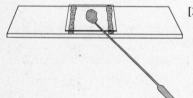

[3] The inoculated agar medium is introduced between the slide and the cover slip. The space is only partially filled.

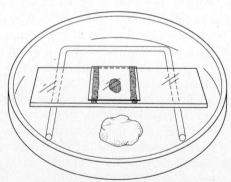

[4] The slide culture is placed in a Petri dish containing glass-rod support and a wet cotton ball.

FIGURE 52.1
Preparation of a slide culture for the examination of mold morphology.

c Select a clean cover slip, pass it through the flame, and then place it upon the paraffin-petrolatum ridges on the slide.

d Inoculate the tube of melted, cooled (45 to 50°C) Sabouraud's agar with one of the mold cultures. (In this instance inclusion of some mold spores will be advantageous since they will be more easily distributed throughout the medium than just the hyphae alone.) Roll each tube between the palms of your hands to distribute the inoculum.

e With a sterile Pasteur pipette, transfer a small amount of the inoculated or seeded agar to the space between the cover slip and slide. Allow the medium to fill approximately one-half of this space.

f Place the slide culture in a Petri dish containing a sterile glass support and a small ball of cotton soaked in water. (The wet cotton provides moisture to prevent drying of the medium.) Incubate the preparation for 2 to 5 days at room temperature.

4 Prepare slide cultures for each of the four species provided. FIGURE 52.2 (on the following page) illustrates structures you will observe.

NOTE: It is advisable to observe the inoculated slant and slide cultures daily to determine when the growth best exhibits the morphological structures.

FIGURE 52.2
Morphological structures of molds to be observed.

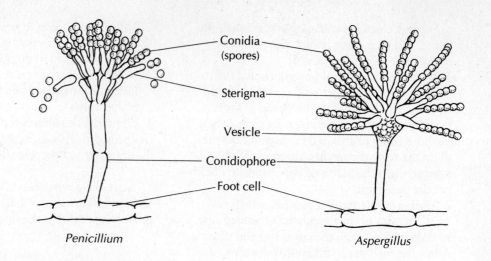
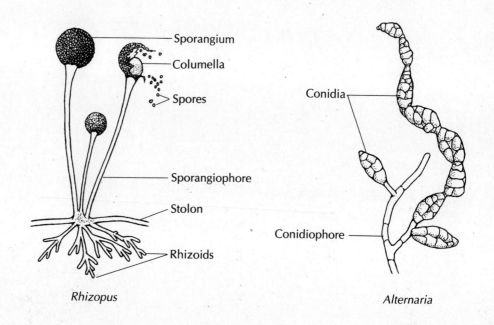

NAME _____

Morphological and Cultural Characteristics of Molds

RESULTS

1 Examine the slant cultures microscopically. Record the gross appearance of the growth of each mold.

2 Examine each of the slide cultures microscopically, using the low-power and high-power objectives only. Make drawings of typical structures found in each culture, identifying such structures as the following: mycelia, septate and nonseptate (coenocytic) hyphae, conidia and conidiophores, sporangia and sporangiophores, and rhizoids. Refer to FIGURE 52.2 for aid in identification of structures seen microscopically.

Penicillium notatum

Macroscopic appearance of cultural growth:

Microscopic appearance of slide culture:

Low-power objective   High-power objective

323

Aspergillus niger

Macroscopic appearance of cultural growth:

Microscopic appearance of slide culture:

Low-power objective   High-power objective

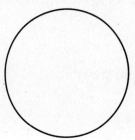 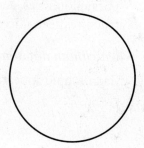

Rhizopus stolonifer

Macroscopic appearance of cultural growth:

Microscopic appearance of slide culture:

Low-power objective   High-power objective

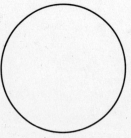 

52 CHARACTERISTICS OF MOLDS (Continued)

Alternaria sp.

Macroscopic appearance of cultural growth:

Microscopic appearance of slide culture:

Low-power objective   High-power objective

QUESTIONS 1 Explain why morphological characteristics are more important in the classification of molds than of bacteria.

2 How does the pH of Sabouraud's agar differ from the pH of media usually used for the cultivation of bacteria?

3 Besides a low pH, give two other means that can be used to keep down bacterial contamination of mold cultures.

4 List two distinguishing characteristics of each of the four molds studied in this exercise.

EXERCISE 53

Sexual Reproduction of Molds

OBJECTIVE
To observe zygospore formation in *Mucor hiemalis*.

OVERVIEW
Molds produce different kinds of sexual spores; for example, zygomycetes produce *zygospores* and ascomycetes produce *ascospores*. In this exercise, you will observe zygospore formation by *Mucor hiemalis*. Zygospores are produced where two mycelia meet to copulate, usually after incubation for about 3 days. In 5 days, a dark line of mature zygospores can be seen. In this process there is a fusion of two protoplasts from the union (conjugation) of cells of two compatible hyphae to form a zygote with exchange of nuclear material.

The stages in zygospore development are shown in FIGURE 53.1. The colonial morphology of the two compatible mold strains (+ and −) is shown in FIGURE 53.2. FIGURE 53.3 shows the microscopic appearance where (+) and (−) strains have sexual reproduction forming zygospores (dark round bodies).

REFERENCES
MICROBIOLOGY, Chap. 4, "Procaryotic and Eucaryotic Cell Structures."
MICROBIOLOGY, Chap. 10, " The Major Groups of Eucaryotic Microorganisms: Fungi, Algae, and Protozoa."

MATERIALS
3 plates Sabouraud's agar
Mucor hiemalis (+) strain and (−) strain

PROCEDURE
1 Draw a line across the center of the outside bottom part of each plate of Sabouraud's agar. Inoculate the plates in two locations, A and B, at opposite sides of each dish, as shown in FIGURE 53.4, with the *Mucor hiemalis* strains as follows:
Plate 1: Strain (+) inoculated at A; strain (−) inoculated at B
Plate 2: Strain (+) inoculated at A and B
Plate 3: Strain (−) inoculated at A and B

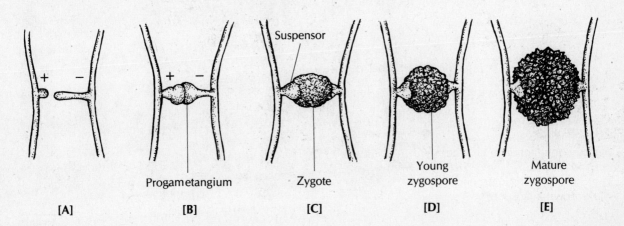

FIGURE 53.1
Stages in the development of a zygospore.

XI EUCARYOTIC MICROORGANISMS

FIGURE 53.2

Appearance of plates inoculated at two sites with (+) and (−) strains of *Mucor hiemalis*. Vertical dark line in bottom plate is composed of zygospores.

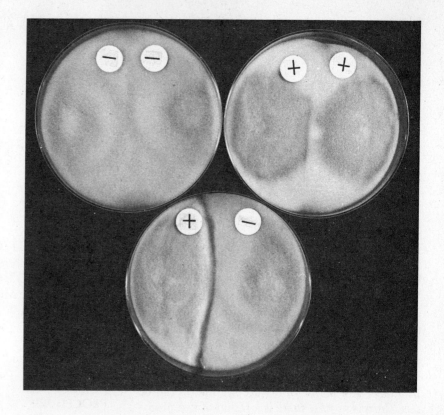

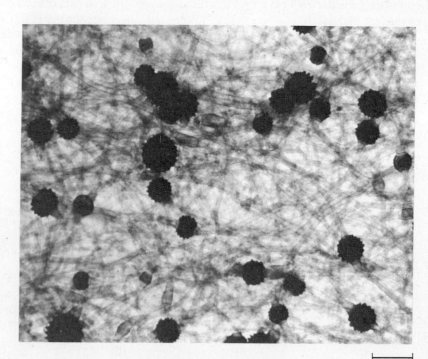

FIGURE 53.3

Microscopic appearance where (+) and (−) strains have sexual reproduction forming round, dark brown zygospores.

80 μm

2 Incubate the plates at room temperature in a container that can be humidified, for example, a large enclosed can or jar containing a beaker of water.

3 Observe the plates at each subsequent laboratory period. Use the low-power objective to view the morphological structures, particularly at the converging margins of growth. It will be better if you can use a stereomicroscope to observe the morphology of the cultures.

4 Using a stiff wire needle, remove some zygospores along the dark line where the mycelia of opposite strains meet, and prepare them in a wet mount.

5 Examine the wet mount under the low-power and the high-power objectives.

FIGURE 53.4
Sites of inoculation on Sabouraud's agar plate.

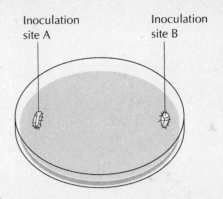

NAME _____

Sexual Reproduction of Molds

RESULTS 1 Describe the appearance of growth at each time of observation in the chart provided.

\multicolumn{3}{c}{GROWTH AND ZYGOSPORE DEVELOPMENT IN *MUCOR HIEMALIS*}		
Date	Amount of growth	Description of mold structures

2 Observe the structures associated with stages in the production of a zygospore with the aid of FIGURE 53.1.

QUESTIONS 1 Differentiate between the following sexual spores: zygospore, ascospore, oospore, and basidiospore.

2 Why do we consider zygospore formation in molds as a form of sexual reproduction?

3 How does *Mucor hiemalis* reproduce asexually?

EXERCISE 54

Dimorphism of *Mucor rouxii*

OBJECTIVE
To observe the phenomenon of dimorphism in the fungi.

OVERVIEW
Some pathogenic fungi exhibit **dimorphism**, a phenomenon in which a species of fungus has two growth forms. In pathogenic fungi, the yeast form usually exists in tissue while the filamentous mold form exists in vitro on laboratory medium.

This exercise demonstrates the ability of a fungus to grow as either a filamentous mold or a yeast. Its development ranges from filamentous mycelium to spherical yeasts.

REFERENCES
MICROBIOLOGY, Chap. 4, "Procaryotic and Eucaryotic Cell Structures."
MICROBIOLOGY, Chap. 10, "The Major Groups of Eucaryotic Microorganisms: Fungi, Algae, and Protozoa."

MATERIALS
Slant culture of *Mucor rouxii* on yeast-extract-peptone-glucose (YPG) medium
5 ml of sterile tap water
1 5-ml pipette
1 1-ml pipette
Water-bath set at 40 to 45°C
10 ml of YPG medium
1 thin plastic cup (5-ml polystyrene disposable beakers available from Fisher Scientific Co., cat. no. 02-544-30)
1 single-edged razor blade

PROCEDURE
1 Obtain a slant culture of *Mucor rouxii* on yeast-extract-peptone-glucose (YPG) medium.
2 Pipette 3 ml of sterile tap water into the slant culture.
3 Scrape the surface of the fungal mat with a stiff straight needle. Suspend the spores evenly by shaking the tube.
4 Melt 10 ml of YPG medium in a test tube and cool down to 40 to 45°C in a water-bath.
5 Inoculate the molten medium with 1 ml of the spore suspension, and mix quickly but gently to avoid forming bubbles.
6 Rapidly pour the inoculated medium into a thin plastic cup (it need not be sterile) to the brim. Allow the agar to solidify.
7 Incubate the cup with inoculated medium for 2 days at room temperature.
8 Cut open the plastic vessel with a single-edge razor blade as shown in FIGURE 54.1.
9 Make a thin (1-mm) vertical slice of the agar medium with the razor blade, and mount it between a microscope slide and cover slip.
10 Under the microscope, using the low-power and high-power objectives, examine the morphology of the fungus from the top to the bottom of the mount (vertical slice) corresponding to the top and the bottom of the culture vessel. FIGURE 54.2 shows the appearance of the dimorphic gradient that should be observed.

Exercise adapted from and courtesy of S. Bartnicki-Garcia, "The Dimorphic Gradient of *Mucor rouxii:* A Laboratory Exercise," *ASM News* 38(9):486–488, 1972.

FIGURE 54.1

[A] Cultivation of *Mucor rouxii* in disposable polystyrene beaker. [B] The beaker can be readily cut open to prepare agar slices for microscopic observation.

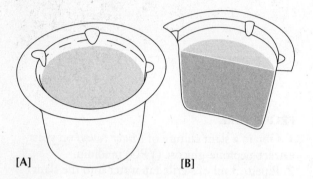

FIGURE 54.2

Dimorphic gradient of *Mucor rouxii*. Area A, mycelial development. Area B, transition forms. Area C, yeast cells. Note the tubular sporangiophores and globose sporangia developing above the agar surface; this type of morphological differentiation occurs only in open air.

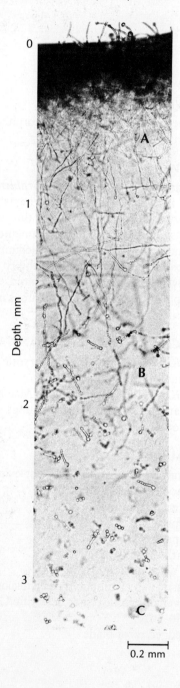

NAME _____

Dimorphism of *Mucor rouxii*

RESULTS Describe the morphology of the fungus in the chart below. Note also the presence of tubular sporangiophores and globose sporangia developing above the agar surface.

MORPHOLOGY OF *MUCOR ROUXII*	
Location in mount or culture vessel	**Morphology of fungus**
Agar surface (air)	
Top portion	
Middle portion	
Bottom portion	

QUESTIONS 1 What is the explanation for getting a dimorphic gradient in the exercise performed?

2 What other factors might be responsible for dimorphism in pathogenic fungi?

EXERCISE 55

Morphology of Yeasts

OBJECTIVE
To examine the morphology of some yeast cells.

OVERVIEW
Yeasts are nonfilamentous fungi; that is, they do not form hyphae, although some species form pseudohyphae. (A *pseudohypha* is a series of cells adhering end to end to form a short chain.) Yeasts are therefore unicellular fungi, typically round or ovoid in shape. They are larger than bacteria but are less complex morphologically than molds. Laboratory study of yeasts is performed in essentially the same manner as for bacteria.

Yeasts are widely distributed in nature. Some are found frequently on fruits and leaves, for example *Saccharomyces cerevisiae*. Others are found as commensals on the human body and form part of the normal microbiota, for example, *Candida albicans*, although they can become opportunistic pathogens.

Yeasts are facultative anaerobes. Their metabolic activities are harnessed to many industrial fermentation processes, for example, the production of alcoholic beverages like wine and beer. They are also used in the preparation of many foods, such as bread.

Biochemical activities (e.g., carbon assimilation tests) are also used in the characterization and identification of yeasts because of the paucity of morphological features.

The life cycle of two common yeasts exemplifying two modes of asexual reproduction are shown in FIGURES 55.1 and 55.2. The species are the budding yeast *Saccharomyces cerevisiae* and the fission yeast *Schizosaccharomyces octosporus*.

REFERENCES
MICROBIOLOGY, Chap. 4, "Procaryotic and Eucaryotic Cell Structures."

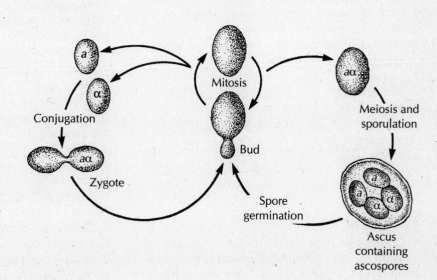

FIGURE 55.1
The life cycle of *Saccharomyces cerevisiae*. The *a* and *α* refer to mating-type alleles. (In a diploid cell two alleles of a given gene occupy corresponding loci on a pair of homologous chromosomes.) Unlabeled cells may be *a*, *α*, or *aα*.

FIGURE 55.2
The life cycle of *Schizosaccharomyces octosporus*, a fission yeast. It reproduces asexually by binary fission. Sexual reproduction is by conjugation of compatible cells with the subsequent formation of ascospores.

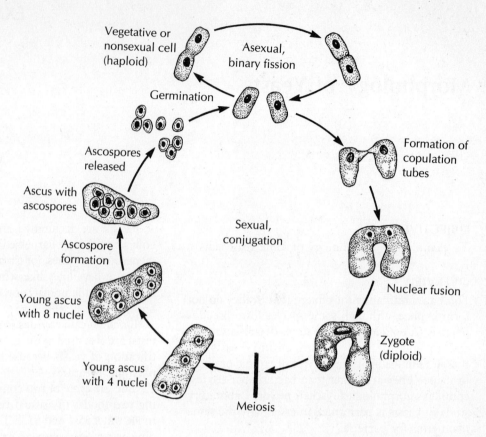

MICROBIOLOGY, Chap. 10, "The Major Groups of Eucaryotic Microorganisms: Fungi, Algae, and Protozoa."

MATERIALS

4 Petri plates with Sabouraud's agar
1 Petri plate with 0.5% sodium acetate agar
1 Petri plate with cornmeal agar
4 fermentation tubes of each carbohydrate broth (glucose, maltose, lactose)
1 stoppered test tube with approximately 1 ml of sterile water
Sabouraud's agar slant cultures of *Saccharomyces cerevisiae*, *Candida albicans*, and *Schizosaccharomyces octosporus*
Compressed yeast cake
Water-iodine solution (3 volumes water to 1 volume Gram's iodine)

PROCEDURE

1 Obtain four Petri plates with Sabouraud's agar.
2 With the aid of the inoculating loop, transfer a small piece of yeast cake to the tube with sterile water. Agitate the tube vigorously.
3 Prepare a streak plate (see Exercise 17) of this yeast-cake suspension on Sabouraud's agar medium. Prepare streak plates of each of the pure cultures of yeast provided on Sabouraud's agar medium. Make a streak plate on the sodium-acetate-agar medium with *Saccharomyces cerevisiae*. Make a streak plate on the cornmeal-agar medium with *Candida albicans*.

Incubate the Sabouraud's agar plates at room temperature (25°C) for 2 to 4 days. Inoculate the carbohydrate-broth fermentation tubes with the yeast cultures and the yeast-cake suspension, and incubate them in a similar manner. Incubate the streaked plate of sodium-acetate-agar medium and the cornmeal-agar medium for 7 days at room temperature.

4 Make a wet-mount preparation of each pure culture and yeast-cake suspension. Mix a drop of methylene blue with the yeast cells to make the wet mounts. Examine these using the high-power and oil-immersion objectives.
5 Prepare a wet mount of *Saccharomyces cerevisiae* in a water-iodine solution. Examine it using the high-power and oil-immersion objectives.
6 Make drawings to illustrate the structures observed in the wet-mount preparations, using FIGURES 55.3, 55.4, and 55.5 for reference.

7 From the yeast growth on the sodium-acetate-agar medium (after incubation of 1 week at room temperature), make one stained preparation of the culture by the malachite green spore-staining technique (see Exercise 10.A).

From the yeast growth on the cornmeal-agar medium (after 1 week's incubation), make a wet mount with a drop of methylene blue stain. Examine it using the high-power and oil-immersion objectives.

FIGURE 55.4
Morphology of *Schizosaccharomyces octosporus*. Cells undergoing fission are shown by filled arrow. White arrow shows ascospores within an ascus. *(Courtesy of L. Kapica and E. C. S. Chan, McGill University.)*

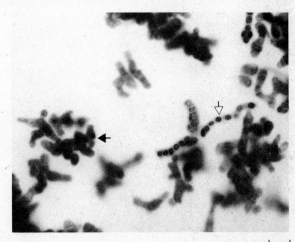

FIGURE 55.3
Morphology of *Saccharomyces cerevisiae*. Budding cell is shown by filled arrow. Ascospores within an ascus are shown by a white arrow. *(Courtesy of L. Kapica and E. C. S. Chan, McGill University.)*

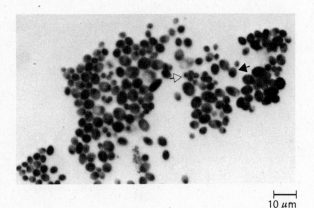

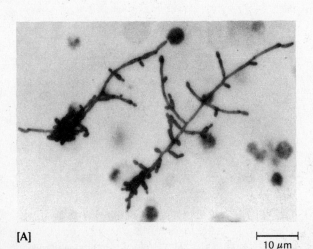

[A]

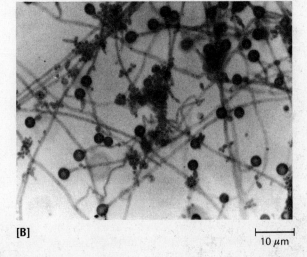

[B]

FIGURE 55.5
[A] *Candida albicans*, pathogenic phase in urine sample of infected patient. Shown are pseudomycelia with blastospores. [B] Culture phase. Chlamydospores seen with pseudomycelia and blastospores.

NAME _____

55

Morphology of Yeasts

RESULTS

1 Make drawings of a few cells from each of the methylene blue wet-mount preparations. Observe the shape of the individual cells and the characteristic arrangement of cells. FIGURES 55.3, 55.4, and 55.5 will aid in your observations.

Specimen

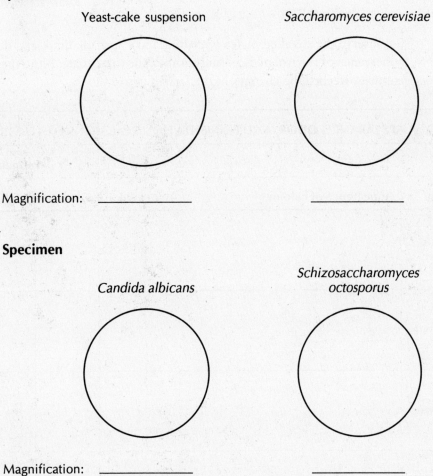

Yeast-cake suspension

Saccharomyces cerevisiae

Magnification: _____   _____

Specimen

Candida albicans

Schizosaccharomyces octosporus

Magnification: _____   _____

341

2 Describe the appearance (including color) of any intracellular structures observed in the water-iodine wet mount of *S. cerevisiae*.

3 Observe the streaked plates on Sabouraud's agar medium, and describe the appearance of the isolated colonies in the following chart. Note the odor of the cultures. Record the changes in the fermentation tubes.

APPEARANCE, ODOR, AND FERMENTATION REACTIONS OF YEASTS					
			Fermentation		
Specimen	Appearance of colonies	Odor	Glucose	Maltose	Lactose
Yeast-cake suspension					
S. cerevisiae					
C. albicans					
S. octosporus					

NAME _____ 55 MORPHOLOGY OF YEASTS *(Continued)*

4 Draw and describe the appearance of the ascospores in the spore-stained preparation. Similarly, draw and describe the appearance of the yeast culture grown on cornmeal-agar medium.

Saccharomyces cerevisiae
on sodium-acetate medium

Candida albicans
on cornmeal-agar medium

Magnification: _____   _____

Description:

QUESTIONS **1** Compare in general terms the morphological characteristics of bacteria, molds, and yeasts.

2 How did the cells from the culture of *Saccharomyces cerevisiae* compare with those from the yeast cake?

3a What was the color of the intracellular granules observed in the water-iodine wet-mount preparation?

3b What is the most probable chemical nature of these granules?

3c What probable function do these granules serve for the cells?

MICROORGANISMS AND DISEASE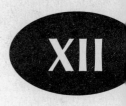

The procedures for the study of pathogenic microorganisms (which actually constitute only a small proportion of all living microorganisms), particularly their characterization and identification, are basically very similar to those used for other microorganisms. For example, many of the exercises performed so far in this manual are applicable in the field of medical diagnostic microbiology—the isolation and identification of unknown cultures, the staining of bacterial cells, the use of differential and selective media, various biochemical and physiological tests, and so on.

Of course, in handling pathogenic microbes, precaution and due respect for specimens must be exercised. All necessary safeguards must be taken so as to prevent infection of oneself or others. Never forget that clinical specimens are ordered for culture by the physician because infectious disease is suspected. Therefore, any specimen should be presumed to contain viable pathogenic microorganisms. The Centers for Disease Control (CDC) has introduced a strategy referred to as **universal precautions** to address concerns about transmission of human immunodeficiency virus (HIV, the etiologic agent of AIDS, that is, acquired immunodeficiency syndrome) in the health care environment. This strategy assumes that **"all patients should be assumed to be infectious for HIV and other blood-borne pathogens."** This means that the careful microbiologist working in the diagnostic laboratory, exposed to specimens that may contain infected body fluids, wears protective gloves, mask, and appropriate clothing. Precautionary practice includes avoidance of needle-stick injuries and thorough washing of hands after contact with specimen material or other contaminated articles.

> **CAUTION:** The exercises in this section, as far as possible, make use of alternate nonpathogenic microorganisms, which exhibit cell properties similar to those of real pathogens, as well as attenuated pathogenic microbes. Nevertheless, the student should consider all these microorganisms as potential pathogens and handle them with care.

In medical microbiology diagnostic work, which is practiced in hospitals and private laboratories, laboratory procedures are determined by the nature of the specimens and the pathogens suspected. For example, if a sputum specimen from a person suspected of having tuberculosis is submitted to the diagnostic laboratory, one of the first tests would be preparation and examination of an acid-fast smear. Additional tests would include inoculation of special media, such as Lowenstein-Jensen or Petragnani media. On the other hand, if a fecal specimen from a suspected case of typhoid fever were received, the first test would be the inoculation of a selective medium. Following incubation, suspected colonies would be isolated to yield pure cultures for subsequent biochemical and serological characterization. Generally, following the isolation and identification of the etiologic agent, a determination of which antimicrobial agents might be effective in the treatment of the infection is carried out. However, regardless of the nature of the specimen and the suspected pathogen, there are certain basic steps involved in the process of microbiological identification and testing in diagnostic microbiology.

1 Collection of the specimen to be processed
2 Processing of the specimen for culture
3 Identification of the medically important bacteria in the specimen
4 Antibiotic susceptibility testing of the identified bacteria

When it is not possible to isolate the pathogen from the specimen, one can use an alternative approach, namely, an immunological procedure. During the course of many infections, antibodies are produced by the infected individual against antigens of the etiologic agent. These antibodies may or may not aid the patient in resisting the infection. However, their presence in the serum indicates that the infection has occurred. Serological tests are performed to determine the presence of such antibodies in the patient's serum. Such antibodies are of great value in diagnosing the occurrence of certain diseases, especially if there is a ***rising antibody titer,*** or concentration, to the specific antigens of an infectious agent. Such an approach is particularly useful in viral infections.

Finally, it should be mentioned that many laboratories in large city hospitals have adopted commercial multitest systems that can yield final identification of etiologic agents in a few hours. Most of these systems are partially or totally automated. Data are fed into a computer that analyzes them and provides the necessary results, such as a definitive identification. However, all this sophisticated instrumentation is based on sound microbiological principles, which the beginning microbiology student is obliged to learn and remember.

EXERCISE 56

Normal Flora of the Human Body

OBJECTIVE
To examine some microorganisms on the human skin as an introduction to the normal microflora on the human body.

OVERVIEW
Everyone of us harbors and carries around with us billions of microorganisms all the time. They constitute the *normal microflora* of our body. This microflora is composed of *resident microorganisms* since they reside as permanent guests in a particular anatomical site. Microbes that are present for brief periods on any anatomical site because they have accidentally landed there are called *transient microorganisms* and are not considered part of the normal microflora.

Resident microorganisms generally cause us no harm but instead contribute to our well-being by synthesizing useful metabolic substances we may need, as in the intestinal tract, by clearing dead cellular debris or by preventing transient microorganisms from infecting us through sheer competition. However, in some instances, resident microorganisms can work against us as *opportunistic pathogens,* or deadly invaders of our weakened system of defense. (Witness the opportunistic infections that kill AIDS patients!)

Certain resident microorganisms are regularly found at specific anatomical sites due to the selection pressures of environmental factors of the particular site. Such factors include pH, oxygen concentration, amount of moisture, and presence of specific body secretions (e.g., hormones or perspiration). These factors interact with the physiology of the microorganisms to give the latter the capacity to adapt to the specific anatomical site and achieve residency status. Resident microbes colonize only *superficial* body tissues that include not only the skin but also mucous membranes that extend from the body surface inward into the body cavity, such as those that line the respiratory, intestinal, and genital tracts. Interior mucous membrane surfaces offer warmth, moisture, and nutrient secretions.

In this exercise we shall examine some microorganisms isolated from the human skin. Microorganisms that normally colonize the skin, or any other anatomical site, are called *saprophytes* because they live on inanimate organic material and do not invade living cells and tissues. They are also called *commensals* because they do "dine together" with their host, in a state of harmony.

Resident microorganisms on the human skin include the following: *Staphylococcus* species, such as *S. epidermidis* and *S. aureus*; *Streptococcus* species that are α-hemolytic, β-hemolytic, or nonhemolytic; *Propionibacterium acnes*, a Gram-positive, pleomorphic rod; *Corynebacterium* species, that is, diphtheroids with pleomorphic morphology; *Pityrosporum* species, small budding yeasts; and perhaps even some fungi.

REFERENCES
MICROBIOLOGY, Chap. 17, "Normal Flora of the Human Body,"
MICROBIOLOGY, Chap. 18, "Host-Parasite Interactions: Nonspecific Host Resistance."

MATERIALS
1 blood-agar plate
1 mannitol-salt-agar plate
1 sterile cotton swab
1 tube containing 1 ml of sterile normal saline (0.85% NaCl)
Hydrogen peroxide, 3%, stored in a brown bottle under refrigeration

PROCEDURE
1 Aseptically remove one sterile cotton swab from its dispenser and moisten it with sterile normal

saline. Squeeze out excess diluent from the cotton tip by rolling it against the inside wall of the tube containing the diluent.

2 Swab your skin over an area of about 5 in^2.

3 Return the swab to the sterile normal saline tube and agitate it vigorously in the saline solution to suspend any adhering microorganisms. Discard the swab.

4 Use an inoculating needle, after sterilizing it in the flame, to streak a blood-agar plate and a mannitol-salt-agar plate for isolated colonies with inocula from the saline solution (see Exercise 17).

5 Incubate the plates in the usual inverted position at 35°C for 48 h.

6 Following the incubation period, examine the blood-agar plate, and note the numbers of *different* kinds of colonies. Note the kind of zones of blood hemolysis, that is α-, β-, or nonhemolytic zones.

7 Make Gram stains of each colony type.

8 Examine the mannitol-salt-agar plate and look for colonies surrounded by yellow-colored zones. This medium is selective for staphylococci. Avirulent commensal staphylococci are differentiated from potentially pathogenic *Staphylococcus aureus* because the latter is able to ferment mannitol, causing yellow coloration of this medium around the colony.

9 From either medium plate, select several colonies that are composed of either *streptococcal* or *staphylococcal* cells (identified from your Gram-stain examination).

10 Carry out a *catalase test* with cells from these colonies using the slide method in the following manner:

    a Transfer cells from the center of a well-isolated colony with a flame-sterilized looped inoculating needle to the surface of a clean glass slide.

    b Add 1 or 2 drops of 3% hydrogen peroxide to the cells. Bubbles of oxygen will be seen if the cells are staphylococci. Streptococci are catalase-negative. (The alternate procedure is to add a few drops of hydrogen peroxide reagent directly to the surface growth on an agar plate or slant. We cannot use this procedure here because blood agar has erythrocytes that possess catalase activity. Catalase is a hemoprotein enzyme that decomposes hydrogen peroxide to oxygen and water.) [See PLATE 24.]

Normal Flora of the Human Body

RESULTS Record your observations in the chart below.

	BACTERIAL COLONIES OBSERVED FROM SKIN SWAB				
Colony designation	Medium: blood or mannitol	Type of hemolytic zones	Gram reaction	Catalase activity	Colony description
A					
B					
C					
D					
E					

QUESTIONS **1** What are the most probable species of bacteria in the several colonial types examined?

Colony A:

Colony B:

Colony C:

Colony D:

Colony E:

2 For each colonial type examined, explain why you consider the cells of each to be either *resident* or *transient microorganisms*.

EXERCISE 57

Precautions and Methodology of Specimen Collection and Processing

OBJECTIVE
To demonstrate the importance of proper precaution in the transport of microbial specimens to the laboratory.

OBJECTIVES
The first step involved in diagnostic microbiology is the collection of an *appropriate* specimen from the patient. This procedure, which is usually the responsibility of the physician or the nurse, must be properly done because the results obtained in the laboratory are only as good as the specimen collected. Three concerns should be borne in mind when a specimen is being collected:

1 The specimen must be obtained from the proper body site to permit recovery of the etiologic agent of the infection.
2 The specimen must be obtained in a manner that avoids contamination with extraneous microorganisms.
3 The specimen must be obtained in a manner that maximizes chances for the recovery of the etiologic agent.

Direct sampling, with a swab, of an infected site should be done whenever possible, such as when the infection is on or near an external surface of the body like the throat, eye, or genital tract. However, infected sites, such as the lungs, gastrointestinal tract, and urinary tract, may be inaccessible to direct sampling. Fortunately, in such cases, discharges, secretions, and excretions eliminated from these sites can be collected for processing and culture. There are still other infected sites, such as in the brain, blood, and joints, that are not accessible to direct sampling and that do not produce discharges and other fluids. They require special aseptic procedures for collecting the specimen, for example, a sterile needle to obtain spinal fluid, blood, or joint fluid.

The primary goal of specimen collection is to obtain infected material from the site that contains the causative agent.

In collecting a specimen, care must be exercised to achieve minimal contamination of the specimen with extraneous microorganisms, especially that of the normal microbiota. (Of course the infection can sometimes be due to opportunistic microbes of the normal microbiota; this is usually the case in compromised or debilitated patients.) For example, the bladder and kidney are sterile in the healthy state. However, if infection is suspected in these sites, urine must be collected. It is voided though the external opening of the genitourinary tract, which harbors a normal microbiota. To avoid contamination of the urine, the external genitalia are cleaned with soap and water and the first few milliliters of urine are discarded before the urine is collected for microbiological culture.

To maximize chances of recovery of the etiologic agent in specimen collection, certain steps should be exercised to protect medical specimens. Appropriate containers must be used; they should be sterile and disposable. Once the specimen is obtained, it must be placed in an environment that will maintain the viability of the pathogen. Some pathogens are very vulnerable to drying, so specimens must be kept in a moist environment. For example, the use of special ***transport media*** provides sufficient moisture if the specimen is on a dry swab. Specimens suspected of containing obligate anaerobes must be kept in an oxygen-free or reduced environment. Devices to exercise these precautions are available commercially under various trade names, for example, Culturette and Anaerobic Culturette.

Time delays between sampling (specimen collection) and transportation to the laboratory must be kept to a minimum. Many pathogens die if they are not placed in an environment suitable for growth within a short time, even though they may be deadly killers inside the human body. Delays in transporting specimens to the laboratory may also allow overgrowth of the specimen with contaminant or extraneous bacteria.

If immediate processing of the sample in the laboratory is not possible, some specimens can be refrigerated. These include citrated whole blood, urine, stool, water, food, and milk. However, other types of samples cannot be refrigerated and should be kept warm; these include cough plates for *Bordetella pertussis*, which causes whooping cough, and specimens for *Neisseria meningitidis*, which causes infectious meningitis, and *N. gonorrhoeae*, the cause of gonorrhea. It is best of course, to process the specimens in the laboratory as soon as they are received.

Processing of the specimens includes an initial direct examination of the sample for the purpose of assessing the acceptability of the sample and of determining the probable identity of the etiologic agent. For example, the specimen container used for collection must be sterile. It should not be leaking or contaminated on the outside. The specimen must have been properly collected for the particular purpose and promptly transported. Microscopic examination is done to provide information that will help to identify rapidly the infective agent. This usually provides only presumptive identification of the etiologic agent and must be confirmed by further tests. Some pathogens have distinct morphology and/or staining reaction that permits them to be recognized easily even in the presence of extraneous bacteria. For example, spirochetes (*Treponema pallidum*) in a scraping from a chancre under dark-field microscopy suggest syphilis; Gram-negative diplococci (*Neisseria gonorrhoeae*) in a Gram stain of urethral exudate from a male suggest gonorrhea [FIGURE 57.1]; acid-fast bacilli (*Mycobacterium tuberculosis*) in sputum suggest tuberculosis.

Once a specimen has been collected and processed as described above, it is ready for inoculation into a series of media for *primary culture*. This series is usually suggested by the nature of the specimen and the nature of the infection as diagnosed by the examining physician requesting the analysis.

REFERENCES

MICROBIOLOGY, Chap. 18, "Host-Parasite Interactions: Nonspecific Host Resistance."
Isenberg, H. D., J. A. Washington, A. Balows, and A. C. Sonnenwirth, "Collection, Handling, and Processing of Specimens," in E. H. Lennette,

FIGURE 57.1
Gram stain of urethral exudate showing *Neisseria gonorrhoeae* within leukocytes.

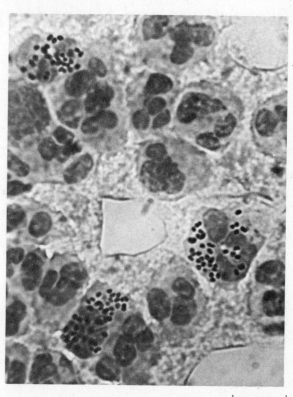

editor-in-chief, *Manual of Clinical Microbiology*, 4th ed., American Society for Microbiology, Washington, D.C., 1985.
Isenberg, H. D., editor-in-chief, *Clinical Microbiology Procedures Handbook*, American Society for Microbiology, Washington, D.C., 1992. A comprehensive and detailed, daily-use procedural handbook for the microbiological and immunological diagnosis of infectious diseases.

MATERIALS

3 blood-agar plates
3 sterile calcium alginate (or other synthetic fiber) swabs (The common cotton swab is said to contain inhibitory fatty acids for some fastidious microbes.)
1 sterile test tube with cap
1 tube with anaerobic transport medium from commercial source (e.g., Port-A-Cul tube from Becton Dickinson Microbiology Systems)
1 thioglycolate-broth culture of *Bacteroides fragilis*, highly diluted

> **CAUTION:** Exercise due care in handling this microbial species.

PROCEDURE

1. Dip a sterile swab aseptically into the *highly diluted* culture of *B. fragilis*. Roll the swab against the side of the tube until it no longer drips.
2. Put the swab into a sterile test tube.
3. Repeat step 1 with a fresh swab, and place it into a tube of anaerobic medium.
4. Repeat step 1 with another fresh swab and spread it on a blood-agar plate to get isolated colonies. Discard the swab in a receptacle containing a disinfectant. Incubate the inoculated agar plate for 48 h at 37°C under anaerobic condition (such as in an anaerobic chamber or an anaerobic jar as shown in Exercise 21).
5. Keep the test tube containing the first swab and the tube with transport medium plus swab for 24 h at room temperature under aerobic condition.
6. Inoculate each stored swab onto separate blood-agar plates to obtain isolated colonies. Discard the used swabs into a disinfectant. Incubate the plates anaerobically for 48 h at 37°C.
7. Record the numbers of colonies and describe the colonial morphology in the Results section.
8. Do a Gram stain on a typical colony and describe the Gram reaction and cell morphology of *B. fragilis*.

Precautions and Methodology of Specimen Collecting and Processing

RESULTS 1 Complete the following chart:

SURVIVAL OF *BACTEROIDES FRAGILIS* DURING TRANSPORT		
Swab	No. of colonies	Colonial morphology
Fresh		
No transport medium		
Transport medium		

2 Complete the following:

Gram reaction of cells:

Cell morphology:

QUESTIONS 1 Name at least two ingredients that might make up the composition of an anaerobic transport medium and give reasons why they are used.

2 List three reasons why a clinical specimen should be transported to the diagnostic laboratory as soon as possible.

EXERCISE 58

Immunology and Serology: Immunoprecipitation

OBJECTIVE
To perform and interpret serological tests in microbial infections.

OVERVIEW
Immunology is the study of properties of the host that confer resistance against infectious agents. The immune system is composed of a single integrated cellular system producing effector products of two types: serum ***antibodies*** that form part of the humoral immunity and sensitized cells called ***lymphocytes*** that constitute cell-mediated immunity. Antibodies are effective in ***opsonizing*** bacteria (rendering them more vulnerable to phagocytosis) and in neutralizing toxins and viruses. The in vitro study of antibodies in serum and their interaction with antigens is called ***serology***. Lymphocytes are important in eliminating intracellular parasites and viruses and rejecting tumors and transplants. Thus the immune response elicits reactive antibodies and cells in response to antigens. It not only forms the principal means of defense in vertebrates against infection by pathogenic microorganisms and larger parasites but it also acts as a surveillance mechanism against the transformation of host cells into cancer cells. The various types of acquired specific immunity are shown in FIGURE 58.1.

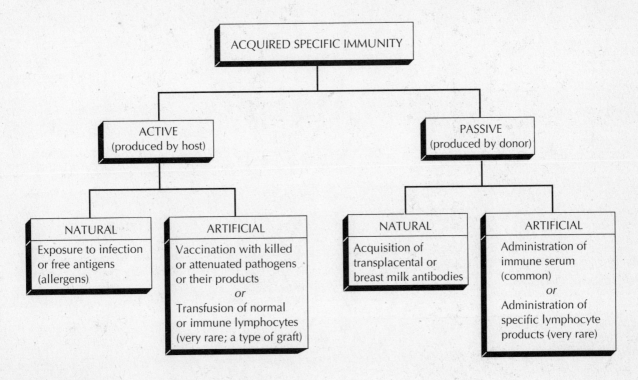

FIGURE 58.1
Various types of acquired specific immunity.

Antibodies are *immunoglobulins*, that is, a type of protein molecule, and can be measured by various serological tests. In clinical laboratories the three major tests used to measure antibody are precipitation, agglutination, and complement-fixation. *Precipitation* is a reaction that occurs between a *soluble antigen* and its specific antibody. It can be demonstrated in several ways, one of the simplest being the ring test. A small amount of antigen is carefully layered over the antibody. If the antibody is specific for the antigen, a fine line, or ring, of precipitate forms at the interface between the two within a few minutes. The ring test is very useful, particularly when only small amounts of either reagent are available, for example, in medicolegal identification of bloodstains. In carrying out the precipitin reaction, one uses a constant dilution of antibody and varying dilutions of antigen because the bulk of the precipitate is formed by the antibody.

Agglutination, on the other hand, is the reaction that occurs between a *particulate antigen* and its specific antibody. Examples of such antigens include bacteria and erythrocytes. Agglutination reactions may be done in the test tube or on a slide, where varying dilutions of antiserum are used with a constant amount of antigen. The test-tube method gives a more quantitative measure of the amount of antibody present, whereas the slide gives a rapid answer to such questions as whether there is antibody in the serum to a particular antigen. On the other hand, if the antibody is known, the slide method can be used to identify an unknown antigen. In agglutination reactions constant antigen and varying antibody are used because the bulk of the agglutinated material is formed by the antigen.

A discussion of complement-fixation will be found in Exercise 61.

REFERENCES

MICROBIOLOGY, Chap. 19, "The Immune Response: Specific Host Resistance."

MICROBIOLOGY, Chap. 20, "Practical Aspects of Immunity."

EXERCISE 58.A

Immunology and Serology: Immunoprecipitation
58.A Precipitin Ring Test

OBJECTIVE
To perform a precipitin ring test.

OVERVIEW
This form of the precipitin test is used in the identification of bacteria, for example, streptococci grouping using cell extract as the antigen in a precipitin reaction with known antisera. It is also widely used in forensic medicine in identifying bloodstains and in detecting food adulteration. The suspected material is dissolved or extracted in saline (0.85% sodium chloride), which is then serially diluted and tested with a variety of antisera. The appearance of a white ring of precipitate at the interface of antigen and antibody indicates a positive test.

REFERENCE
MICROBIOLOGY, Chap. 20, "Practical Aspects of Immunity."

MATERIALS
1.0 ml of bloodstain extract A
1.0 ml of bloodstain extract B
2.0 ml of antihuman rabbit serum (dispensed in two narrow tubes, about 7 mm internal diameter)
2 Pasteur pipettes
Rack for narrow tubes

PROCEDURE
1 You are provided with two tubes labeled "Extract A" and "Extract B," each containing a saline extract made from a different bloodstain.
2 To determine whether either extract came from a human source, you are provided with two narrow tubes containing antihuman rabbit serum (the blood serum of a rabbit immunized with human blood). Label these tubes "A" and "B."
3 Using a fresh Pasteur pipette each time, carefully layer an equal amount of extract A over the antiserum in tube A and similarly an equal volume of extract B in tube B. (Avoid mixing by allowing the extracts to run down the side.)
4 Carefully place tubes in a rack without mixing, and observe periodically for at least 30 min.

NAME _____

Precipitin Ring Test

RESULTS Record the reaction of each tube with a + or −, that is, ring or no ring.

_____ Tube A (with extract A)

_____ Tube B (with extract B)

QUESTIONS 1 Which extract was from a human source?

2 What disadvantages are there in precipitation tests carried out in solutions?

EXERCISE 58.B

Immunology and Serology: Immunoprecipitation
58.B Ouchterlony Immunodiffusion

OBJECTIVE
To perform a precipitation test in gel.

OVERVIEW
Immunoprecipitation usually occurs when antibody (Ab) combines with a polyvalent-soluble antigen (Ag). At the equivalence point, antigen and antibody combine in optimal proportions (equivalently) to form an insoluble lattice or immune precipitate. Immune complexes are soluble in antigen excess or antibody excess, so a precipitate is only seen at the equivalence point.

Precipitation in liquid phase forms quickly and can be quantitated gravimetrically. However, precipitation in gel allows both quantitative and qualitative assessment of antigen concentration and specificity. Single-dimensional diffusion from agar (Oudin technique) allows the qualitative determination of the number of different antigens in a foreign material, as they will form different precipitin bands between single sources of antigen and antibody. By allowing two-dimensional diffusion in agar monolayers between multiple wells (Ouchterlony technique), the relationship between different antigens can be visualized (identity, nonidentity, or partial identity or cross-reactivity). This technique can also be crudely quantitative, as the square of the distance that the band is situated from the antigen source well is proportional to the antigen concentration. A more precise quantitative method allows the antigen to diffuse from wells into a gel containing antiserum, thus forming a ring precipitate (Mancini technique or radial immunodiffusion). The concentration of antigen is proportional to the area of the ring or, more simply, the square of the diameter of the ring. A recent adaptation of this technique employs electrophoresis to "pull" the antigen from the well into the gel in one direction only. This again forms a precipitin band, which now appears as an elongated peak or "rocket." The antigen concentration is simply proportional to the height of the peak. Conventional immunoelectrophoresis is a qualitative assay in which a sample of antigen is electrophoretically separated in agar gel, after which an immune serum is applied in a trough parallel to the electrophoretic separation. As the antibody and separated antigen varieties diffuse together, precipitin arcs are formed. The location of the arcs can identify various antigens, depending upon the specificity of the antiserum and the isoelectric point of the antigen and pH of the assay buffer.

All these assays involve immunoprecipitation and immunodiffusion to quantitate antigen and determine some of its qualitative properties. Similarly, in a reversed format, antibodies can be quantitated. The lattice theory of precipitate formation is essential to the understanding of these reactions.

THE OUCHTERLONY TECHNIQUE WITH PRECIPITIN REACTION PATTERNS
The Ouchterlony two-dimensional double-diffusion assay is the most commonly used qualitative test for antigen specificity. A wide variety of antigenic extracts or culture products can be compared to known pure antigens and their identity or dissimilarity graphically demonstrated. The combination of an easily visible result with a simplicity of operation makes this test very popular.

Antigen and antibody (antiserum) are placed in separate wells cut into agar in a Petri dish as shown in FIGURE 58.2. Homologous reactants produce a line of precipitate where they meet in agar. Like antigens produce bands that meet exactly as in FIGURE 58.2A, indicating a reaction of identity. Unlike antigens produce bands which cross as in FIGURE 58.2B, a reaction of partial identity where there is incomplete homogeneity of antigens (XY and X) by the development of one spur in the band of precipitation,

Exercise courtesy of Dr. Malcolm Baines, McGill University.

FIGURE 58.2
Precipitin reaction patterns in the Ouchterlony immunodiffusion test.

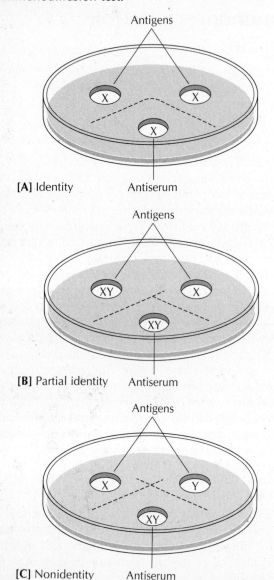

[A] Identity

[B] Partial identity

[C] Nonidentity

and as in FIGURE 58.2C a reaction of nonidentity where there is a lack of homogeneity between the antigens (X and Y) by the formation of two separate bands of precipitation which cross each other forming a double spur.

REFERENCE
MICROBIOLOGY, Chap. 20, "Practical Aspects of Immunity."

MATERIALS
10 ml of 1% agarose in 0.85% saline containing 0.01% sodium azide

1 60-mm plastic Petri plate
8 Pasteur pipettes or micropipettes with disposable tips
1 ruler with fine subdivisions or vernier caliper
1 3-mm punching tool
Vacuum aspirator
1 beaker for boiling water and beaker stand
1 Bunsen burner and striker
1 plastic storage box (8 by 8 in)
0.2 ml of rabbit anti-normal human serum (aNHS) in tube
0.2 ml of 1% human serum albumin (HSA) in tube
0.2 ml of 1% human gamma globulin (HGG) in tube

PROCEDURE
1 Heat agarose in boiling water.
2 Dispense 10 ml of agarose into the Petri plate.
3 Let stand at room temperature for 30 min.
4 When agarose is gelled, cut seven holes with a 3-mm punch in a hexagonal pattern, leaving about 5 mm between wells, using the pattern shown below.

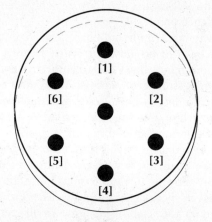

5 Suck out plugs with a Pasteur pipette and vacuum aspirator.
6 Fill the peripheral wells of the plate in clockwise order with 50 µl of [1] NaCl, [2] 1% HGG, [3] 1% HSA, [4] 1% normal or whole human serum (NHS), [5] 1% HGG, and [6] 1% HSA. Put 50 µl of rabbit aNHS in the center well of the plate. NOTE: Remember to mark the rim of the Petri plate to indicate where well [1] is located.
7 Leave the plate uncovered at room temperature for 30 min.
8 Cover the plate, invert quickly, and place in the plastic humid container provided. Wet the towel in the bottom.
9 Place in a cold room (4°C) and inspect after 24 to 48 h for precipitin bands.

NAME _____

Ouchterlony Immunodiffusion

RESULTS

1 Trace the pattern of precipitation in the diagram below.

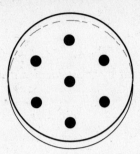

2 Discuss the patterns of precipitation with respect to the identity, partial identity, and nonidentity of the antigens tested.

58.B

365

QUESTIONS 1 What advantages does the Ouchterlony immunodiffusion technique have over the ring test performed in Exercise 58.A?

2 What is meant by the *equivalence point* in antigen-antibody reactions?

EXERCISE 59

Immunoagglutination: Bacterial Agglutination Tests

OBJECTIVE
To perform immunoagglutination tests with bacteria.

OVERVIEW
The agglutination test is one of the most widely used serological tests for the identification of microorganisms. The outer surfaces of bacterial cells contain antigens that can be used in agglutination reactions. Such reactions occur between particulate antigens, such as whole cells or cell walls, flagella, and capsules bound to cells, and antibodies. The procedure entails mixing a suspension of microorganisms (agglutinogen) with serum containing antibody (agglutinin), incubating the mixture, and then observing for clumping of bacterial cells (or antigen). This procedure can be performed on a slide or in test tubes. A tube-agglutination test is more quantitative because it can be used to estimate the concentration (titer) of antibody in serum. A tube-agglutination test may be used to determine whether a particular microorganism is causing the disease in a patient. The endpoint is the greatest dilution of serum showing an agglutination reaction. If an increase in titer is shown in successive days, this is evidence for a rising antibody titer and is an indication that the patient has an infection caused by the specific microbe used in the test. Serological reactions are highly specific and are used extensively for identification of bacterial species or strains within a species. (A strain of bacterium based on specific serological reaction is called a *serovar*.)

Practical use of agglutination reactions has spread from the areas of infectious disease and immunohematology to other studies embracing autoimmune disease, endocrine assay, and so on. The conventional procedures of direct agglutination have given rise to a variety of techniques including coating of carrier particles with antigen or antibody.

REFERENCE
MICROBIOLOGY, Chap. 20, "Practical Aspects of Immunity."

MATERIALS
Antiserum diluted for rapid slide-agglutination test
Antiserum for tube-agglutination technique
Normal serum
Heat-killed or phenolized-saline (normal saline + 0.5% phenol) suspension of bacterial cells against which antisera were prepared
Physiological salt solution (0.85% sodium chloride)
10 1-ml serological pipettes
10 serological test tubes

PROCEDURE
SLIDE AGGLUTINATION
1 Using a glass-marking pencil, draw two circles about the size of a dime at opposite ends of a clean slide as shown in FIGURE 59.1.

2 Within the area of one circle, add one drop of antiserum (for slide test), and within the other, add one drop of normal serum. To each of these drops, add one drop of antigen [FIGURE 59.1].

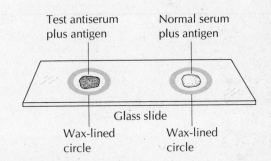

FIGURE 59.1
Procedure for slide-agglutination test.

3 Grasp the slide between the thumb and forefinger, and rock it gently back and forth. Observe the mixtures against a white background and with a good light source. Look for clumping (agglutination) of cells.

TUBE AGGLUTINATION

1 Arrange 10 serological test tubes in a row in a test-tube rack and label them "1" to "10."
2 Add 0.8 ml of physiological saline to the first tube and 0.5 ml to all the rest. Refer to TABLE 59.1 as you proceed.
3 Pipette 0.2 ml of antiserum into the first tube. Using a clean pipette, mix the contents of the first tube by alternately sucking up and blowing back the mixture. (Alternatively, a vortex apparatus may be used to mix the contents.) Following this, transfer 0.5 ml from the first tube to the second tube. Mix the contents of tube 2 in a like manner, using a *clean* pipette, and then transfer 0.5 ml to the third tube. Continue this procedure through the ninth tube. The 0.5 ml removed from the ninth tube is discarded since the tenth tube serves as a control for the antigen suspension. This procedure of dilution, the twofold dilution series, is shown in FIGURE 59.2.
4 Add 0.5 ml of antigen suspension to each tube. Now grasp the rack containing the tubes with both hands and shake it vigorously back and forth in short abrupt strokes. Place the rack, with tubes, in a constant-temperature water-bath set at 45°C for 2 to 4 h.
5 Observe each tube individually for evidence of clumping.

NOTE: More pronounced reactions can be obtained by placing the tubes in a refrigerator overnight after the water-bath incubation.

TABLE 59.1 Protocol of Tube-Agglutination Test

Volumes shown in milliliters	Tube no.									
	1	2	3	4	5	6	7	8	9	10*
Saline	0.8	0.5	0.5	0.5	0.5	0.5	0.5	0.5	0.5	0.5
Antiserum	0.2									
Transfer to next tube	0.5	0.5	0.5	0.5	0.5	0.5	0.5	0.5	**	
Antigen	0.5	0.5	0.5	0.5	0.5	0.5	0.5	0.5	0.5	0.5
Reciprocal of final serum dilution	10	20	40	80	160	320	640	1,280	2,560	

*Antigen control.
**0.5ml discarded from tube 9.

FIGURE 59.2
Procedure for twofold dilution series.

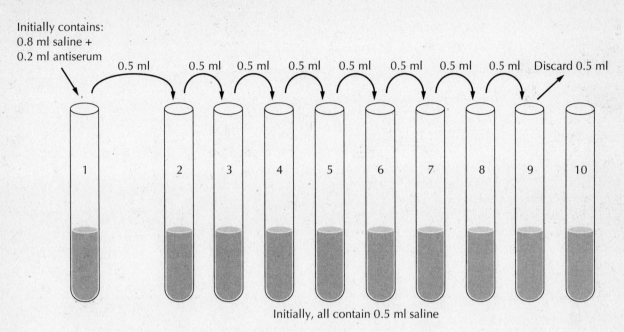

NAME _____

Immunoagglutination: Bacterial Agglutination Tests

RESULTS

1 Describe what occurred to the cells in each of the drops in the slide-agglutination test.

2 On the following chart, record the reaction observed in each tube by the following scheme:

++++ = complete clumping of all cells, fluid clear
+++ = nearly complete clumping of all cells, fluid very slightly turbid
++ = moderate degree of clumping of cells, fluid moderately turbid
+ = slight degree of clumping of cells, fluid turbid
0 = no clumping of cells, fluid reveals even suspension of cells

	TUBE-AGGLUTINATION REACTION									
	Tube no.									
	1	2	3	4	5	6	7	8	9	10
Rating										

QUESTIONS 1 Explain why agglutination reactions may be used for the identification of microorganisms.

2 What precaution should be observed in the interpretation of agglutination reactions for microbial identification?

3 What is the titer of the antiserum used in the tube-agglutination technique?

4 How do you prepare an antiserum against a bacterial antigen?

EXERCISE 60

Streptolysin O Neutralization

OBJECTIVE
To demonstrate the neutralization of a toxin by antibody.

OVERVIEW
Neutralization of activity of various substances can occur by both direct or indirect mechanisms by utilizing specific or nonspecific reagents or procedures. Specific antibodies can react with the antigenic determinants on soluble molecules (such as toxins) and particulate materials (such as viruses, thereby masking their specific attachment structures and neutralizing their functional activity). Antibody can directly attach to the active site or may indirectly impair its function through combination at a nearby site, thereby inhibiting activity through steric hindrance or alteration to the tertiary structure of the antigen. Nonspecific reagents such as enzymes and complement can also neutralize activity.

An example of toxin neutralization is demonstrated in this exercise using the streptococcal hemolysin that causes β hemolysis of erythrocytes in blood-agar plates. Streptolysin O (STO) is a product of Lancefield group A streptococci; it is labile to heat and oxygen and is thus active only in its reduced form. In cases of rheumatic fever, the presence of antibodies to STO is quantitated by assaying the ability of the antibodies to neutralize streptolysin O activity. Titers of less than 50 Todd units/ml of anti-STO (ASTO) activity are taken as evidence that there has been no recent infection with β-hemolytic streptococci.

REFERENCES
MICROBIOLOGY, Chap. 19, "The Immune Response: Specific Host Resistance."

Exercise courtesy of Dr. Malcolm Baines, McGill University.

Todd, E. W., "Antigenic Streptococcal Hemolysin," *J. Exptl. Med.* 55:267, 1932.

MATERIALS
5 ml of STO (reduced)
20 ml of phosphate-buffered saline (PBS), pH 6.5
6 ml of ASTO antiserum (1 unit/ml)
15 ml of 5% rabbit red blood cells (R-RBC)
5 1-ml pipettes
10 5-ml pipettes
24 test tubes (17 × 100 mm)
Test-tube rack
Water-bath at 37°C
Centrifuge with rotor for 17- × 100-mm tubes
1 ml of serum containing ASTO activity

PROCEDURE
Preparation of Reduced STO
1. 1 ml of alkaline PBS
2. 6 mg of cysteine hydrochloride
3. 3 ml of stock STO
4. Mix gently for 10 min.

Preparation of PBS Stock (pH 6.5)
1. 4.2 g of NaCl
2. 3.17 g of KH_2PO_4
3. 3.58 g of $Na_2HPO_4 \cdot 12H_2O$
4. 1.0 liter of distilled water

Preparation of Alkaline PBS
1. 1.0 liter of PBS stock, pH 6.5
2. 1.6 g of NaOH

STREPTOLYSIN O TITRATION
1. Place a row of 12 test tubes in the test-tube rack. Add the reagents to the tubes as follows:

Tube no.	1	2	3	4	5	6*	7	8	9	10	11	12*
STO, ml	0.5	0.4	0.3	0.2	0.1	0	0.5	0.4	0.3	0.2	0.1	0
PBS pH 6.5, ml	0	0.1	0.2	0.3	0.4	0.5	1.0	1.1	1.2	1.3	1.4	1.5
ASTO, ml	1.0	1.0	1.0	1.0	1.0	1.0	—	—	—	—	—	—

*Tubes 6 and 12 are negative controls, and both should show no lysis.

2 Mix thoroughly and incubate at 37°C for 15 min.
3 Add 0.5 ml of 5% R-RBC in PBS to each tube.
4 Mix thoroughly and incubate at 37°C for 45 min, resuspending the R-RBC at 15-min intervals.
5 Centrifuge the tubes at $1000 \times g$ on the bench centrifuge for 5 min.
6 Observe the tubes and choose the tube that just fails to show hemolysis. Use the chart provided in the result section to record and calculate your results. At this point the unit of ASTO has neutralized 1 hemolytic unit of STO. Calculate how many hemolytic units of STO are contained in 1 ml of stock STO solution. You will need this value for the next part of this exercise.

ANTISTREPTOLYSIN O TITRATION

1 Place a row of 10 tubes in the test-tube rack.
2 Pipette 1 ml of PBS pH 6.5 into each tube.
3 Add 0.5 ml of ASTO serum to tube 1 and mix thoroughly. You now have a dilution 2:3 in tube 1. Transfer 0.5 ml from tube 1 to tube 2 and mix thoroughly. You now have a dilution of 2:9 in tube 2. Continue for tubes 3 to 8, making dilutions 2:27, 2:81, 2:243, 2:729, 2:2187, and 2:6561. Discard 0.5 ml from tube 8.
4 Prepare positive and negative control tubes as follows:
  Tube 9 (positive lysis): no additions
  Tube 10 (negative lysis): add 0.5 ml of PBS
5 Dilute the STO solution with alkaline PBS to give 2 lytic units/ml. Add 0.5 ml of diluted STO solution to tubes 1 through 9.
6 Mix thoroughly and incubate at 37°C for 15 min.
7 Add 0.5 ml of 5% R-RBC to all tubes (1 through 10).
8 Mix thoroughly and incubate at 37°C for 45 min, resuspending the RBC at 15-min intervals.
9 Centrifuge at $1000 \times g$ for 15 min and read tubes for lysis.
  NOTE: Tube 9 should show complete lysis, and tube 10 should show no lysis.
10 The ASTO titer is expressed in Todd units of antibody activity, which is the reciprocal of the highest dilution showing no hemolysis or complete neutralization of STO activity, for example, no lysis at tube 4, reciprocal dilution = 81:2 = 40.5 Todd units/ml as shown in the following table:

Tube no.	1	2	3	4	5	6	7	8
Todd units/ml	1.5	4.5	13.5	40.5	121.5	364.5	1093.5	3280.5

Streptolysin O Neutralization

RESULTS

Streptolysin O Titration

1 Rate each tube using the following scale, and enter the result in the chart provided.

```
   0  = no lysis
   ±  = trace lysis
   +  = distinct lysis
  ++  = 50% lysis
 +++  = almost complete lysis
++++  = complete lysis
```

NOTE: Tubes 6 and 12 are negative controls and should show no lysis (no STO). Tube 7 is a positive control and should show complete lysis (no ASTO).

TITRATION OF STREPTOLYSIN O												
Tube no.	1	2	3	4	5	6	7	8	9	10	11	12
Rating												

2 Calculation of STO concentration:

$$\frac{1 \text{ ml}}{\text{STO added, ml}} = \frac{1 \text{ ml}}{\underline{\qquad} \text{ ml}} = \underline{\qquad} \text{ units STO/ml}$$

Antistreptolysin O Titration

1 Dilution of STO to give 5 ml of 2 units/ml:

X ml × stock STO units/ml = 5 ml × 2 STO units/ml

Therefore,

$$X \text{ ml} = \frac{5 \text{ ml} \times 2 \text{ STO units/ml}}{\underline{\qquad} \text{ STO units/ml}} = \underline{\qquad} \text{ ml}$$

2 Enter the results of reading the tubes in the chart provided.

TITRATION OF ANTISTREPTOLYSIN O										
Tube no.	1	2	3	4	5	6	7	8	9	10
Rating										

$$\text{Titer of ASTO} = \frac{1}{\text{serum dilution}} = \underline{} \text{ Todd units/ml}$$

QUESTIONS 1 What chemical was used to keep the stock solution of STO active (reduced)?

2 What is the name of the species of streptococcus that causes rheumatic fever?

EXERCISE 61

Complement-Fixation

OBJECTIVE
To learn the principles of complement-fixation and appreciate the many controls used in its performance.

OVERVIEW
With some antigen-antibody systems, no visible reaction occurs; in such cases complement-fixation may be used. **Complement,** which occurs in the serum of most warmblooded animals, is a thermolabile complex made up of 20 protein components. It takes part in various serological reactions, and whenever an antigen-antibody combination occurs in the presence of complement, the complement is adsorbed, or fixed. The fixation of complement (abbreviated C') to an antigen-antibody complex is not necessarily visible and may have to be demonstrated by an indicator system. The most widely used indicator system is a combination of red blood cells and their specific antibody (*hemolysin*); this combination is known as a *sensitized red-cell suspension.* The sensitized cells, representing a specific antigen-antibody system, also fix complement, and in this case the combination of antigen-antibody and complement leads to **hemolysis,** or the dissolution of red blood cells with escape of the hemoglobin from the cell. The gross appearance changes from a turbid suspension of red blood cells showing a clear, colorless supernatant fluid after centrifuging to a red transparent fluid when hemolysis has taken place with no change in color after centrifuging. In a complement-fixation test, the antigen and antibody are allowed to react in the presence of C'. After the initial reaction period the indicator system is added and the mixture allowed to react. If the original antigen and antibody are *homologous,* that is, specific for each other, no hemolysis is seen since the C' is bound. If the original antigen and antibody are *heterologous,* that is, not specific, then hemolysis occurs since the C' is not fixed by the first system and is free to act on the sensitized cells of the indicator system. Thus, *failure* to obtain lysis denotes a *positive* reaction, while complete *hemolysis* indicates a *negative* reaction. The principle of complement-fixation is shown in FIGURE 61.1

Complement-fixation tests may be used in the diagnosis of infectious diseases, especially those caused by viruses, rickettsiae, and fungi. Complement-fixation tests are considerably more elaborate and difficult to perform than agglutination or precipitation tests. But they have a wide range of application and can be carried out with antigens in soluble or particulate form and under conditions in which the antigen-antibody combination does not lead to visible reaction by any other method.

Careful standardization and titration of reagents used in the test are essential for valid results. Four of the five components used in complement-fixation tests must be known, and only one component may represent the unknown factor. Either antigen or antibody is identified in the common complement-fixation tests.

REFERENCES
MICROBIOLOGY, Chap. 18, "Host-Parasite Interactions: Nonspecific Host Resistance."
MICROBIOLOGY, Chap. 20, "Practical Aspects of Immunity."

MATERIALS
HEMOLYSIN TITRATION
1 ml of hemolysin diluted 1:100 with 0.85% NaCl or saline (The saline used throughout the complement-fixation test contains 1.0 g magnesium sulfate per liter.)
5 ml of complement diluted 1:10 with saline (*kept on ice*)
5 ml of 2% sheep red blood cells
15 ml of 0.85% saline
4 1-ml pipettes

FIGURE 61.1
Principle of the complement-fixation test. If complement has been fixed by the antigen-antibody reaction in the test system (stage 1), then it is no longer available for hemolysis of the added sensitized sheep red blood cells in the indicator system (stage 2).

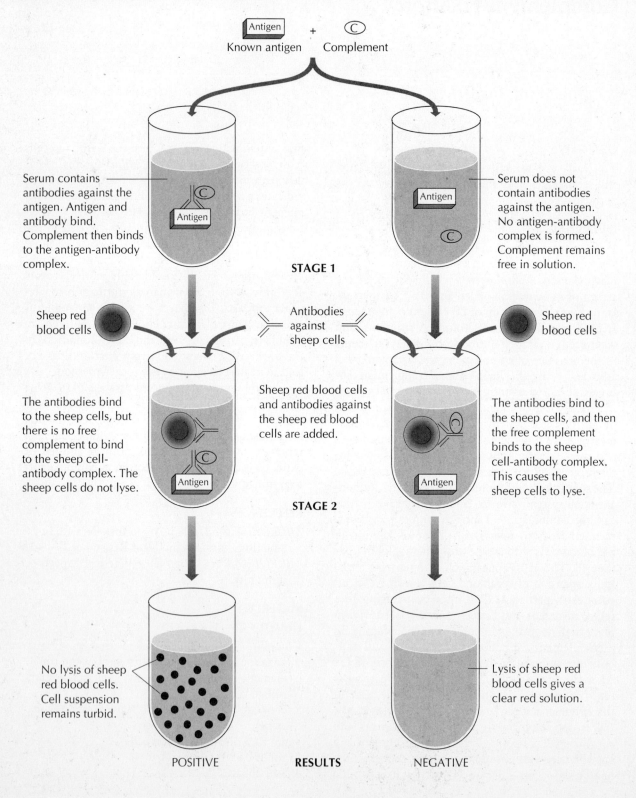

Rack with 9 serological test tubes
Water-bath at 37°C

TITRATION OF COMPLEMENT

6 ml of hemolysin (2 MHD/ml) (See results section for the definition of MHD.)
15 ml of 0.85% saline
3.5 ml of complement (1:10) *(kept on ice)*
6 ml of 2% sheep red blood cells
Rack with 10 serological test tubes

DETERMINATION OF ANTIGENIC DOSE

6.5 ml of antigen solution in initial dilution
5 ml of antiserum (1:5 dilution)
20 ml of complement (4 MHD/ml) *(kept on ice)*
17 ml of hemolysin (4 MHD/ml)
Rack with 33 serological test tubes
5 ml of saline

COMPLEMENT-FIXATION TEST

1.5 ml of test serum heated at 56°C for 30 min to inactivate complement
1.5 ml of normal serum
5 ml of complement (appropriately diluted) *(kept on ice)*
2 ml of antigen
8 ml of 0.85% saline
5 ml of hemolysin (against sheep red blood cells, appropriately diluted)
6 ml of 2% sheep red blood cells

10 1.0-ml pipettes (Use each only for its specific reagent.)
Rack with 9 serological test tubes

PROCEDURE

Titrations of hemolysin and complement are preliminary to any complement-fixation reaction and in practice are set up each day that complement-fixation tests are performed. The antigen used in any complement-fixation test must also be standardized and titrated. The antigen must be tested for anticomplement and hemolytic activity, and the optimal concentration to be used must be determined. However, all these tests need to be done only once for each batch of antigen, provided the antigen is stable.

HEMOLYSIN TITRATION

1 Add 0.5 ml of saline to each of nine tubes as indicated in TABLE 61.1.
2 Add 0.5 ml of hemolysin to tube 1, mix, and make halving dilutions of the hemolysin up to and including tube 8.
3 Add 0.5 ml of complement to each tube.
4 Add 0.5 ml of 2% red blood cells (RBC) to each tube.
5 Add 1 ml of saline to each tube. (Saline is added here to bring the volume to 2.5 ml, the standard volume adopted for our purposes. A constant volume must be maintained in all tests.)

TABLE 61.1 Protocol for Hemolysin Titration

	Tube no.								
	1	2	3	4	5	6	7	8	9 (control)
Saline	0.5	0.5	0.5	0.5	0.5	0.5	0.5	0.5	0.5
Hemolysin, 1:100	0.5	Make	halving	dilutions	till	tube	8		
Hemolysin dilution	1:200	1:400	1:800	1:1600	1:3200	1:6400	1:12,800	1:25,600	
Complement, 1:10	0.5	0.5	0.5	0.5	0.5	0.5	0.5	0.5	0.5
RBC, 2%	0.5	0.5	0.5	0.5	0.5	0.5	0.5	0.5	0.5
Saline	1	1	1	1	1	1	1	1	1
Final hemolysin dilution	1:1000	1:2000	1:4000	1:8000	1:16,000	1:32,000	1:64,000	1:128,000	

6 Shake the rack and place it in the water-bath at 37°C for 1 h.

NOTE: The highest dilution of hemolysin giving complete (100%) hemolysis represents 1 unit of hemolysis under the conditions of the test (volume of 2.5 ml, complement 1:10, 2% red blood cells, etc.). This unit is termed the *minimal hemolytic dose (MHD)*. (A more accurate endpoint is obtained by determining the amount or dilution that will give 50 % lysis. However, such determinations require more complex titrations and equipment and are not essential in the routine performance of complement-fixation tests.) In the actual test, 2 to 3 MHD of hemolysin is used per volume unit; for example, 1 MHD in 0.5 ml of 1:32,000 dilution; 0.5 ml of 1:16,000 = 2 MHD.

TITRATION OF COMPLEMENT

1 Follow the protocol in TABLE 61.2 after setting up 10 serological test tubes in a rack.
2 Shake the rack, incubate for 1 h in a 37°C water-bath, and read the results.

NOTE: The *smallest* amount of complement giving hemolysis equals 1 MHD (2 to 3 MHD of complement is used in the actual test).

DETERMINATION OF ANTIGENIC DOSE

1 Prepare a master titration of the antigen provided through six doubling dilutions (1:40 through 1:1280) in 3-ml quantities as follows.
2 Arrange five rows of serological tubes with six tubes in each row.
3 Pipette the antigen dilutions as follows:
  Tube 1 of each row: 0.5 ml of 1:40 antigen
  Tube 2 of each row: 0.5 ml of 1:80 antigen
  Tube 3 of each row: 0.5 ml of 1:160 antigen
  Tube 4 of each row: 0.5 ml of 1:320 antigen
  Tube 5 of each row: 0.5 ml of 1:640 antigen
  Tube 6 of each row: 0.5 ml of 1:1280 antigen
4 Prepare a master titration of a strongly positive antiserum through five doubling dilutions (1:5 through 1:80) in 4-ml quantities as follows.
5 Pipette serum dilutions as follows:
  Add 0.5 ml of 1:5 serum to each tube in first row.
  Add 0.5 ml of 1:10 serum to each tube in second row.
  Add 0.5 ml of 1:20 serum to each tube in third row.
  Add 0.5 ml of 1:40 serum to each tube in fourth row.
  Add 0.5 ml of 1:80 serum to each tube in fifth row.
6 Add 0.5 ml of complement (2 units) to each tube.
7 Prepare these controls:
  a *Serum control*: 0.5 ml of 1:5 serum
  0.5 ml of complement
  (2 units)
  0.5 ml of saline
  b *Antigen control*: 0.5 ml of 1:40 antigen
  0.5 ml of complement
  (2 units)
  0.5 ml of saline
  c *Hemolytic system control*: 0.5 ml of complement (2 units)
  1.0 ml of saline
8 Incubate in a 37°C water-bath for 30 to 60 min.
9 After incubation, add 0.5 ml of hemolysin (2 units) and 0.5 ml of 2% red blood cells to each tube.

TABLE 61.2 Protocol for Complement Titration

	Tube no.									
	1	2	3	4	5	6	7	8	9	10 (control)
Saline	0.45	.4	.35	0.3	.25	0.2	.15	0.1	.05	0.5
Complement, 1:10	0.05	0.1	0.15	0.2	0.25	0.3	0.35	0.4	0.45	
Hemolysin, 2 MHD/ml	0.5	0.5	0.5	0.5	0.5	0.5	0.5	0.5	0.5	0.5
RBC, 2%	0.5	0.5	0.5	0.5	0.5	0.5	0.5	0.5	0.5	0.5
Saline	1	1	1	1	1	1	1	1	1	1

10 Mix the contents by shaking the rack, and incubate at 37°C for 30 min.
11 Read the results.

NOTE: The dose of antigen to be employed in the complement-fixation test is the largest amount giving complete hemolysis with the smallest amount of serum.

PERFORMANCE OF THE COMPLEMENT-FIXATION TEST

1 Put ingredients into each tube as shown in the protocol of TABLE 61.3.
2 Shake until all the ingredients are mixed and place the tubes in the water-bath at 37°C for 1 h, after which add the ingredients as shown in TABLE 61.4.
3 Replace the tubes in the water-bath for 1 h, shaking again halfway through this period. At the end of the hour read your results *without further shaking*.

NOTE: The following are the contents of each tube and the results that should be obtained:

Tube 1: Hemolysin + RBC = no hemolysis (complement lacking)
Tube 2: Complement + RBC = no hemolysis (hemolysin lacking)
Tube 3: Complement + hemolysin + RBC = hemolysis
Tube 4: Antigen + complement + hemolysin + RBC = hemolysis (no antibody)
Tube 5: Positive serum + complement + hemolysin + RBC = hemolysis (no antigen)
Tube 6: Positive serum + antigen + complement + hemolysin + RBC = no hemolysis (This is a *positive* test for complement-fixation because there is specific antibody present that reacted with the antigen and fixed the complement.)
Tube 7: Normal serum + complement + hemolysin + RBC = hemolysis (no antigen, no antibody)
Tube 8: Normal serum + antigen + complement + hemolysin + RBC = hemolysis (This is a *negative* test for complement-fixation because there is no specific antibody present to react with the antigen and thus fix the complement.)
Tube 9: Complement inactivated by heat + hemolysin + RBC = no hemolysis

For the results to be reliable, it is necessary to be sure that the following results are observed in the controls:

Hemolysin (f) alone does not lyse the RBC (tube 1).
Complement (c) alone does not lyse the RBC (tube 2).
Complement (c) + hemolysin (f) + RBC (g) give complete hemolysis (tube 3).
Antigen (d) alone does not fix the complement (tube 4).
Positive serum (a) alone does not fix the complement (tube 5).
Normal serum (b) alone does not fix the complement (tube 7).

TABLE 61.3 Protocol for the Test System in the Complement-Fixation Test

Pipette	Ingredient	Code	Tube no.								
			1	2	3	4	5	6	7	8	9*
A	Test serum, ml	a	0	0	0	0	0.5	0.5	0	0	0
B	Normal serum control, ml	b	0	0	0	0	0	0	0.5	0.5	0
C	Complement, ml	c	0	0.5	0.5	0.5	0.5	0.5	0.5	0.5	0.5
D	Antigen, ml	d	0	0	0	0.5	0	0.5	0	0.5	0
E	Saline, ml	e	1.5	1.5	1.0	0.5	0.5	0.0	0.5	0.0	1.0

*Tube 9 is to be heated at 56°C for 30 min instead of incubating to show inactivation of complement. Replace in your series.

TABLE 61.4 Protocol for the Indicator System in the Complement-Fixation Test

			\multicolumn{9}{c}{Tube no.}
Pipette	Ingredient	Code	1	2	3	4	5	6	7	8	9*
F	Hemolysin, ml	f	0.5	0	0.5	0.5	0.5	0.5	0.5	0.5	0.5
G	RBC, ml	g	0.5	0.5	0.5	0.5	0.5	0.5	0.5	0.5	0.5

*Tube 9 is to be heated at 56°C for 30 min instead of incubating to show inactivation of complement. Replace in your series.

NAME _____

Complement-Fixation

RESULTS **Hemolysin Titration**

1 Record your readings in the chart provided as follows: 4+ = complete hemolysis (clear red fluid); 2+ = partial hemolysis (red fluid but also cells in suspension); 0 = no hemolysis (turbid cell suspension).

	RESULTS OF HEMOLYSIN TITRATION								
	Tube no.								
	1	2	3	4	5	6	7	8	9
Hemolysin dilution	1:1000	1:2000	1:4000	1:8000	1:16,000	1:32,000	1:64,000	1:128,000	
Reading									

Titration of Complement

2 Record your readings of hemolysis in the chart provided using the same scale as for hemolysin titration: 4+ = complete hemolysis; 2+ = partial hemolysis; 0 = no hemolysis.

	RESULTS OF COMPLEMENT TITRATION									
	Tube no.									
	1	2	3	4	5	6	7	8	9	10 (control)
C', ml	0.05	0.1	0.15	0.2	0.25	0.3	0.35	0.4	0.45	
Reading										

Determination of Antigenic Dose

3 Record your results in the chart provided. Be sure to inspect the controls first.

	RESULTS OF ANTIGENIC DOSE DETERMINATION					
	Antigen dilutions					
Serum dilutions	1:40	1:80	1:160	1:320	1:640	1:1280
1:80						
1:40						
1:20						
1:10						
1:5						

Performance of the Complement-Fixation Test

4 Read the results and enter them into the chart provided as follows:
+ = hemolysis; − = no hemolysis.

	RESULTS OF THE COMPLEMENT-FIXATION TEST								
	Tube no.								
	1	2	3	4	5	6	7	8	9
Reading									

5 Interpret your results as described on page 381.

NAME _____ 61 COMPLEMENT-FIXATION *(Continued)*

QUESTIONS 1 When is complement-fixation useful in antigen-antibody interaction?

2 What is hemolysin and how is it prepared?

3 Why were complement solutions always kept on ice in laboratory manipulations?

4 Explain why the indicator system is necessary in the complement-fixation test.

EXERCISE 62

Complement-Mediated Lysis in Agar

OBJECTIVE

To show the relationship between complement and antibody concentration and their interdependence in producing antibody-dependent complement-mediated lysis of sheep erythrocytes.

OVERVIEW

Since the complement-fixation assay is a complex quantitative assay requiring scrupulous attention to accurate preparative procedures (see Exercise 61), the purpose of this exercise is to demonstrate graphically in a simpler manner the antibody dependency of complement (C')-mediated lysis and its generally quantitative properties.

If erythrocytes are used as "targets" for C'-mediated lysis in an agar matrix, then areas of hemolysis can be measured at the end of the test. The diameter of the areas of lysis will be proportional to the original concentration of the limiting factor, either antibody or complement:

$$\text{Diameter} = a + b \sqrt{x(\text{concentration})}$$

where a equals the diameter of the well and b is a diffusion and reaction constant characteristic of the assay. Therefore, the square root of the concentration of the limiting variable x, antibody or complement, can be plotted against the diameter of the area of lysis. The specific objective of this exercise is to show the relationship between complement and antibody concentration and their interdependence in producing antibody-dependent complement-mediated lysis of sheep erythrocytes.

REFERENCES

MICROBIOLOGY, Chap. 18, "Host-Parasite Interactions: Nonspecific Host Resistance."
MICROBIOLOGY, Chap. 20, "Practical Aspects of Immunity."

Exercise courtesy of Dr. Malcolm Baines, McGill University.

MATERIALS

2 100-mm Petri plates containing a thin layer (1 mm) of 1% agar containing 5% sheep red blood cells (SRBC) in 2.5 ml over an underlayer of 20 ml of 1% agar in phosphate-buffered saline (PBS)
1 3-mm gel punching tool
Vacuum aspirator and Pasteur pipettes
10 1-ml pipettes (or a 50- to 200-μl micropipette and tips)
10 glass test tubes (12 × 75 mm)
2 ml of diluted (about 1:10) anti-SRBC hemolysin, tube A (for plate 1)
0.2 ml of stock anti-SRBC hemolysin, tube B (for plate 2)
50 ml of PBS, pH 7.2 (Dutton's modification):
Solution A (10× concentrate), pH 7.2
Dextrose 10 g
KH_2PO_4 0.6 g
$Na_2HPO_4 \cdot 12H_2O$ 4.8 g
Distilled water 1 liter
Solution B (10× concentrate)
$CaCl_2 \cdot 2H_2O$ 1.9 g
KCl 4 g
NaCl 80 g
$MgCl_2 \cdot 6H_2O$ 2 g
$MgSO_4 \cdot 7H_2O$ 2 g
Distilled water 1 liter
For use, take 1 volume of solution A and 1 volume of solution B and add to 8 volumes of distilled water.
1 plastic sandwich box (8 × 8 in)
0.2 ml stock guinea pig complement, tube C (*on ice*)
2 ml of diluted (about 1:10) guinea pig complement, tube D (*on ice*)
1 ruler with fine subdivisions or vernier caliper
2 sheets of graph paper
1 marking pen
1 100-μl heparinized capillary tube (commercially available)
0.2 ml of fresh human serum (*on ice*)

PROCEDURES

1 Using the well-punching tool provided, cut seven wells in the plates using the pattern shown:

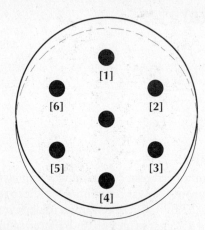

2 Suck out the plugs using a Pasteur pipette and vacuum aspirator.
3 Label the plates on the top and edge as 1 and 2.
4 Add 2 ml of the solution of diluted antibody of SRBC to plate 1 (provided in tube A) to flood the whole plate.
5 Prepare a 1:2 dilution series of five tubes of the antibody in the following manner. Put 0.1 ml of PBS in each of four tubes labeled "B2" to "B5." Starting with 0.1 ml of anti-SRBC antibody from tube B, add to 0.1 ml of PBS in B2. Mix and serially transfer 0.1 ml from tube to tube to make five tubes ranging from the undiluted stock (tube B) to a 1:16 dilution in tube B5. Use the same 1-ml pipette for each dilution.
6 Starting with the lowest dilution and using the same pipette, place 50 μl of the antibody dilutions in the wells of plate 2 in the following order: well 1 = undiluted, 2 = 1:2, 3 = 1:4, 4 = 1:8, 5 = 1:16, and 6 = PBS only. (If a micropipette is not available, simply fill the wells using a Pasteur pipette, being careful not to overfill.)
7 Place both plates on a moistened paper towel in your plastic box and incubate at 37°C for 1 h.
8 While the plates are incubating, make a 1:2 dilution series of five tubes starting with 0.1 ml of fresh guinea pig serum as a source of complement from tube C in 0.1 ml of PBS as you did for antibody in step 5. (**NOTE:** This must be done on ice as complement is very heat-labile.)
9 When the incubation period is completed, thoroughly empty the liquid from the plates by gently aspirating the contents of the wells using the vacuum aspirator and a Pasteur pipette. The plates should be rinsed twice with 2 ml of PBS which is then reaspirated. Invert over absorbent paper to allow complete drainage of rinsing buffer.
10 To plate 1 add the 1:2 dilutions of complement in the following order: well 1 = undiluted stock, 2 = 1:2, 3 = 1:4, 4 = 1:8, 5 = 1:16, 6 = PBS.
11 To plate 2 add 2 ml of a 1:10 dilution of guinea pig complement (tube D) to flood the whole plate.
12 Replace the plates in the plastic box and incubate for 60 min at 37°C.

NAME _____

Complement-Mediated Lysis in Agar

RESULTS

1 Observe the plates and record the diameters of the rings (zones) of hemolysis to the nearest 0.1 mm, if possible.

Plate 1

Well 1 _____   Well 2 _____   Well 3 _____

Well 4 _____   Well 5 _____   Well 6 _____

Plate 2

Well 1 _____   Well 2 _____   Well 3 _____

Well 4 _____   Well 5 _____   Well 6 _____

2 Plot the square root of antiserum or complement concentrations on the y ordinate against the diameter of the rings of hemolysis on the x axis. If you consider this as an estimate of standard activity for complement (plate 1) and for antibody (plate 2), respectively, then unknowns could be crudely related to these values by using the additional wells for their assay.

As an example of this test, you can place 50 μl of human serum in the central well of each assay plate at step 6. You can use some of your own whole blood obtained by pricking your finger with the lancet provided and collecting your blood in the heparinized 100-μl capillary tube. (Or you can ask your instructor for a specimen of fresh human serum he or she will have prepared.) Using a dispensing tube/capillary holder, gently blow out half (50 μl) of the contents of the capillary tube (100 μl) into each central well of each plate.

3 Assuming that guinea pig serum has a complement activity of 350 CH50 units/ml (50% complement hemolysis), extrapolate from the graph plotted and read (obtain) the approximate activity of human serum. If you used whole blood in your assay, multiply the result by 2 to adjust for the cellular component of the blood.

QUESTIONS **1** Besides lysis of red blood cells, can complement also mediate lysis of bacterial cells?

2 What is the visible manifestation of lysis of a bacterial culture, and how can this be measured quantitatively?

EXERCISE 63

Bacterial Infection of a Plant: Demonstration of Koch's Postulates

OBJECTIVE
To demonstrate Koch's postulates using bacterial infection of a plant as a model.

OVERVIEW
Crown gall is a disease of plants characterized by the formation of tumors or galls. Tumors may be produced anywhere on the plant, on the roots, or in the stems. The size of the tumors ranges from less than 1 cm to 30 cm or more in diameter. Tumors may enlarge to the extent that they exert pressure on surrounding xylem tissue to reduce the flow of water and photosynthetic products to healthy tissue. Consequently, there is a reduction in plant size and flower as well as fruit yield.

The etiologic agent is a bacterium called *Agrobacterium tumefaciens*. It is an aerobic Gram-negative rod and is motile with one to six peritrichous flagella. It survives for a long time in soil and attacks the plants by binding to the cells of root or stem tissues that have been abraded by handling, grafting, and insects. The bacterium possesses a plasmid (pTi) that carries a sequence of genes, called T-DNA, that is the cause of pathogenesis. Some or all of the plasmid appears, by an as yet unknown process, in the plant cell while the bacterium remains extracellular. Part of the plasmid, the T-DNA, becomes inserted into the plant-cell chromosomal material. It then induces an imbalance of plant hormones (auxins such as indoleacetic acid; cytokinins such as zeatin) in the infected cell, leading to the unregulated proliferation of cells (hyperplasia) and the formation of tumors. The inserted T-DNA apparently replicates with the plant-cell chromosome during mitosis and is inherited in a mendelian fashion.

In this exercise, you will use *A. tumefaciens* to infect a plant and to fulfill Koch's postulates. Fulfillment of Koch's postulates helps identify a specific microorganism as the causative agent for a particular disease. This is not always easy to do because many microorganisms can be isolated from a diseased tissue, but their very presence does not prove that any or all of them caused the disease. A microorganism that is present in a diseased site may be a secondary invader, part of the normal flora selected by the diseased condition, or a member of a transient flora of the affected site. That is, a microorganism may be in a diseased site because of the disease and may not have caused it. This is why Robert Koch (1843–1910) established four criteria that are embodied in his postulates:

1 The same microorganism must be present in every case of the disease.
2 The microorganism must be isolated from the diseased tissue and grown in pure culture in the laboratory.
3 The microorganism from the pure culture must cause the disease when inoculated into healthy, susceptible laboratory animals (or plants).
4 The microorganism must again be isolated from the inoculated animals and must be shown to be the same pathogen as the original microorganism.

However, it should be remembered that these criteria, albeit useful, cannot be fulfilled for all infectious diseases. For instance, viruses cannot be cultivated on artificial media, and they are not readily observable in a host; some diseases affect only humans and not laboratory animals.

REFERENCES
MICROBIOLOGY, "Prologue: Discovering the Microbial World."
MICROBIOLOGY, Chap. 18, "Host-Parasite Interactions: Nonspecific Host Resistance."

MATERIALS
2 potted tomato plants approximately 6 weeks old
2 sterile disposable 1- or 2-ml syringes

1 Petri plate of nutrient agar
1 tube of sterile nutrient broth (5 ml)
Broth culture of *A. tumefaciens*
Mortar and pestle
1 sterile razor blade
1 bottle with 50 ml of sterile water

PROCEDURE
1 Streak a nutrient-agar plate with *A. tumefaciens* to produce isolated colonies. Incubate for 48 h at room temperature. Then characterize the colonies and make and examine a Gram-stained preparation.
2 Draw a small amount of the broth culture into the sterile syringe and use it to inoculate the tomato plant in the following manner: puncture the tissue near a stem node and inject a small amount of the culture into the "cut." Repeat in two or three different locations as indicated in FIGURE 63.1. Using a second syringe in which you have drawn sterile nutrient broth, make similar punctures and inoculations into a second tomato plant (control).

NOTE: An incubation period of 2 or 3 weeks is required for symptoms (tumors) to become evident on the plant. Accordingly, the plants must be maintained in a suitable environment, for example, light, temperature, and humidity, for this period of time.
3 When tumors form on the plant, cut one off, using a sterile blade. Place this specimen in a sterile 250-ml flask and wash it with several changes of sterile water. Following the last washing, grind the tumor in a mortar with a pestle. After the tissue is adequately macerated, remove a loopful and streak a nutrient-agar plate; incubate the plate for 24 to 48 h. Prepare and examine a Gram-stained specimen from this macerated tissue.

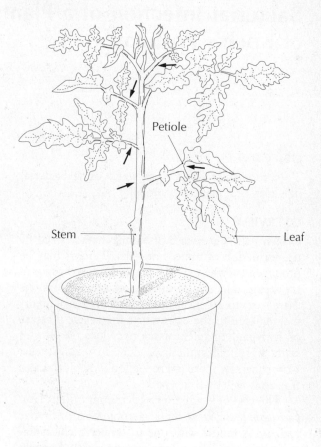

FIGURE 63.1
Inoculation sites (arrows) of *Agrobacterium tumefaciens* on a tomato plant.

NAME _____

Bacterial Infection of a Plant: Demonstration of Koch's Postulates

RESULTS

1 Describe the characteristics of the colonies and the organisms from the original culture and from those obtained from the tumor.

2 Sketch the plant, showing the location and size of the tumor(s) at the time the experiment was terminated.

QUESTIONS 1 What is the morphology and Gram reaction of most bacterial plant pathogens?

2 How are plant diseases usually transmitted in nature?

3 What human malady can benefit from research on the disease caused by *A. tumefaciens?*

EXERCISE 64.A

Airborne Infections
64.A The Corynebacteria and the Gram-Positive Cocci

OBJECTIVE
To become acquainted with the characteristics of some pathogenic microorganisms that are transmitted by air.

> **CAUTION:** Students are again reminded to exercise special care in handling any pathogens used in this exercise.

OVERVIEW
In the following three exercises in this part of the manual, pathogenic bacteria will be introduced according to their *route of infection*, for example, airborne infections.

From the initial examination of the primary cultures from clinical specimens (see Exercise 57), potentially pathogenic microorganisms are selected for further identification tests. The degree to which a microorganism is identified varies among laboratories and is determined also by the microbe itself. The major types of identification tests have been introduced already in previous exercises in this manual. These include physiological, biochemical, tolerance (e.g., salt tolerance), and immunologic tests.

The genus *Corynebacterium* comprises non-spore-forming Gram-positive rods. There are many species in this genus, including species pathogenic to humans, animals, and plants. A very important human bacterial pathogen of the respiratory tract is *C. diphtheriae*, the causative agent of diphtheria. *C. diphtheriae* produces a potent exotoxin that is antigenic and specifically neutralized by antitoxin.

The streptococci are chain-forming Gram-positive cocci. They are for the most part commensal bacteria living on the mucous membranes of humans and lower animals. Some species invade and cause infection only when the host resistance is low. Others, like *Streptococcus pyogenes*, are very virulent, and, except in a small percentage of the population who are carriers, their very presence means infection. They are the major cause of bacterial sore throats (acute pharyngitis). *S. pyogenes* is a pus-producing, or pyogenic, microorganism. On the other hand, there are some streptococci that are of very considerable importance in the dairy industry.

Streptococci are identified by biochemical characteristics, including hemolytic reactions, and antigenic characteristics based upon cell-wall specific antigens (giving rise to Lancefield groups). Hemolytic reactions are based on hemolysins that are produced by streptococci while growing on blood-enriched agar. The two types of hemolysis on blood-agar plates are:

1 *Beta (β) hemolysis:* Complete hemolysis, giving a clear zone around the colony
2 *Alpha (α) hemolysis:* Incomplete hemolysis, producing methemoglobin and a green, cloudy zone around the colony

No hemolysis, or no change in the blood agar, around the colony is sometimes referred to as ***gamma (γ) hemolysis***. Alpha- and gamma-hemolytic streptococci are usually members of the normal flora, while beta-hemolytic streptococci are frequently pathogens. Over 90 percent of human streptococcal infections are caused by Lancefield group A beta-hemolytic streptococci. [See PLATES 7 and 8.]

Streptococcus pneumoniae is characteristically a somewhat elongated Gram-positive coccus that occurs in pairs with flattened, opposed surfaces and pointed distal ends. Each pair is surrounded by a capsule. The bacterium may form short chains, and the capsule may be lost on artificial cultivation, and since colonies on blood agar are surrounded by a zone of α hemolysis, it may be difficult to distinguish it from other α-hemolytic streptococci. Two tests make the distinction as shown in the following table:

Bacterium	Bile solubility	Inulin fermentation
S. pneumoniae	+	+
α-hemolytic streptococci, e.g., S. salivarius	−	−

REFERENCE
MICROBIOLOGY, Chap. 24, "Airborne Diseases."

MATERIALS
1 culture of *Corynebacterium xerosis* on Loeffler's slant
1 blood-agar plate with a mixture of *S. pyogenes* and *S. faecalis*
1 blood-agar plate with a mixture of *S. salivarius* and *S. pneumoniae*
Brain-heart-infusion-broth culture of *S. salivarius*
1 Todd-Hewitt-broth culture of *S. salivarius* (use only 1 ml)
1 Todd-Hewitt-broth culture of *S. pneumoniae* (use only 1 ml)
Gram stains
Loeffler's alkaline methylene blue stain
1 tube of cystine-trypticase-agar or cystine-trypticagar (CTA) medium with glucose
1 tube of CTA medium with maltose
1 tube of CTA medium with sucrose
2 tubes of CTA medium with inulin
1.5 ml of sodium desoxycholate solution (10%)
Reagents for capsule stain

PROCEDURE
THE CORYNEBACTERIA

1 A culture of *C. xerosis* is provided on a slant of Loeffler's medium. Make two smears and stain with Gram's stain and Loeffler's alkaline methylene blue stain. After staining with Loeffler's methylene blue, metachromatic granules can be seen in some cells of corynebacteria [FIGURE 64.1].

2 Inoculate *C. xerosis* into the following set of CTA medium sugars: glucose, maltose, and sucrose.

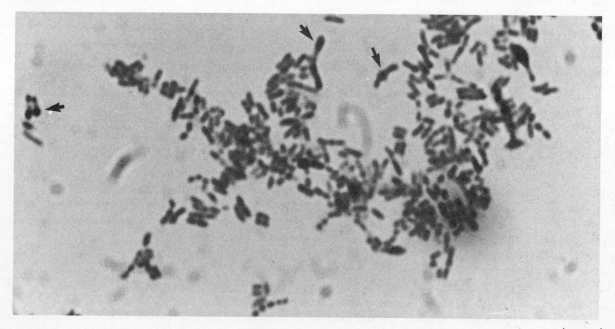

FIGURE 64.1

Corynebacterium diphtheriae. Stained with Loeffler's methylene blue to show metachromatic granules (arrows). *(Courtesy of Liliane Therrien and E. C. S. Chan, McGill University.)*

3 Incubate at 37°C for 24 h and observe the results.

THE GRAM-POSITIVE COCCI

1 You are provided with two blood-agar plates. One has been inoculated with a mixture of two species of the genus *Streptococcus*: *S. pyogenes* (β-hemolytic) and *S. faecalis*, (nonhemolytic, or γ, type). The other has been inoculated with two species of the chain-forming cocci, *S. salivarius* and *S. pneumoniae*, both of which produce the same change in blood agar, that is, α hemolysis but which can be differentiated nevertheless by subtle differences in colony form. Also, *S. pneumoniae* has a smaller zone of blood hemolysis than *S. salivarius*.

2 Use transmitted light (daylight provides the best results) to observe. Note particularly any alterations that the various species have produced in the blood agar.

3 Stain *S. pneumoniae* and other cultures of streptococci from the blood-agar plates for capsules. (Refer to Exercise 10.B for the procedure to do this.)

4 You are provided with a brain-heart-infusion-broth culture of *S. salivarius*. Make a Gram-stain preparation to study particularly the arrangement of the cells. Chain formation in the streptococci is poorly developed when the organisms are grown on solid medium but well developed in a fluid medium. FIGURE 64.2 shows the typical cell arrangement of the streptococci.

5 You are provided with two tubes of inulin CTA medium.

6 Inoculate one tube with *S. pneumoniae*, the other with *S. salivarius*. Incubate at 37°C and observe the results after 24 h.

7 You are given cultures in Todd-Hewitt broth of *S. salivarius* and *S. pneumoniae*, with 1 ml of culture per tube. To each of these cultures add 0.5 ml of

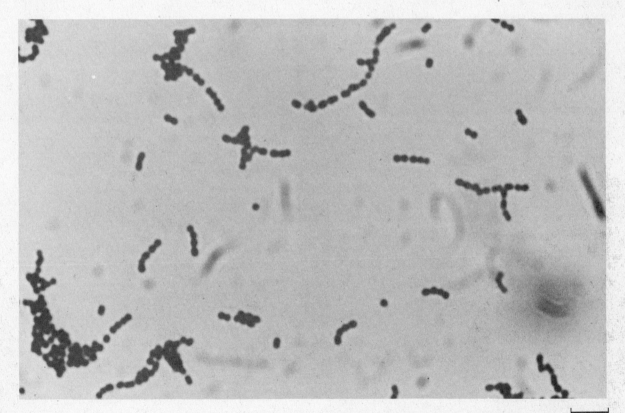

FIGURE 64.2
Gram-stained specimen of *Streptococcus pyogenes*. (Courtesy of Liliane Therrien and E. C. S. Chan, McGill University.)

sodium desoxycholate, shake, and note the tube in which the turbidity of the cell suspension is cleared. This is the bile-solubility test.

NOTE: A buffered medium like the Todd-Hewitt broth maintains the pH of the culture in the range of optimal activity of autolytic enzymes that are responsible for the lysis of *S. pneumoniae*. The bile salts play their part in this test in that they speed up a process, for diagnostic convenience, that would eventually occur spontaneously under prolonged incubation of the cultures.

NAME _____

The Corynebacteria and the Gram-Positive Cocci

RESULTS

1 Describe the appearance of *C. xerosis* in the Gram-stained and Loeffler's alkaline methylene-blue-stained preparations.

2 Record the sugar reactions of *C. xerosis* with + for acid production and – for no reaction. Fill in the blanks in the chart provided using a reference for sugar reactions of *C. diphtheriae*, and compare them with those of *C. xerosis*.

	COMPARATIVE SUGAR REACTIONS OF CORYNEBACTERIA		
	CTA medium with:		
Species	Glucose	Sucrose	Maltose
C. diphtheriae			
C. xerosis			

+ = acid (yellow); – = no acid produced (red).

3 Describe the hemolytic reactions of:

S. salivarius:

S. pyogenes:

S. faecalis:

64.A

4 Note the results of the capsule stains of *S. pneumoniae* and other species of streptococci on the blood-agar plates.

5 Describe the cell arrangement of *S. salivarius*.

6 Give the results of inulin fermentation by:
S. pneumoniae:

S. salivarius:

7 Describe the results of the bile-solubility test on:
S. pneumoniae:

S. salivarius:

8 How do these results correlate with the inulin-fermentation test?

NAME _____ 64.A THE CORYNEBACTERIA AND THE GRAM-POSITIVE COCCI *(Continued)*

QUESTIONS 1 Describe how a blood-agar medium is prepared.

2 Why is blood-agar medium commonly used in the clinical microbiology laboratory to study the streptococci?

3 What is the significance of knowing the Lancefield groupings of the streptococci?

4 Describe what is characteristic about the morphology of the cells of the corynebacteria.

EXERCISE 64.B

Airborne Infections
64.B The Mycobacteria, Neisserias, and Gram-Negative Aerobic Rods

OBJECTIVE
To become acquainted with the characteristics of more pathogenic microorganisms that are transmitted by air.

OVERVIEW
The genus *Mycobacterium* consists of acid-fast bacteria, including *M. tuberculosis* that causes tuberculosis primarily in humans.

Members of the *Neisseria* group are Gram-negative ellipsoidal cells occurring characteristically as paired cocci with the division line found along the axis of the paired cells. Some of the species are important human pathogens. For example, *Neisseria meningitidis* causes meningitis (an airborne disease) and *N. gonorrhoeae* causes gonorrhea (a contact disease) in humans.

A group of Gram-negative rods smaller in size than the bacteria you have seen during the first part of this course will be introduced in this exercise. They are grouped together in *Bergey's Manual of Systematic Bacteriology*, vol. 1, as genera not assigned to any family in Section 4, "Gram-Negative Aerobic Rods and Cocci." Representative genera are *Brucella*, *Bordetella*, and *Francisella*. Although they are smaller than microorganisms like *Escherichia coli*, there is a marked degree of pleomorphism characteristic of this group. Long, straight, and curved filaments, balloon and stalked forms, and bipolar staining are seen in both motile and nonmotile species. The group is predominantly and obligately parasitic and includes highly pathogenic microorganisms for human and other warmblooded animals. The causative agents of whooping cough, brucellosis, and tularemia, as well as other serious infections, are found in this group of small Gram-negative rods. They are generally parasitic in the respiratory tract or enter the body by inhalation or through some injured part of the skin. The majority of the species require body fluid enrichment or the incorporation of specific growth factors in a medium for isolation and subculture. Identification of genera and species depends upon cultural, biochemical, and serological characterization and, for the most important pathogenic species, upon a pathogenicity test in a suitable laboratory animal.

REFERENCE
MICROBIOLOGY, Chap. 24, "Airborne Diseases."

MATERIALS
Tube of tuberculosis sputum (*autoclaved*) and/or culture of *Mycobacterium smegmatis* on slant of nutrient agar (plus 1.8% glycerol to enhance acid-fastness)
1 18-h culture of *Neisseria subflava* on a brain-heart-infusion-agar plate
1 nutrient-agar plate of *Bordetella bronchiseptica* colonies
Gram-stain reagents
Acid-fast-stain reagents
Methylene blue stain
Freshly prepared tube of 1% tetramethyl-*p*-phenylenediamine dihydrochloride (also commercially available in sealed tubes)

PROCEDURE
THE MYCOBACTERIA
Prepare and examine Gram-stained and acid-fast-stained films from the tuberculosis sputum and/or *M. smegmatis* culture provided [see PLATE 3].

THE NEISSERIAS
1 You are provided with an 18-h culture of *N. subflava* grown on a brain-heart-infusion-agar plate.
2 Note the cultural characteristics of the species.
3 Make and study Gram-stained films of the species. Note the staining reaction, size, shape, and arrangement of the cells. A Gram-stained specimen of *N. gonorrhoeae* is shown in FIGURE 64.3.

4 Add 2 to 3 drops of reagent (1% tetramethyl-*p*-phenylenediamine dihydrochloride) directly to each of several isolated bacterial colonies of *N. subflava*. Neisseria colonies should turn bright purple (deep blue) within 10 s.

The **oxidase test** is useful to detect the presence of *Neisseria* on plates inoculated from specimens that may contain a variety of contaminating organisms [see PLATE 25]. In the reduced state, the dye is colorless. However, in the presence of cytochrome c and cytochrome oxidase (which *Neisseria* colonies have), the reagent is oxidized developing a purple coloration due to the formation of indophenol blue. Biochemically, electrons are transferred from the reagent to cytochrome c and thence, via cytochrome c oxidase, to molecular oxygen. Thus, the test is most helpful in screening out colonies of bacteria that lack the cytochromes or are obligate anaerobes (because the cytochrome system is found in aerobic, or microaerophilic, and facultatively anaerobic microorganisms). (It is important to remember too that other genera of bacteria besides members of the genus *Neisseria*, such as *Aeromonas*, *Pseudomonas*, *Campylobacter*, and *Pasteurella*, are oxidase-positive.)

BORDETELLA

1 A nutrient-agar plate containing colonies of *B. bronchiseptica* is provided for examination of colonial and microscopic morphology.
2 Examine and describe the colonial morphology of *B. bronchiseptica*.
3 Make a Gram-stained preparation and observe the Gram reaction, size, and shape of the cells [FIGURE 64.4].
4 Carry out a methylene blue stain on the bacterium and look for characteristic bipolar staining.

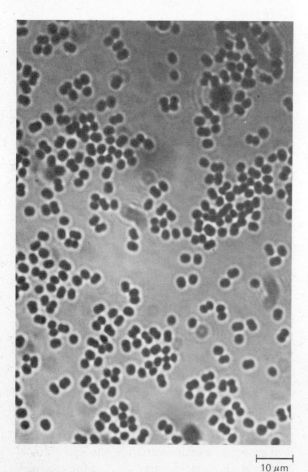

FIGURE 64.3
Neisseria gonorrhoeae. Gram-stained. Note the diplococcal cell arrangement. *(Courtesy of Liliane Therrien and E. C. S. Chan, McGill University.)*

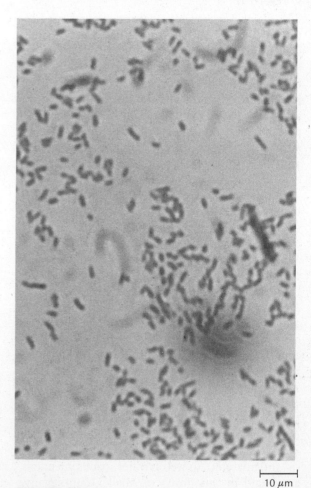

FIGURE 64.4
Bordetella parapertussis. Gram-stained. *(Courtesy of Liliane Therrien and E. C. S. Chan, McGill University.)*

NAME _____

The Mycobacteria, Neisseria, and Gram-Negative Aerobic Rods

RESULTS

1 What are the Gram-stain and acid-fast reactions of *Mycobacterium* species?

2 Describe the cultural characteristics of *N. subflava*.

3 Illustrate the shape and arrangement of cells of *N. subflava*:

Gram reaction: _____

4 What is the result of the oxidase test on *N. subflava*?

5 Describe the colonial morphology of *B. bronchiseptica*.

405

6 Draw the appearance of the cells of *B. bronchiseptica* after methylene blue staining.

QUESTIONS 1 What is meant by *drug-resistant tuberculosis*?

2 Why do you think the acid-fast stain is so useful?

3 What are the species of bacteria that cause the following diseases?
Whooping cough:

Brucellosis:

Tularemia:

4 Explain the biochemical basis of the oxidase test for *Neisseria*.

5 How specific is the oxidase test for *Neisseria* species?

EXERCISE 65

Foodborne and Waterborne Diseases

OBJECTIVE

To become acquainted with the characteristics of some pathogenic microorganisms transmitted by food and by water.

OVERVIEW

In humans, microbial numbers in the large intestine are very high: each gram of feces contains about 10^{11} bacterial cells. Fifty or sixty percent, dry weight, of fecal material may consist of bacteria and other microorganisms. On the other hand, the stomach and small intestine have relatively few microorganisms due to the hydrochloric acid produced by the stomach and the rapid movement of food through the small intestine.

Most intestinal bacteria are commensal microorganisms, and many live in mutualistic association with the human host. For example, some intestinal bacteria synthesize vitamins, such as folic acid and vitamin K. Also, the normal microbiota prevent colonization of extraneous and pathogenic bacteria by producing antimicrobial substances and by successful competition.

Intestinal bacteria are mostly anaerobes belonging to the genera *Bacteroides*, *Bifidobacterium*, *Lactobacillus*, *Peptostreptococcus*, and *Fusobacterium* and facultative bacteria of the genera *Escherichia*, *Enterobacter*, *Citrobacter*, and *Proteus*. *Escherichia coli* is, of course, the well-known species and is indispensable to microbiologists, geneticists, and biochemists as a research subject.

Most diseases of the gastrointestinal tract result from the ingestion of food or water contaminated with pathogenic microorganisms. Among these are species of *Salmonella*, which are the causative agents of typhoid and paratyphoid fevers, and of *Shigella*, which are the causative agents of bacillary dysentery [FIGURE 65.1]. An important genus causing intestinal discomfort is *Staphylococcus*, which forms clusters of cells in grapelike bunches. There are two species, *Staphylococcus aureus*, which is pathogenic, and *S. epidermidis*, which is usually not pathogenic but can be an opportunistic pathogen. FIGURE 65.2 shows a photomicrograph of *S. aureus*.

Many types of differential and selective media have been developed to differentiate between the lactose-fermenting enteric bacteria and the non-lactose-fermenting enteric bacteria. The lactose fermenters are called *coliforms* and are generally not pathogenic although some strains may cause diarrhea. Members of the non-lactose-fermenting group, such as the salmonellas and the shigellas, are pathogenic. Such media have been invaluable in public health microbiology; they are used routinely in control laboratories to help in breaking the fecal-oral cycle of disease. Such solid media include eosin-methylene-blue (EMB) agar, Endo agar, and MacConkey agar. A simplified key for the identification of enterics based on major physiological characteristics is given in FIGURE 65.3.

One set of fluid biochemical tests that is most useful in differentiating fecal *Escherichia coli* from *Enterobacter aerogenes* (which may not be of fecal origin) constitutes the so-called IMViC tests. IMViC is an acronym for the following tests: indole, methyl red, Voges-Proskauer, and citrate; "i" is used for euphony. [See PLATES 18, 26, 27, and 28.] The indole test determines the ability of an organism to cleave the amino acid tryptophane into indole, ammonia, and pyruvic acid. The indole is detected by the addition of Kovacs' reagent and the subsequent appearance of a cherry-red-colored layer (see Exercise 27.B). The methyl-red test measures the acidity of the medium after growth of the organisms in glucose broth. Methyl red is an acid-base indicator that turns red below pH 4.5. *Escherichia* ferments glucose to form a considerable amount of acid and is methyl-red positive. *Enterobacter* carries out a

FIGURE 65.1
Gram-stained preparation of a species of *Shigella*. *(Courtesy of Liliane Therrien and E. C. S. Chan, McGill University.)*

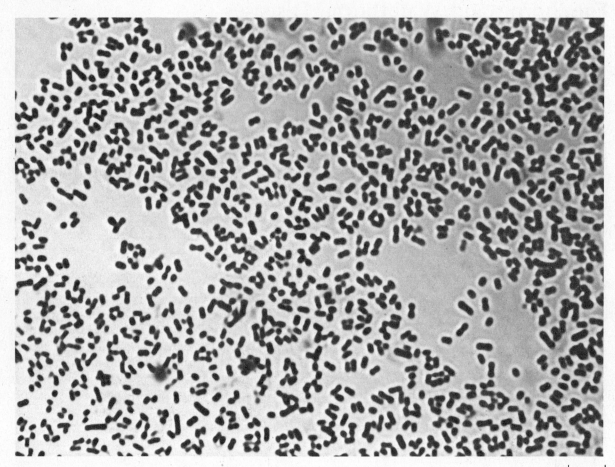

10 μm

butylene glycol fermentation and hence does not produce sufficient acid from glucose to be methyl-red positive. Voges-Proskauer is the name of a test used to detect the presence of acetoin (acetylmethylcarbinol), the immediate precursor of 2,3-butylene glycol. Thus it is positive for *Enterobacter* and negative for *Escherichia*. A positive reaction is indicated by the formation of a cherry red color upon the addition of a potassium hydroxide–creatine solution. (Acetoin is oxidized to diacetyl in the presence of excess alkali, and this compound reacts with creatine to form a red compound.) The citrate test determines the ability of microorganisms to utilize citrate as the sole carbon source. Citrate is not used by *Escherichia*, which lacks the permease to transport citrate into the cell. *Enterobacter* can both transport citrate and use it as a sole carbon source.

REFERENCE
MICROBIOLOGY, Chap. 25, "Foodborne and Waterborne Diseases."

MATERIALS
MacConkey agar plate with *Escherichia coli* and *Salmonella typhimurium* (mixed-culture plate)
Nutrient-agar slant cultures of *E. coli*, *S. typhimurium*, *Staphylococcus aureus*, and *S. epidermidis*
Nutrient-broth cultures of *E. coli* and *Ent. aerogenes*
Blood-agar plate containing a mixture of *S. aureus* and *S. epidermidis*
Tube of physiological saline solution
2 tubes of lactose tryptone water with Durham tube
2 tubes of glucose tryptone water with Durham tube

65 FOODBORNE AND WATERBORNE DISEASES

FIGURE 65.2
Gram-stained preparation of *Staphylococcus aureus*. (Courtesy of Liliane Therrien and E. C. S. Chan, McGill University.)

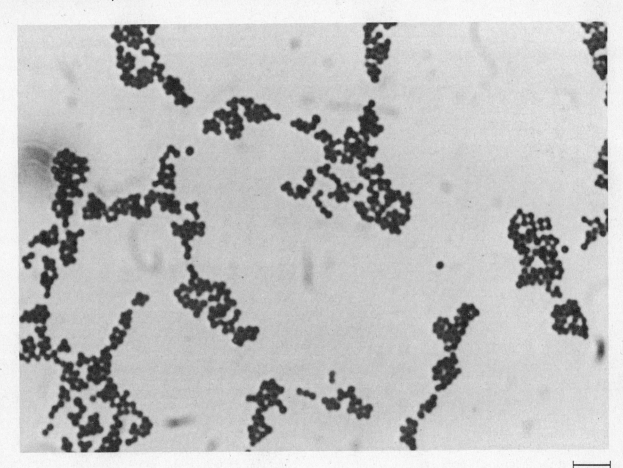

10 μm

4 tubes of mannitol tryptone water with Durham tube
2 tubes of iron agar
2 tubes of tryptone broth
4 tubes of MR-VP broth
2 tubes of Simmon's citrate medium
Gram-stain reagents
2 tubes of nutrient gelatin
2 tubes of oxalated rabbit plasma
37°C water-bath
Kovacs' reagent
Methyl-red indicator
Voges-Proskauer reagent
3 1-ml pipettes
Pasteur pipettes or plastic transfer pipettes (commercially available; sterile in individual wraps) for dispensing drops of reagents
2 small test tubes

PROCEDURE

1 You are provided with a plate of MacConkey agar containing colonies of *E. coli* and *S. typhimurium*. Note and describe the growth characteristics of these organisms on the medium.

2 Cultures of the above bacteria are also provided on nutrient-agar slants. Make Gram-stained films of the two cultures and do hanging-drop motility tests in physiological saline solution.

3 Use each of the agar slant cultures in step 2 above to inoculate the following media: lactose, glucose, and mannitol in tryptone water, and the tube of iron agar. Incubate at 37°C.

IMViC TESTS FOR ENTERICS

1 With the use of an inoculating loop, inoculate nutrient-broth cultures of *E. coli* and *Ent. aerogenes* into one tube of tryptone broth, two tubes of

FIGURE 65.3

Simplified key for the identification of the enteric bacilli.

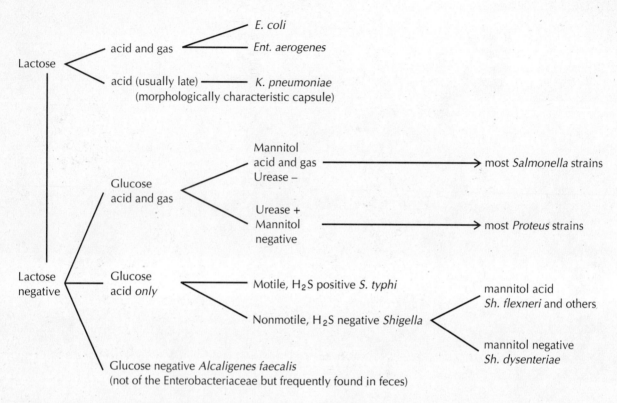

MR-VP broth, and one tube of Simmon's citrate medium.

2 Incubate all tubes at 37°C for 24 h and then carry out the following procedures.

3 *Indole test*: Add 10 drops of Kovacs' reagent to each tube. The development of a red color is a positive test.

4 *Methyl-red test*: Add 4 to 5 drops of methyl-red indicator to one tube of MR-VP broth for each culture. The development of a red color is a positive test. A yellow color is a negative test.

5 *Voges-Proskauer test*: Pipette 1 ml of each culture in MR-VP broth into two small test tubes; add 1 ml reagent to each culture; shake the tubes to aerate them for 1 min. A positive test is indicated by the development of a pink color.

6 *Citrate test*: The presence of a blue color is a positive reaction.

THE STAPHYLOCOCCI

1 You are provided with a blood-agar plate that has been inoculated with a mixture of *S. aureus* and *S. epidermidis*.

2 Study and record the colony form, pigmentation, and so on of the two species as they occur as discrete colonies on the blood-agar plate and as confluent growth on the nutrient-agar slants.

3 Prepare Gram-stained films of each of the two species. Examine the films for Gram reaction and the characteristic grouping of the cells (make your films from the slant cultures). Inoculate each of the two species into (a) tryptone water containing mannitol and (b) a tube containing nutrient gelatin. Incubate at 37°C for 24 h.

4 You are supplied with two tubes of oxalated rabbit plasma. Add a full loopful of *S. aureus* culture to one tube. Add a full loopful of *S. epidermidis* to the other tube. Distribute the growth in the plasma by shaking, and place both tubes in the water-bath at 37°C. Examine the tubes after 2 h. The pathogenic *S. aureus* has the ability to clot oxalated or citrated blood plasma, whereas the nonpathogenic species does not. Organisms that have the capacity to clot are called *coagulase-positive*; those that cannot clot such plasma are *coagulase-negative*. [See PLATE 29.]

NAME _____

Foodborne and Waterborne Diseases

RESULTS The Enterics

1 Describe the growth characteristics of the following microorganisms on MacConkey agar:

E. coli:

S. typhimurium:

2 Note the results of Gram staining, motility, fermentation, and biochemical reactions in the chart provided:

GRAM-STAIN, MOTILITY, FERMENTATION AND BIOCHEMICAL REACTIONS OF THE ENTERICS							
				Tryptone water with:			
Species	Gram	Motility	H_2S*	Lactose	Glucose	Mannitol
Escherichia coli						
Salmonella typhimurium						

*Iron-agar medium turns black with production of hydrogen sulfide.

3 Fill in the results of the IMViC reactions in the following chart:

IMViC REACTIONS				
Species	I	M	Vi	C
Escherichia coli				
Enterobacter aerogenes				

The Staphylococci

1 Describe the colonial growth characteristics (margin, elevation, pigmentation, etc.) on blood agar of the following staphylococci:

S. aureus:

S. epidermidis:

2 Report the Gram-stain reaction and cell arrangement of:

S. aureus:

S. epidermidis:

NAME _____   65 FOODBORNE AND WATERBORNE DISEASES *(Continued)*

3 Note the biochemical activities of *S. aureus* and *S. epidermidis* in the chart provided:

BIOCHEMICAL ACTIVITIES OF THE STAPHYLOCOCCI			
Species	Mannitol in tryptone water	Nutrient gelatin	Coagulase test
Staphylococcus aureus			
Staphylococcus epidermidis			

QUESTIONS 1 What is the biochemical basis of the coagulase test?

2 Why is MacConkey agar medium considered to be both a differential and a selective medium?

3 How is the genus *Staphylococcus* differentiated from the genus *Micrococcus*?

4 Is the coagulase test a test for virulence of the staphylococcal strain?

EXERCISE 66

Contact Diseases

OBJECTIVE
To become acquainted with the characteristics of some pathogenic microorganisms that are transmitted by contact.

OVERVIEW
In this exercise examples of microorganisms causing contact diseases will be studied. The term *contact diseases* is used in a broad sense and includes not only diseases transmitted by direct contact with an infected individual but also infections that occur through trauma or injury or diseases transmitted by means of a vector, usually an arthropod.

A large number of pathogenic microorganisms are capable of causing contact diseases in humans. Some you have already studied because they can also be transmitted in other ways, and some will be new to you.

One group of bacteria, the spirochetes, is responsible for a number of diseases in this category. The bacteria in this group are flexible and spiral; most cannot be readily cultivated and are difficult to stain with routinely used stains. Most are parasitic in vertebrates, and some are pathogenic. There are three genera pathogenic for humans: *Borrelia*, *Leptospira*, and *Treponema*. One of the prime examples of a direct-contact disease, syphilis, a sexually transmitted disease (STD), is caused by *Treponema pallidum*. This is a tightly coiled bacterium with pointed ends and 8 to 14 spirals. Another example of a contact disease, transmitted by the bite of a tick (*Ixodes dammini*), is Lyme disease; it is caused by a spirochete named *Borrelia burgdorferi* [FIGURE 66.1].

Gonorrhea is also a direct-contact disease like syphilis. It too is an STD and is caused by *Neisseria gonorrhoeae*, a Gram-negative coccus.

Many microorganisms can enter the body and cause infection through wounds or abrasions in the skin and mucous membranes. They include the hemolytic streptococci, the staphylococci, and the clostridia. The latter, normally found in the soil, are anaerobes. When wounds are contaminated with species of the clostridia, such conditions as gas gangrene and tetanus may result.

REFERENCES
MICROBIOLOGY, Chap. 23, "'Sexually Transmitted Diseases."
MICROBIOLOGY, Chap. 26, "Arthropod-Borne Diseases."
MICROBIOLOGY, Chap. 27, "Wound and Skin Infections Acquired by Direct Contact."

MATERIALS
Chocolate-agar slant culture of *Neisseria gonorrhoeae*
1 each brain-heart-infusion-agar slant culture of *Clostridium tetani* and *C. perfringens*
Starch-broth culture of *C. perfringens*
5 toothpicks or packet of Stim-U-Dent sticks (Johnson & Johnson; available from a drug store)
Hollande's stain reagents
Dilute carbol fuchsin
Nigrosin solution
Gram-stain reagents
Spore-stain reagents
Capsule-stain reagents

PROCEDURE
SPIROCHETES
1 Due to the difficulty of culturing these bacteria routinely, you will work with spirochetes found in the mouth. The numbers of these spirochetes present in the mouth of healthy individuals vary, but most people have some of them at the gum margins. Make at least three films (as described below) with scrapings removed from between your teeth near, or preferably just below, the gums. Use the

FIGURE 66.1

The Lyme disease spirochete, *Borrelia burgdorferi*, as seen by dark-field microscopy. *(Courtesy of Antonia Klitorinos, R. Siboo, and E. C. S. Chan, McGill University.)*

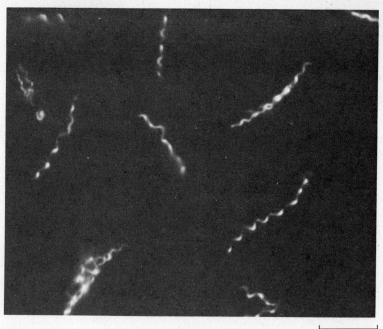

10 µm

toothpicks or Stim-U-Dent sticks provided to obtain the scrapings.
2 Stain one film by Hollande's method.
  a Fix with absolute alcohol for 1 to 2 min.
  b Drain off the alcohol, but do not let the preparation dry and do not wash it off.
  c Flood with the mordant, warming it to the steaming point for 1 min.
  d Wash thoroughly with tap water, holding the slide horizontally.
  e Stain with pyridine-silver solution, warming gently for 2 to 3 min. (Repeat this step, if necessary.)
  f Wash thoroughly and air-dry.
3 Stain one film with dilute carbol fuchsin, warming for 30 s.
4 Emulsify scrapings from between your teeth in nigrosin, and spread the mixture in a thin film; dry rapidly.
5 Examine all stained preparations.

NEISSERIA GONORRHOEAE

You are provided with a chocolate-agar slant of *N. gonorrhoeae*. Make a Gram stain and observe the morphology of this organism.

THE CLOSTRIDIA

1 You are provided with brain-heart-infusion-agar slant cultures of *C. tetani* and *C. perfringens* and a starch-broth culture of *C. perfringens*, which have been grown anaerobically. Do a Gram stain and a spore stain from each culture. Note that *C. perfringens* forms very few, if any, spores when grown on brain-heart-infusion-agar slants and requires special cultural conditions, such as starch broth, to sporulate.
  2 Using the capsule-staining reagents provided, stain *C. perfringens* for capsules. Examine. (This is the only encapsulated member of the pathogenic clostridia.)

NAME _____

Contact Diseases

RESULTS 1 Draw the appearance of any spirochetes from the teeth scrapings stained by each of the following methods:

Hollande's method

Dilute carbol fuchsin

Nigrosin

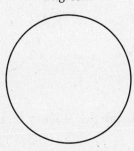

2 Draw the morphology of *N. gonorrhoeae*:

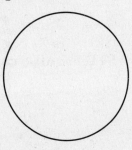

Gram reaction: _____

3 Draw the Gram-stain morphology of the following clostridia:

Clostridium tetani *Clostridium perfringens*

Gram reaction: _____   _____

4 Draw the appearance of *C. perfringens* grown in starch broth after spore staining. Label the cell parts and indicate their color reaction:

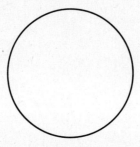

5 Draw the cells of *C. perfringens* after capsule staining:

QUESTIONS **1** Explain how Lyme disease may be avoided.

2 Explain how an STD like AIDS can be avoided.

EXERCISE 67

Bacteriological Analysis of Urine

OBJECTIVE
To carry out a bacteriological analysis of urine.

OVERVIEW
In a healthy person, the kidney, urinary bladder, and ureters are free of microorganisms. However, bacteria are commonly found in the lower portion of the urethra of both males and females. The upper portion of the urethra, near the bladder, has few microorganisms, apparently because of some antibacterial effect exerted by the urethral mucosa and because of the mechanical removal of microorganisms by the frequent flushing of the urethral epithelium by urine.

Despite these defense mechanisms, urinary tract infections can sometimes occur. Such infections range from severe to mild and may affect the kidneys, ureters, bladder, or urethra. Usually the sources of the causative microorganisms are the external skin surfaces and the lower portion of the urethra. In the case of a urinary tract infection, urine may have millions of bacteria per milliliter. The bacterium most often encountered as a cause of urinary tract infections is *Escherichia coli*; other frequently encountered organisms are *Proteus* spp., *Pseudomonas* spp., enterococci such as *Streptococcus faecalis*, *Acinetobacter* spp., staphylococci, and *Alcaligenes* spp. One factor that seems to favor the establishment of urinary tract infections is obstruction of the urinary flow, such as caused by a twisted ureter, kidney or bladder calculi (stones), an enlarged prostate gland, or cancer of the bladder or urethra. In a hospital, another factor that can lead to urinary tract infection is the practice of introducing a sterile catheter through the urethra into the bladder in order to obtain samples of urine from the bladder; this procedure may sometimes transfer bacteria from the distal portion of the urethra to the bladder and thus lead to infection of the bladder.

As indicated previously, bacteria commonly colonize the lower portion of the urethra, even in a healthy person. During voiding, sterile urine from the bladder may acquire these bacteria as it passes from the bladder along the urethra to the outside of the body. Urine may also acquire organisms from the external surface of the body (tip of the penis, labial folds, vulva). Thus, in collecting a urine sample for bacteriological analysis, these external surfaces are first thoroughly cleansed. The first portion of the voiding is discarded because it is usually heavily contaminated with organisms from the lower portion of the urethra, and the subsequent urine is collected in a sterile container. Despite these precautions, a voided urine sample from a healthy person may contain up to 1000 bacteria per milliliter. However, in cases of urinary tract infections the population of bacteria usually exceeds 100,000 cells per milliliter.

The first step in laboratory detection of a urinary tract infection is to determine the viable bacterial count per milliliter of a urine sample from the patient. The pour-plate method can be used for this purpose; however, the calibrated loop–direct streak method (see below) is also accurate and is more convenient and rapid.

It should be mentioned that several types of reliable devices for detecting genitourinary tract infections are available from commercial sources. They provide a simple and convenient means of culturing freshly voided urine (and urogenital-discharge specimens) as soon as it is collected.

REFERENCES
MICROBIOLOGY, Chap. 22, "Nosocomial Infections."

MICROBIOLOGY, Chap. 27, "Wound and Skin Infections Acquired by Direct Contact."

MATERIALS

2 blood-agar plates (prepared 1 day in advance and allowed to dry overnight to remove water of syneresis on surface)

Sterile disposable plastic calibrated 1.0-μl (0.001-ml) loop needle (available from a scientific supply company like Fisher or Canlab)

Regular wire loop needle

Urine samples A and B (one normal urine, one infected urine) to be tested, in sterile containers

PROCEDURE

1 Immerse the 1.0-μl plastic disposable loop needle vertically just below the surface of the first urine sample. Withdraw the needle vertically to remove 1.0 μl.

2 Streak the needle in a straight line all the way across the center of one blood-agar plate; then discard the needle in a jar of disinfectant.

3 Use a regular wire loop needle to make a series of streaks at 90° to the original streak on the plate [FIGURE 67.1]. This spreads out the 1.0-μl inoculum over the plate so that isolated colonies can be obtained. Sterilize the needle by flaming when finished.

FIGURE 67.1
Method of streaking a urine sample on a blood-agar plate.

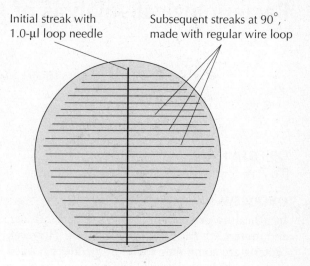

4 Repeat steps 1 to 3 using the second urine sample and the second blood-agar plate.

5 Incubate the two blood-agar plates aerobically at 35°C for 18 to 24 h.

NAME _____

Bacteriological Analysis of Urine

RESULTS Count the colonies on both plates. Multiply the number of colonies per plate by 1000 to obtain the colony count per milliliter of urine.

(NOTE: If > 100 colonies occur on the plate, this represents a colony count that is > 100,000 per milliliter of urine, indicative of a urinary tract infection.)

Colony count per milliliter of urine sample A: _____

Colony count per milliliter of urine sample B: _____

QUESTIONS

1 From the results obtained, which sample of urine, A or B, would be considered to be from a patient with a urinary tract infection?

2 Assume that characterization tests of the predominant organism in the infected urine sample indicated that the bacterium was *Escherichia coli*. What is the normal habitat of this species? How might this species be able to cause a urinary tract infection?

3 Suppose that an anaerobic bacterium was causing a urinary tract infection. If you were to analyze a urine sample from the patient, would the procedure you used in this exercise be likely to indicate that an infection was present? What modification(s) might you need to make in the procedure?

EXERCISE 68

Commercially Marketed Rapid and Convenient Bacterial Identification Kits

OBJECTIVE
To become familiar with miniaturized multitest methods for identifying bacteria.

OVERVIEW
Conventionally, the identity of a species of enteric bacteria is determined by a series of sugar fermentations and biochemical tests performed in tubed and plated media such as those carried out in Exercise 65. Such tests have been well worked out for the differentiation and identification of the enteric bacterial species [see FIGURE 65.3] and lend themselves to commercial development into identification kits in the form of rapid and convenient miniature *multitest systems*. Such systems use microtechniques that incorporate a number of media into a single unit. At least six multitest systems are commercially available.

Obvious advantages of these kits are minimal storage space for tubed and plated media, use of less media, cost-effectiveness, rapidity in obtaining results, and the possibility of analyzing the data by a computerized code system to identify a species. However, there are some disadvantages of minitest systems: accuracy of some tests is said to be questionable by some workers, interpretation of reactions can be difficult, obtaining the proper inoculum size is difficult, there is a possibility of medium carryover from one compartment to another, and some identification tests are inadequate for the characterization of certain species (e.g., those tests that employ serology or phage-typing). Nevertheless, such miniaturized identification kits are used routinely by many diagnostic microbiology laboratories for the identification of members of the family Enterobacteriaceae.

Two commonly used kits are the Enterotube II Multitest System (Roche Diagnostics, Hoffman-La Roche Inc., Nutley, NJ 07110) and the API 20E System (Analytab Products, Inc., Division of Sherwood Medical, Plainview, NY 11803). Both kits can be refined by using computer-assisted programs. Accuracy is increased because each test is given a point value. Tests that are considered more important are given more points. Point values obtained are fed into a computer for identification of the culture.

ENTEROTUBE II MULTITEST SYSTEM
This system consists of a single tube with 12 compartments, each of which contains a different medium, as well as a self-enclosed wire inoculating needle [FIGURE 68.1 and PLATE 30]. This needle is used to touch a single isolated colony from a differential and selective medium plate (e.g., MacConkey agar, EMB agar, Hektoen enteric agar) inoculated with fecal specimen material or rectal swab. The needle is then drawn through all 12 compartments in one operation to inoculate all the test media. Fifteen standard biochemical tests can thus be performed in one inoculating procedure. Following incubation, the color changes that take place in each of the compartments are interpreted according to the manufacturer's instructions to identify the microorganisms. That is, the enteric culture can be identified by conventional patterns of positive and negative fermentation and biochemical reactions. Alternatively, the identification method has been further refined by means of a computer-assisted system called ENCISE (Enterobacteriaceae numerical coding and identification system for Enterotube). That is, the combination of biochemical reactions is reduced to an identification (ID) value of a five- or seven-digit profile number using binary mathematics. These numbers are then used with computer software to achieve the best fit of numerical values to make an identification.

API 20E SYSTEM
The functional unit of this kit is a stiff plastic strip containing 20 individual microtubes, each of which

423

FIGURE 68.1

Main components of the Enterotube II.

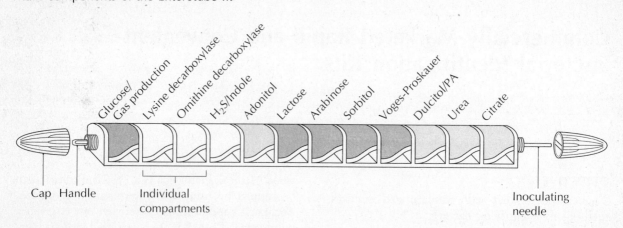

has an upper cupule, or depression [FIGURE 68.2]. Each microtube contains a dehydrated medium in its bottom. The dried media are reconstituted into aqueous form when the inoculum of the test organism is added. Following inoculation, the strip is then incubated in a plastic-covered tray to prevent dehydration of the media by evaporation. Even though there are only 20 different media in the same number of microtubes, 22 biochemical tests are performed. After incubation, results are read by color changes in the media or obtained with the use of specific reagents added to the media (see PLATE 31]. Identification of the microbe is then made using the manufacturer's differential charts. Some laboratories may prefer to use the manufacturer's rapid service for computer-assisted identification by means of a database system called API Profile Recognition System (PRS) and Analytical Profile Indexes.

In the following exercise you will inoculate an Enterotube and an API strip with an unknown species of enteric bacteria. After incubation of the kits, you will identify the unknown by the conventional method, noting the characteristic color changes and interpreting them according to the manufacturer's instructions.

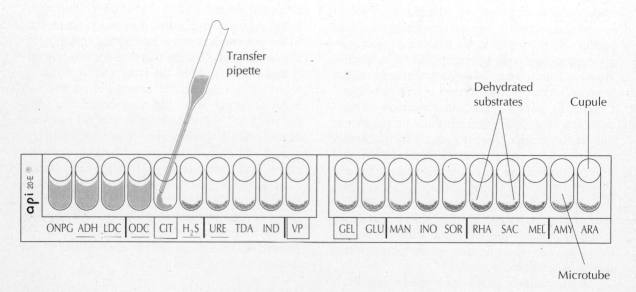

FIGURE 68.2

Main components of the API 20E System.

MATERIALS

24-h trypticase soy-agar plate culture, number-coded and with isolated colonies of one of the following species: *Enterobacter aerogenes, Escherichia coli, Klebsiella pneumoniae, Proteus vulgaris, Salmonella typhimurium,* or *Shigella dysenteriae*
1 Enterotube II
1 API 20E strip
1 5-ml tube of 0.85% sterile saline
1 tube of sterile mineral oil
Enterotube reagents
API 20E reagents
1 5-ml pipette
Sterile Pasteur pipettes or plastic transfer pipettes
Enterotube color reaction charts and ENCISE pads (from the manufacturer)
API Profile Recognition System and differential identification charts (from the manufacturer)

PROCEDURE
ENTEROTUBE SYSTEM

1 Familiarize yourself with the Enterotube (see FIGURE 68.1 and literature from the manufacturer).
2 Label the Enterotube with your name and the coded number of the culture supplied by the instructor. Use FIGURE 68.3 as a guide for the following steps.
3 Remove the screw caps from both ends of the Enterotube. Using the inoculating needle contained in the Enterotube, touch the straight end to an isolated colony to pick up inoculum cells. The bent end is the handle [FIGURE 68.3A].
4 Inoculate the Enterotube as follows:
  a Twist the needle in a rotating motion and withdraw it slowly through all 12 compartments of the Enterotube [FIGURE 68.3B].
  b Reinsert the needle in the Enterotube with a rotating motion through all 12 compartments. Then withdraw the needle to the fourth compartment (H_2S/IND). (The presence of the needle remaining in the first 3 compartments maintains anaerobiosis for dextrose fermentation, carbon dioxide production, and the decarboxylation of lysine and ornithine.)
  c Break the needle at the exposed notch by bending, discard the broken-off needle, and replace the caps loosely at both ends of the Enterotube [FIGURE 68.3C].
5 Remove the blue tape covering the ADO, LAC, ARB, SOR, VP, DUL/PA, URE, and CIT compartments. This uncovers the tiny air vents that provide aerobic conditions in these compartments [FIGURE 68.3D].
6 Place the clear plastic slide band over the GLU compartment to contain the wax, which may escape by the excessive production of gas by some microorganisms.
7 Incubate the tube on a flat surface for 24 to 48 h at 37°C.
8 Record and interpret all reactions in the chart provided in the results section using the manufacturer's instructions, as summarized in TABLE 68.1. Read all the other tests before the indole and Voges-Proskauer tests, which require the following additional steps:
  a *Indoletest:* Melt a small hole in the plastic film covering the H_2S/indole compartment using a hot inoculating loop. Add 1 to 2 drops

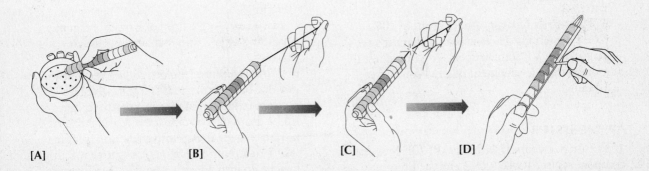

FIGURE 68.3
Main steps in the procedure for inoculating the Enterotube II.

TABLE 68.1 Enterotube Biochemical Reactions

Test	Initial color	Final color	Comments
GLU	Red	Yellow	Acid from glucose
GAS			Gas from glucose fermentation causes separation of the wax
LYS	Yellow	Purple	Lysine decarboxylase
ORN	Yellow	Purple	Ornithine decarboxylase
H_2S			Ferrous ions react with sulfide ions, forming black precipitate
IND	Beige	Red	Kovacs' reagent added to H_2S/IND compartment for detection
ADON	Red	Yellow	Adonitol fermentation
LAC	Red	Yellow	Lactose fermentation
ARAB	Red	Yellow	Arabinose fermentation
SORB	Red	Yellow	Sorbitol fermentation
VP	Beige	Red	Voges-Proskauer reagents detect acetoin
DUL	Green	Yellow	Dulcitol fermentation
PA		Black precipit.	Phenylpyruvic acid released from phenylalanine combines with iron salts
UREA	Yellow	Pink	Ammonia changes pH of medium
CIT	Green	Blue	Citric acid used as carbon source

of Kovacs' reagent, and allow the reagent to touch the agar surface. A positive test is indicated by development of a red color within 1 min.

b *Voges-Proskauer test:* Add 2 drops of 20% KOH containing 5% α-naphthol (Barritt's reagent) to the VP compartment. A positive test is indicated by development of a red color after 20 min.

API 20E SYSTEM

1 Familiarize yourself with the API 20E System components (see FIGURE 68.2 and the literature from the manufacturer).

2 Label the elongated flap on the incubation tray with your name and the coded number of the unknown culture as given by your instructor. Use FIGURE 68.4 as a guide for the following steps.

3 Add approximately 5 ml tap water to the incubation tray with a pipette.

4 With a sterilized loop, transfer cells from an isolated colony on a streaked plate of the unknown culture to a 5-ml tube of sterile normal saline [FIGURE 68.4A]. Shake the suspension well (or vortex) to distribute the cells uniformly.

5 Remove the API strip from its sterile envelope and place it in the incubation tray.

6 Slant the incubation tray. With a sterile plastic transfer pipette (or Pasteur pipette) containing the bacterial suspension, fill the tube section of each compartment by placing the tip of the pipette against the side of the cupule. Fill the cupules of the CIT, VP, and GEL microtubes with the bacterial suspension [FIGURE 68.4B].

7 With a second sterile plastic transfer pipette, fill the cupules of the ADH, LDC, ODC, H_2S, and URE microtubes with sterile mineral oil to create an anaerobic environment [FIGURE 68.4C].

8 Cover the inoculated strip with the tray lid and incubate from 24 to 48 h at 37°C.

9 Record and interpret all reactions in the chart provided in the results section using the manufacturer's instructions, as summarized in TABLE 68.2. Read all the other tests before the TDA, VP, and IND tests which require the following additional steps:

a *TDA test:* Add 1 drop of 10% ferric chloride. A positive test is brown-red. Indole-positive microorganisms may produce an orange color, which is considered a negative reaction.

b *VP test:* Add 1 drop of VP reagent II, then 1 drop of VP reagent I. A positive reaction produces a red color (not pale pink) after 10 min.

c *Indole test:* Add 1 drop of Kovacs' reagent. A red ring after 2 min indicates a positive reaction.

d *Nitrate reduction:* Before adding reagents, look for bubbles in the GLU tube. Bubbles indicate reduction of nitrate to nitrogen gas. Add 2 drops of nitrate reagent A and 2 drops of nitrate reagent B. A positive reaction (red) may take 2 to 3 min to develop. A negative reaction may be confirmed with zinc dust (see Exercise 27.C).

68 COMMERCIALLY MARKETED RAPID AND CONVENIENT BACTERIAL IDENTIFICATION KITS

FIGURE 68.4
Main steps in the procedure for inoculating the API 20E System tray.

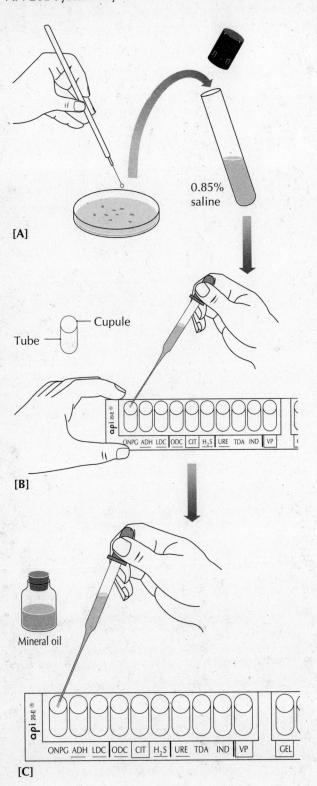

TABLE 68.2 API 20E Biochemical Reactions

Test	Positive	Negative	Comments
ONPG	Yellow	Colorless	O-nitrophenyl-beta-D-galactopyranoside
ADH	Red	Yellow	Ornithine, CO_2, NH_3 from arginine
LDC	Red	Yellow	Cadaverine liberated from lysine
ODC	Red	Yellow	Putrescine produced
CIT	Dark blue	Green	Citric acid as sole carbon source
H_2S	Black	None	Thiosulfate reduced to H_2S
URE	Red	Yellow	Urea hydrolyzed to NH_3 and CO_2
TDA	Brown	Yellow	Indole and pyruvic acid produced
IND	Red ring	Yellow	Indole detected by Kovacs' reagent
VP	Red	Colorless	Acetoin detected by reagents
GEL	Diffusion	None	Hydrolysis of gelatin diffuses pigment
GLU	Yellow	Blue	Glucose fermentation
MAN	Yellow	Blue	Mannitol fermentation
INO	Yellow	Blue	Inositol fermentation
SOR	Yellow	Blue	Sorbitol fermentation
RHA	Yellow	Blue	Rhamnose fermentation
SAC	Yellow	Blue	Sucrose fermentation
MEL	Yellow	Blue	Melibiose fermentation
AMY	Yellow	Blue	Amygdalin fermentation
ARA	Yellow	Blue	Arabinose fermentation
NO_2	Red	Yellow	Nitrate reduction
N_2	Bubbles	None	Nitrate reduction (may have to use zinc test)

NAME _____

Commercially Marketed Rapid and Convenient Bacterial Identification Kits

RESULTS As instructed previously, enter all reaction observations in the chart provided.

\	\	DATA OF OBSERVATIONS WITH THE ENTEROTUBE AND API STRIP	
		Results, + or –, of:	
Code	Name of test	Enterotube	API 20E
ONPG	Beta-galactosidase		
ADH	Arginine dihydrolase		
LDC	Lysine decarboxylase		
ODC	Ornithine decarboxy.		
CIT	Citrate		
H_2S	Hydrogen sulfide		
URE	Urease		
TDA	Trypt. deaminase		
IND	Indole		
VP	Acetoin		
GEL	Gelatin		
GLU	Glucose		
MAN	Mannitol		

		Results, + or −, of:	
Code	Name of test	Enterotube	API 20E
INO	Inositol		
SOR	Sorbitol		
RHA	Rhamnose		
SAC	Sucrose		
MEL	Melibiose		
AMY	Amygdalin		
ARA	Arabinose		
OXI	Oxidase		
NO_3	Nitrate reduction		
GAS	Gas production		
PHE	Phenylalanine		
LAC	Lactose		
DUL	Dulcitol		
Species			

DATA OF OBSERVATIONS WITH THE ENTEROTUBE AND API STRIP (Continued)

NAME _____   68 BACTERIAL IDENTIFICATION KITS (Continued)

Enterotube II System

1 Based on your data, identify your unknown microorganism using the differential identification chart, and enter it in the chart on pages 429–430.

2 Circle the number corresponding to each positive reaction below the appropriate compartment in the diagram below. Add up the numbers in each bracket, and place the sum into the diamond space below. Determine the five-digit ID value and identify your unknown organism by referring to the computer-coding manual. Does this identification agree with step 1 above?

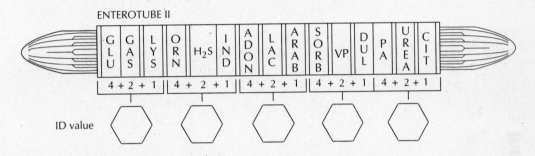

API 20E System

1 Based on your data, identify your unknown microorganism using the differential identification chart, and enter it in the chart on pages 429–430.

2 Indicate + and − in the 24- or 48-h line of the chart below. (**NOTE:** Ignore the 5-h line; you will not be recording these results.) The value for each test is given under its code. Add the points for each positive reaction within each section divided by a bold line. Then determine the seven-digit profile number by reading the seven total numbers across. Identify your unknown organism by referring to the API 20E Analytical Profile Index (from the manufacturer). Does this identification agree with step 1 above?

	ONPG 1	ADH 2	LDC 4	ODC 1	CIT 2	H_2S 4	URE 1	TDA 2	IND 4	VP 1	GEL 2	GLU 4	MAN 1	INO 2	SOR 4	RHA 1	SAC 2	MEL 4	AMY 1	ARA 2	OXI 4
5 h																					
24 h																					
48 h																					
Profile number																					

QUESTIONS **1** Which method, the Enterotube II System or API 20E System, do you prefer to use for identification of a member of the Enterobacteriaceae and why?

2 Explain why some laboratories prefer to use such rapid multitest identification kits over conventional methods of bacterial identification.

ENVIRONMENTAL MICROBIOLOGY

The microbial population of naturally occurring ecosystems is frequently very large and complex. Fertile soil, for example, may contain billions of microorganisms (bacteria, yeasts, molds, algae, and protozoa) per gram. Another feature of the natural microbiota is the extent to which interactions occur among the various species and the effect which this has on their combined physiological activities.

Most, if not all, laboratory procedures used to isolate microorganisms from natural materials (like soil, water, sewage) recover only a very small fraction of the total population. The great diversity of physiological types present precludes the possibility of recovery of the majority of the microbial population under a given physical environment on a single nutritional medium.

Consequently, the microbiological analysis is usually directed toward the enumeration and isolation of a particular physiological group of microorganisms. For example, if our purpose is to isolate spore-forming anaerobic bacteria from soil, we can establish suitable selective procedures and cultural conditions to accomplish this by drawing upon our knowledge of the physiology of the bacteria.

EXERCISE 69.A

Microbiology of Soil
69.A Quantitative Enumeration of Microorganisms

OBJECTIVE
To count the number of viable microorganisms in a 1-g sample of soil.

OVERVIEW
Soil harbors microorganisms of all varieties. It is because of them that life is sustained on this planet. They are essential for the degradation of organic matter that is deposited in the soil as animal wastes and dead plant and animal bodies. By their action, especially in the cycles of the elements, such as the carbon, nitrogen, phosphorus, and sulfur cycles, the soil is replenished with basic elemental nutrients. It follows that plants assimilate these nutrients, transforming them into organic macromolecules which, in turn, become food for animals.

Each gram of soil contains enormous numbers and diverse kinds of microorganisms including algae, protozoa, fungi, and bacteria. The most numerous microbes in soil are bacteria. Special media and conditions of incubation are necessary to isolate the diverse bacterial species. This bacterial flora performs a vast array of biochemical changes in soil that constitute essential links in the cycles of matter in nature. FIGURE 2.1 (page 13) shows molds and bacteria growing on agar plates inoculated with soil. In this exercise, several different experiments will be performed, the results of which will give you an appreciation of the soil microflora and its biochemical potential.

REFERENCE
MICROBIOLOGY, Chap. 28, "Microbiology of the Soil and the Atmosphere."

MATERIALS
3 Petri plates of nutrient agar
3 Petri plates of Sabouraud's agar
(All plates prepared several days before and kept at room temperature.)
3 99-ml physiological saline dilution blanks
6 sterile 1.1-ml pipettes or 12 1-ml pipettes
Balance and spatula
Sterile paper squares in Petri dish
Soil sample: finely pulverized garden soil

PROCEDURE
1 Weigh out 1 g of soil and add it to a 99-ml dilution blank.

2 Prepare and plate the following dilutions by the spread-plate method, using nutrient agar as the plating medium:

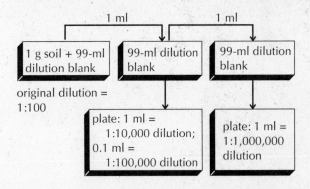

3 Prepare another set of plates of the following dilutions, using Sabouraud's agar as the plating medium: 1:1000, 1:10,000, and 1:100,000. (Use 0.1 ml from the 1:100 dilution for plating the 1:1000 dilution.)

4 Incubate all plates at room temperature for 4 to 7 days.

NAME _____

69.A

Quantitative Enumeration of Microorganisms

RESULTS 1 Count and record the number of colonies on each plate in the chart provided.

COLONY COUNTS OF SOIL DILUTIONS		
Medium	Dilution	Colony counts
Nutrient agar		
Sabouraud's agar		

2 Calculate the plate count per gram of soil on:

Nutrient agar: _____ per gram soil

Sabouraud's agar: _____ per gram soil

3 Describe the predominant colonial types that appear on each of the media.

On nutrient agar:

On Sabouraud's agar:

437

QUESTIONS 1 What types of microorganisms would be missed in the culturing procedure employed?

2 Did you notice any differences between the types of microorganisms recovered on the two plating media?

3 What advantages does a direct microscopic count of microorganisms in soil have over a cultural procedure?

EXERCISE 69.B

Microbiology of Soil
69.B Ammonification: The Liberation of Ammonia from Nitrogenous Compounds

OBJECTIVE
To detect the occurrence of ammonification in the soil.

OVERVIEW
Nitrogen in plants and animals exists mainly in the form of protein. When animals and plants die, the protein is degraded to amino acids which are then deaminated to liberate ammonia. This process of producing ammonia from organic compounds is called *ammonification*. Since most plants can assimilate ammonia, this constitutes a very important step in the nitrogen cycle. Most bacteria found in the soil can carry out ammonification.

REFERENCE
MICROBIOLOGY, Chap. 28, "Microbiology of the Soil and the Atmosphere."

MATERIALS
2 tubes of sterile casein solution
2 tubes of sterile peptone solution
Nessler's reagent (for ammonia test)
Spot plate
Soil sample

PROCEDURE
1 Inoculate one tube of each medium with the soil sample. Make the inoculations by first wetting the loop transfer needle, for example, by touching the sterile transfer needle to the casein solution and then touching the soil sample. The soil adhering to the loop is sufficient inoculum.
2 Incubate the inoculated tubes and an uninoculated tube of each medium for 7 days.
3 After 2 to 4 days' incubation and again at 7 days' incubation, test contents of each tube for ammonia as follows: Place a drop of Nessler's solution in the depression of a spot plate and add a loopful of material from one of the tubes. Observe for any color changes. A faint yellow color is indicative of a small amount of ammonia; deep yellow indicates more ammonia; and a brown color or brownish precipitate signifies a large quantity of ammonia.

NAME _____

69.B

Ammonification: The Liberation of Ammonia from Nitrogenous Compounds

RESULTS Record the results of the ammonia determination in the chart provided.

AMMONIFICATION BY SOIL MICROBES			
		Color with Nessler's reagent	
Substrate	Incubation time	Inoculated tubes	Uninoculated tubes (controls)
Casein			
Peptone			

441

QUESTIONS 1 What is the significance of ammonification in relation to soil fertility?

2 What are the components in casein and peptone that allow ammonification to occur?

Microbiology of Soil
69.C Reduction of Nitrates and Denitrification

OBJECTIVE
To detect the process of denitrification in soil.

OVERVIEW
Denitrification is the reduction of nitrates to nitrites and nitrogen gas. The process is carried out by denitrifying bacteria and may be represented by the following reaction steps:

Nitrate ion (NO_3^-) ⟶ nitrite ion (NO_2^-) ⟶ nitrous oxide (N_2O) ⟶ nitrogen gas (N_2)

The process is also called *anaerobic respiration*. It is carried out under anaerobic conditions by many genera of bacteria including *Pseudomonas* and *Bacillus*. Since nitrates are an important source of nitrogen for plants, it is apparent that denitrification is undesirable for agriculture because the products of denitrification are gaseous. They escape from the soil into the air, leading to a net loss of nitrogen content from the soil. This is why soil is tilled in order to break up large soil clumps where pockets of anaerobiosis, and denitrification, could develop.

MATERIALS
3 Durham fermentation tubes of nitrate-salts medium
Nessler's reagent (for ammonia test)
Trommsdorf's reagent and a 1:3 dilution of sulfuric acid (for nitrite test)
Spot plate and glass rods
Broth culture of *Pseudomonas fluorescens*
Soil sample
Concentrated H_2SO_4 (about 1 ml in small test tube)
Diphenylamine reagent (for nitrate test)

PROCEDURE
1 Inoculate one tube of the nitrate-salts medium with soil, using a wet loop transfer needle. Inoculate another tube with a loopful of *P. fluorescens* culture.
2 Incubate the inoculated tubes and an uninoculated control at room temperature for 7 days.
3 After incubation, observe the tubes for evidence of gas formation. Test the medium in each tube in the following order:
  a *Nitrite test*: Add 1 drop of 1:3 dilution of sulfuric acid solution to a depression in a spot plate. To this add 3 drops of Trommsdorf's reagent. Dip a glass rod in the culture to be tested and touch the end of this to Trommsdorf's reagent mixture. Do not mix. Nitrites are indicated by the immediate appearance of a blue-black color. If the nitrite test is negative, test for nitrates.
  b *Nitrate test*: Add 3 drops of diphenylamine reagent to a depression on the spot plate. Place a drop of the culture to be tested at the surface of this solution. Add 2 drops of concentrated sulfuric acid. If nitrates are present, a light blue color develops within a few minutes. (NOTE: Nitrites also give a blue color with diphenylamine. However, if a negative test was demonstrated with Trommsdorf's reagent, nitrites can be excluded.)
  c *Ammonia test*: Test with Nessler's reagent by placing a drop of Nessler's solution in the depression of the spot plate. Add a drop of culture to this solution. A faint yellow color that develops is indicative of a small amount of ammonia. A deep yellow color indicates more ammonia; and a brown color or brownish precipitate is indicative of a large quantity of ammonia.

NAME _____

Reduction of Nitrates and Denitrification

RESULTS Record the results of the tests carried out in the chart below.

TESTS FOR PRODUCTS OF DENITRIFICATION BY SOIL MICROORGANISMS			
Test	Specimen	Test result, + or –	Gas, + or –
Nitrite	Soil		
Nitrite	*Pseudomonas fluorescens*		
Nitrite	Control		
Nitrate	Soil		
Nitrate	*P. fluorescens*		
Nitrate	Control		
Ammonia	Soil		
Ammonia	*P. fluorescens*		
Ammonia	Control		

QUESTIONS **1** Distinguish between denitrification and nitrification.

2 When the test for nitrite is negative, why is it necessary to retest the inoculated medium (nitrate broth) for presence of nitrate?

3 What soil conditions favor microbial reduction of nitrates?

EXERCISE 69.D

Microbiology of Soil
69.D Symbiotic and Nonsymbiotic Nitrogen Fixation

OBJECTIVE
To learn about the process of nitrogen fixation.

OVERVIEW
Nitrogen fixation is the process by which atmospheric nitrogen can be returned to soil by its conversion into ammonia and amino acids.

Symbiotic bacteria play an important role in nitrogen fixation. Soon after seeds of leguminous plants (such as soybeans, peas, alfalfa, and clover) germinate in the soil, certain bacteria (such as *Rhizobium*) invade the roots and stimulate the formation of root nodules, where the bacteria populate. These root-nodule species fix atmospheric nitrogen in symbiosis with the leguminous partner. Symbiotic nitrogen fixation also occurs in the roots of nonleguminous plants. For example, the actinomycete *Frankia* forms root nodules in alders. Many soils, however, in which legumes are not growing and to which nitrogen is not added in the form of fertilizers, still increase in nitrogen content because of the presence of nonsymbiotic, free-living nitrogen-fixing species of bacteria. There are free nitrogen fixers that are aerobic, for example, *Azotobacter* spp., and free nitrogen fixers that are anaerobic, for example, certain species of *Clostridium*.

MATERIALS
1 125-ml Erlenmeyer flask with 100 ml of nitrogen-free fluid medium
Soil sample
Balance
Petri plate of nitrogen-free *Azotobacter* medium
Gram-stain reagents
2 microscopic slides

PROCEDURE
1 You are provided with a 125-ml Erlenmeyer flask containing 100 ml of sterile nitrogen-free fluid medium.
2 Pour into the medium 1 g of soil (a heavy inoculum is required for the demonstration of nitrogen fixation by the anaerobic nitrogen fixers present in the sample).
3 Incubate the flask at room temperature for 1 week; then observe for the presence and growth of anaerobic nitrogen fixers.
4 Sprinkle soil over the surface of a nitrogen-free agar plate provided.
5 Incubate at room temperature for a week; then examine the surface for the presence of colonies of aerobic free nitrogen fixers.

NAME _____

Symbiotic and Nonsymbiotic Nitrogen Fixation

RESULTS

1 Examine the flask of nitrogen-free fluid medium. Gas bubbles and a rancid odor are evidence of the activity of the bacterium *Clostridium pasteurianum*. Make Gram stains of films from the bottom of the medium and draw the cells in the circles provided. This species is a spore-forming bacillus with a terminal or central bulging spore.

2 Examine the nitrogen-free agar plate. Describe the colonies.

3 Make and examine stained films using Gram staining and draw the cells in the circles provided. (Cells of *Azotobacter* are large ovoid and are Gram-negative.)

QUESTIONS 1 Name one species of bacteria that are symbiotic nitrogen fixers.

2 Name one species of bacteria that are nonsymbiotic nitrogen fixers.

3 Can all legumes be successfully inoculated with the same species of symbiotic nitrogen-fixing bacteria? Explain.

EXERCISE 70

Biological Succession in a Mixed Microbial Flora

OBJECTIVE
To observe the succession of microbial forms in milk.

OVERVIEW
The laboratory study of microorganisms is conveniently done with pure cultures. The initial specimen or sample, for example, soil, water, or sputum, contains a large variety of microbial species. The routine techniques are directed toward isolating certain predetermined types in pure culture; the pure cultures are then characterized. In most natural environments, a multitude of microorganisms representing the numerous species grow in close physical association with each other. Such an association may account for changes that are a direct consequence of the interactions among microbial species. This exercise will serve to illustrate the fluctuation in numbers and the sequence of predominant physiological types of microorganisms that develop in raw milk.

REFERENCE
MICROBIOLOGY, Chap. 31, "Microbiology of Food."

MATERIALS
Quantity or amounts of most materials below are not specified because this depends on the duration of time (incubation of the raw milk) desired for the performance of this exercise. (The student will have practice in calculating these quantities.)
500 ml of raw milk in a 1-liter flask
Milk-protein-hydrolysate agar
Nutrient agar
Sterile Petri dishes
Screw-capped tubes
Gelatin agar
Milk agar
Tributyrin agar
Dilution blanks
Sterile pipettes
Reagents for preparation of stained films of milk (xylol, alcohol, methylene blue)

PROCEDURE
1 Place 500 ml of raw (unpasteurized) milk in a clean, sterile 1-liter Erlenmeyer flask. Place a cotton plug in the mouth of the flask. Store the flask of milk at room temperature for several weeks.
2 Perform the following examinations on the fresh raw milk on the first day of this exercise.
   a Prepare a stained film as follows. Spread a loopful of milk on a clean slide over an area about the size of a dime. Allow it to dry without heating. Cover the milk film with xylol for 1 min to remove fat. Drain off the xylol, allow the slide to air-dry, and then cover the film with alcohol for 1 min. Drain off the alcohol, allow the film to air-dry, and then stain the film with methylene blue for approximately 5 min. Examine the film using the oil-immersion objective. Record the various morphological types and the approximate percentage of each.
   b Perform a standard plate count (see Exercise 19). Plate each dilution in triplicate, and incubate one set of plates at each of the following temperatures: 5, 35, and 50°C. Incubate the plates at 5°C for 7 days and the others for 48 h. At the end of the incubation time, and after plate counts are made, transfer several colonies,

of different types, to nutrient-agar slants. Incubate each slant at the same temperature at which the original colony grew, that is, 5, 35, or 50°C. Subsequently, test each agar slant culture for its ability to hydrolyze fat, protein, and carbohydrate (see Exercise 24).

3 At intervals of approximately 4 to 5 days, over a period of 2 or 3 weeks, repeat all the examinations outlined in step 2. In addition, record the physical appearance of the milk at each of these times.

NAME _____

Biological Succession in a Mixed Microbial Flora

RESULTS   Tabulate results of the various examinations in the following chart.

THE SPOILAGE OF RAW MILK AND ITS MICROBIAL CONTENT						
Date	Appearance of milk	Microscopic examination	5°C	35°C	50°C	Characteristics of cultures isolated

QUESTIONS

1 Of the major substrates of milk, which one was first utilized? What is the experimental evidence in support of your answer?

2 From your results, what generalizations can you make about the succession of (a) morphological types and (b) physiological types of microorganisms during prolonged incubation of milk?

3 What interactions among various microbial species might have occurred during the incubation of the milk?

EXERCISE 71

Enrichment Culture Technique: Isolation of Phenol-Utilizing Microorganisms

OBJECTIVE
To learn the enrichment culture technique for the isolation of a specific type of microorganism.

OVERVIEW
Natural environments are generally inhabited by numerous physiological types of microorganisms; the population of each type may vary considerably. In some instances, species of a type of particular interest to the microbiologist are present in such low numbers that it is difficult, if not impossible, to isolate them by the direct techniques, for example, a streak plate or pour plate of the specimen. When the desired microorganisms are suspected of being few in number, the use of the enrichment technique greatly improves the prospect of their isolation. In principle, this technique provides an environment, that is, media composition and physical conditions, that is preferentially selective for the biotype being sought. Consequently, on successive transfers in such an environment, the particular type will emerge as the predominant, if not exclusive, population.

The enrichment culture technique has been used in a wide range of applications in research as well as in clinical, industrial, and environmental microbiology. In *research*, a microbiologist may wish to isolate for different bacteriophages from raw sewage, soil, or a variety of natural sources. Because the phages are usually found naturally in such sources in low numbers, it is necessary to enrich for them so that they become numerous enough for routine isolation. The enrichment process entails growing the susceptible bacterial host in the sample from the phage source so that the phages (if they are present) can complete their cycles in the host cells and thereby increase in number. The increased number of phages obtained by enrichment can be filtered out and detected by routine cultural techniques (see Exercise 43). In *clinical microbiology*, the technique is used routinely for the isolation of intestinal pathogens from fecal samples where these microorganisms may be present in low numbers. For example, a microbiologist can use an enrichment broth like selenite broth, which inhibits Gram-positive bacteria and enriches for species of *Salmonella* and *Shigella*. In *industrial microbiology*, enrichment methods may be used to isolate for specific soil microorganisms that are able to synthesize economically important products such as steroids, enzymes, and vitamins. The laboratory synthesis of these products by microorganisms can then be scaled up to industrial production. In *environmental microbiology*, the enrichment method may be used to isolate for a petroleum-utilizing microorganism that would be capable of degrading environmentally destructive oil spills in water.

Additionally, in environmental microbiology, the enrichment culture technique can be used to clean up destructive oil spills by enriching for all types of microorganisms that degrade oil at the site of spillage in a process called *bioremediation*. This process has been proved effective in a "real-life" situation. The oil tanker *Exxon Valdez* ran aground on Bligh Reef in March 1989 and spilled approximately 11 million gallons of crude oil into the waters off Alaska's Prince William Sound; it was the worst oil spill in U.S. history. This was an environmental disaster of monumental proportion! Spilled crude oil has the potential of affecting every level of the marine food chain. Floating oil contaminates floating plankton (microscopic algae, invertebrates, etc.). In turn, the small fish that feed on these organisms become contaminated. Larger animals in the food chain become contaminated when they eat these small fish. In addition, marine animals and birds exposed directly to oil in the water ingest the oil and become toxified, or their fur or feathers become "tarred" with oil, followed by obvious consequences. Spilled oil also prevents the germination and growth of marine plants and the reproduction of invertebrates, either by smothering or by toxic effects. However, within one year of the incident, Prince William Sound was essentially clean. There were several reasons for the rapid cleanup of the area.

Petroleum-devouring microorganisms, assisted by a process known as *enhanced bioremediation*, were among the least known and most important factors in the recovery process. Bioremediation is the process in which microorganisms are used to biodegrade oil; enhanced bioremediation is the process in which additional amounts of nitrogen and phosphorus are used to enhance the microbial degradation activity. Prince William Sound is home to naturally occurring single-celled microorganisms that continuously feed on hydrocarbons from natural petroleum seepages. Immediately after the *Valdez* accident, these microorganisms increased their numbers to respond to the oil from the spill. However, their ability to "eat" this oil was limited by the available amounts of nitrogen and phosphorus, the other two major elements (oil supplies the carbon) needed for these microbes to multiply and degrade the oil. These elements were supplied by scientists from the U.S. Environmental Protection Agency and Exxon by the application of special fertilizers to oiled shorelines of Prince William Sound. These fertilizers provided the microbes with added nitrogen and phosphorus, permitting them to reproduce and consume oil at an enhanced rate. In the normal situation without an oil spill, there are about 3 million oil-consuming microorganisms in every ounce of sediment in Prince William Sound. After enhanced bioremediation, their number rose to as high as 30 million per ounce, and their degradative ability was enhanced 10-fold. Bioremediation proved to be both effective and safe in cleaning up oil spills by the enrichment culture technique.

REFERENCE

MICROBIOLOGY, Chap. 31, "Biotechnology: The Industrial Applications of Microbiology."

MATERIALS

Ingredients for mineral-salts medium: $(NH_4)_2SO_4$, K_2HPO_4, KH_2PO_4, $CaCl_2$, $FeSO_4 \cdot 7H_2O$, $MnCl_2 \cdot 4H_2O$, $MgSO_4 \cdot 7H_2O$
Phenol*

*This exercise can be adapted for enrichment culture isolation of other biotypes by substituting the available carbon-energy source, for example, hydrocarbon or cellulose.

Agar
250-ml Erlenmeyer flasks
Sterile Petri dishes
1- and 10-ml sterile pipettes
Soil samples

PROCEDURE

Several different media are required for this exercise. They should be prepared in advance and stored in a refrigerator until required.

MEDIA PREPARATION

1 Phenol-mineral-salts medium:
  a *Basal medium ingredients solution:* Prepare three separate solutions of the basal medium ingredients as indicated below. Sterilize them by autoclaving (15 min at 121°C).

Solution A:
$(NH_4)SO_4$, 1.0 g; K_2HPO_4, 1.0 g; KH_2PO_4, 0.5 g; H_2O, 700 ml

Solution B:
$CaCl_2$, 0.01 g; $FeSO_4 \cdot 7H_2O$, 0.005 g; H_2O, 100 ml

Solution C:
$MgSO_4 \cdot 7H_2O$, 0.2 g; $MnCl_2 \cdot 4H_2O$, 0.10 g; H_2O, 100 ml

b *Phenol solution:* Prepare a solution to contain 7.5 g phenol per 100 ml H_2O.
c *Combination:* The phenol-mineral-salts medium is prepared by combining the above solutions in the following proportions:

7 parts solution A
1 part solution B
1 part solution C
1 part phenol solution

Place 40 ml of this medium into sterile stoppered 250-ml Erlenmeyer flasks; three such flasks will be required for each soil sample. Also dispense 10 ml per tube into 10 sterile test tubes.

2 Double-strength phenol-mineral-salts solution: Prepare as above, but double the concentration of each ingredient. Approximately 100 ml will be needed.

3 Sterile agar solution: Dissolve 3 g of agar in 100 ml of water, and sterilize by autoclaving.

4 Phenol-mineral-salts agar medium: Add the agar solution (100 ml) aseptically to the 100-ml double-strength phenol-mineral-salts solution; mix

thoroughly, and dispense into Petri dishes and sterile test tubes (about 6 each). Slant the tubed medium.

ENRICHMENT OF INOCULUM

Follow the protocol of the exercise as shown in FIGURE 71.1. Inoculate three flasks containing phenol-mineral-salts medium, each with a different sample of soil. Use approximately 0.5 g of soil for inoculum. Incubate these flasks at 25°C for 3 days (preferably on a shaking device); then make a transfer of 0.1 ml from each flask into another flask of phenol-mineral-salts medium. Repeat the same procedure after the second set of flasks has incubated 3 days.

ISOLATION OF PHENOL-UTILIZING MICROORGANISMS

Prepare streak plates on phenol-mineral-salts agar from each of the last series of enrichment culture flasks. Incubate plates at 25°C; then observe them for the presence of any colonies. Transfer representative colonies to phenol-mineral-salts agar slants. Test each pure culture in the fluid phenol-mineral-salts medium to ascertain that it grows in this medium.

For each culture isolated, characterize culturally, morphologically, and biochemically (see Exercise 28).

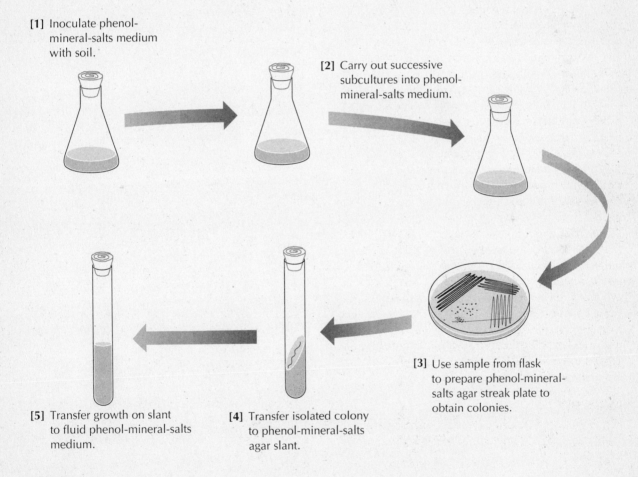

FIGURE 71.1
Enrichment culture procedure for obtaining phenol-utilizing microorganisms.

NAME _____

Enrichment Culture Technique: Isolation of Phenol-Utilizing Microorganisms

RESULTS

1 State why you think that your isolates are phenol-utilizing microorganisms.

2 Summarize the characteristics of phenol-utilizing microorganisms isolated in the chart provided.

CHARACTERISTICS OF PHENOL-UTILIZING MICROORGANISMS		
Culture identification	Source	Characteristics

QUESTIONS

1 Can an organic compound serve both as a carbon source and an energy source for a species of microorganism?

2 If a substance is utilized by a microbe, is it also degraded?

3 How might you account for the following experimental results? A colony is transferred from the phenol-mineral-salts agar into the fluid phenol-mineral salts medium and, after appropriate incubation, no growth occurs.

EXERCISE 72

Standard Method of Water Analysis: Multiple-Tube Fermentation

OBJECTIVE
To carry out a standard analysis of water using the multiple-tube fermentation technique.

OVERVIEW
Natural waters may contain a wide variety of microorganisms. In fact, it is not unlikely that one might find representatives of many of the major categories of microorganisms in a specimen from such sources. A single medium, inoculated and placed under specific conditions of incubation, will reveal only a fraction of the total microbial population. It is necessary, therefore, to employ different procedures for the enumeration and isolation of different microbial types.

Natural waters become polluted when the polluting material upsets the natural balance of microorganisms, plants, and animals living in or near that water or when it makes the waters unsafe for human consumption or recreation. Painful evidence of the pollution of water resources is seen in dead fishes on the shores of a lake or in typhoid or dysentery epidemics. Pollution can be due to increasing industrialization or growth of human habitation. Pollution can be either toxic chemicals or pathogenic microorganisms. Probably the largest single source of potentially pathogenic microbes in water is animal (including human) feces, which contain many billions of bacteria per gram. Some of these bacteria are pathogenic and cause **enteric diseases** (diseases that affect the gastrointestinal tract) such as bacillary dysentery, typhoid fever, cholera, and paratyphoid fever. In addition to intestinal infections caused by bacteria, certain viruses, such as hepatitis and poliomyelitis viruses, are also found in animal feces. Also found are amoebas which can cause amoebic dysentery, as well as other parasites. All these diseases may be spread through drinking polluted water.

Analysis of water samples on a routine basis for each individual type of pathogen is not practical. Therefore, the following principle is universally adopted: If the water supply contains any microorganisms of fecal origin, it follows that this water may also contain pathogens, and therefore, it should not be consumed. The method that has been developed to determine the sanitary quality of drinking water is an example of a procedure designed to detect a certain category of bacteria, namely, the *coliform group* of bacteria. Thus, the basis of water testing for microbiological safety rests on the detection of coliform bacteria. The word *coliform* means having colilike characteristics, that is, like those exhibited by *Escherichia coli*. These characteristics may be summarized as follows:

1 Small Gram-negative rod
2 Nonsporulating
3 Fermentation of lactose with production of acid and gas
4 Production of green "metallic" sheen on eosin-methylene blue (EMB) agar [See PLATE 32.]
5 Aerobic or facultative

The reason *E. coli* is generally used as an *indicator bacterium* (indicating fecal pollution) is because it is found in large numbers in animal feces. It lives longer in water than enteric pathogens. Its population density is directly related to the degree of fecal pollution. It is easily quantified and is relatively harmless to the microbiologist. So if no *E. coli* is present in the water sample, it is likely that there are no intestinal pathogens present. Two places where coliform tests are made routinely are in municipal waterworks (where drinking water is prepared for human consumption) and in sewage plants (where wastewater is treated and tested for its safety before being discharged).

Unfortunately, one complication in testing for fecal coliforms is caused by the bacterium *Enterobacter aerogenes*. It too is found in fecal material, but it is also found in soil and plants. And it shares many of the common characteristics of *E. coli*. Thus, the microbiologist must be able to distinguish between *Ent. aerogenes* and *E. coli*. A group of tests has been devised to do this. These are the well-known IMViC tests, which are, specifically, the indole, the methyl red, the Voges-Proskauer, and the citrate tests (see Exercise 65). However, these tests require the isolation of pure cultures and much time to perform. Instead, an alternative series of tests was devised that could obviate the complication due to the presence of *Ent. aerogenes* without having to isolate and identify pure cultures. This series of tests constitutes the *standard method of water analysis*. These tests are a series of three: the first one is the **presumptive test**, followed by the **confirmed test**, and finishing up with the **completed test**.

One of the procedures in the *presumptive test* is a quantitative method for determining the most probable number (MPN) of coliforms in the water sample. It is basically a statistical approach to the quantitative estimation of bacterial numbers. Samples are serially diluted to the point at which there are few or no viable microorganisms. The detection of this endpoint is based on multiple serial dilutions that are inoculated into a suitable growth medium. Statistical tables are available to estimate the size of the bacterial populations based on the number of replicate tubes (3, 5, or 10) of each dilution [TABLE 72.1].

The microbiologist should be aware of another test for water quality. The amount of organic materials in water or wastewater can be determined by measuring the oxygen consumed by microorganisms, which is proportional to the organic material present. This is the basis for the *biochemical oxygen demand (BOD) test*. In addition to providing information on the load of organic wastes in a water supply, it can determine the effectiveness of a sewage-treatment system. It can also be used to determine whether wastewater is sufficiently stabilized to be discharged into a lake, river, or stream without endangering the oxygen level.

REFERENCES

MICROBIOLOGY, Chap. 29, "Microbiology of Natural Waters, Drinking Water, and Wastewater."

Standard Methods for the Examination of Water and Wastewater, 17th ed., American Public Health Association, 1015 Eighteenth Street, N.W., Washington, D.C., 1989.

MATERIALS

2 tubes of nutrient agar (molten and kept at 45 to 50°C)
2 sterile Petri dishes
11 tubes of lactose broth, single-strength, with Durham tubes
5 tubes of lactose broth, double-strength, with Durham tubes
1 Levine eosin-methylene blue (EMB) agar plate
1 nutrient-agar slant
1 sterile 10-ml pipette
2 sterile 1-ml pipettes
Water sample: Tap water, well water, wastewater, water from lake or pond, and so on (Each student or group of students needs to use only one sample.)

PROCEDURE

Since the bacteriological analysis of water to determine its sanitary quality requires performing certain procedures on successive days, this exercise is outlined in the same fashion. The standard method of water analysis is shown in FIGURE 72.1.

STANDARD PLATE COUNT AND PRESUMPTIVE TEST FOR COLIFORMS

1 Select one of the water samples, and plate 1.0- and 0.1-ml quantities by the pour-plate technique (see Exercise 19). Incubate plates at 35°C for 24 h.

2 With the same sample, inoculate tubes of *single-strength* lactose broth with 1.0 and 0.1 ml (5 tubes each volume) and 5 tubes of *double-strength* lactose broth, each with 10 ml. Incubate these tubes at 35°C from 24 to 48 h. After incubation, proceed as follows.

3 Count and record the number of colonies on the plates.

4 Observe the lactose-broth tubes for evidence of gas production. If no gas is present, incubate the tubes for another 24 h; if gas is still absent at this time, the test for coliforms is considered *negative*. The presence of gas within 48 h in any of the fermentation tubes is a *positive* presumptive test.

NOTE: If the water sample with which you are working gives a negative presumptive test, obtain a lactose-broth tube showing gas production from another member of the class and continue the examination.

CONFIRMED TEST

1 Streak an EMB plate from one of the tubes of lactose broth showing gas formation (choose the tube that received the smallest volume of sample). Incubate this plate at 35°C for 24 h.

2 Observe this plate, after incubation, for colonies of coliform organisms. *Escherichia coli* colonies appear to be bluish black by transmitted light and have a greenish metallic sheen by reflected light.

TABLE 72.1 MPN Index for Various Combinations of Positive and Negative Results When Five 10-ml Portions, Five 1-ml Portions, and Five 0.1-ml Portions Are Used

No. of tubes giving positive reaction out of:				No. of tubes giving positive reaction out of:			
5 of 10 ml each	5 of 1 ml each	5 of 0.1 ml each	MPN index per 100 ml	5 of 10 ml each	5 of 1 ml each	5 of 0.1 ml each	MPN index per 100 ml
0	0	0	<2	4	2	1	26
0	0	1	2	4	3	0	27
0	1	0	2	4	3	1	33
0	2	0	4	4	4	0	34
1	0	0	2	5	0	0	23
1	0	1	4	5	0	1	31
1	1	0	4	5	0	2	43
1	1	1	6	5	1	0	33
1	2	0	6	5	1	1	46
2	0	0	5	5	1	2	63
2	0	1	7	5	2	0	49
2	1	0	7	5	2	1	70
2	1	1	9	5	2	2	94
2	2	0	9	5	3	0	79
2	3	0	12	5	3	1	110
3	0	0	8	5	3	2	140
3	0	1	11	5	3	3	180
3	1	0	11	5	4	0	130
3	1	1	14	5	4	1	170
3	2	0	14	5	4	2	220
3	2	1	17	5	4	3	280
3	3	0	17	5	4	4	350
4	0	0	13	5	5	0	240
4	0	1	17	5	5	1	350
4	1	0	17	5	5	2	540
4	1	1	21	5	5	3	920
4	1	2	26	5	5	4	1600
4	2	0	22	5	5	5	≥2400

Source: Standard Methods for the Examination of Water and Wastewater, 14th ed., American Public Health Association, Inc., N.Y., 1976.

FIGURE 72.1
The standard method of water analysis.

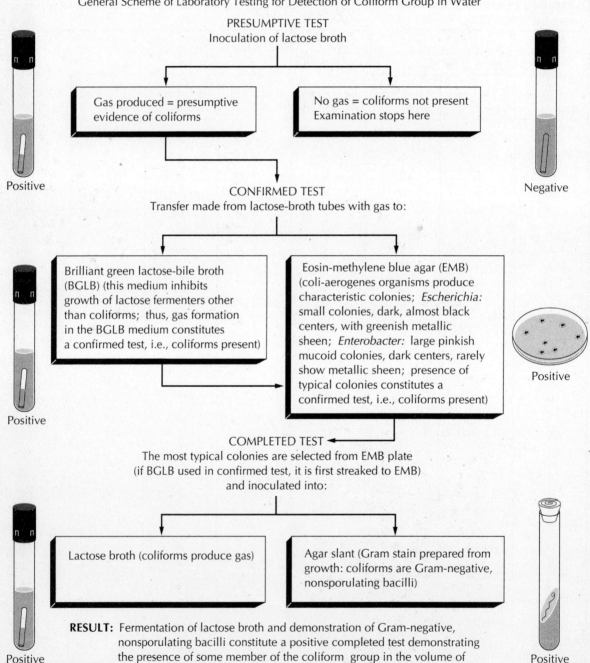

Ent. aerogenes colonies are pinkish and often convex and mucoid; they may tend to coalesce. If typical coliform colonies (the *E. coli*-type colonies) are present on the EMB plate, the result of the confirmed test is considered *positive*.

COMPLETED TEST

1 Inoculate a tube of lactose broth and an agar slant from a typical coliform colony (i.e., one with metallic sheen) on the EMB plate. Incubate these tubes at 35°C for 24 to 48 h.

2 Observe lactose broth for evidence of gas formation. Prepare and examine a Gram-stain preparation from the agar slant culture. If gas appears in the lactose-broth tube and microscopic examination of the smear reveals Gram-negative non-spore-forming bacilli, the test is considered *completed* and the presence of coliform organisms demonstrated.

NAME _____

Standard Method of Water Analysis: Multiple-Tube Fermentation

RESULTS 1 Enter results of this exercise in the chart provided below.

RESULTS OF STANDARD ANALYSIS OF WATER SAMPLE	
Test procedures	Results
Standard plate count per milliliter water sample	
Presumptive test	No. of positive lactose-broth tubes inoculated with: 10 ml: 1 ml: 0.1 ml:
Confirmed test	Appearance of colonies on EMB agar medium:
Completed test	Gas production in lactose broth: Gram reaction and morphology of cells:

2 The number of tubes in the presumptive test showing acid and gas, when referred to a statistical table, gives the most probable number of coliforms per 100 ml of water. Determine the MPN of coliforms per 100 ml of your water sample using TABLE 72.1. For combinations not shown in the table, use the simple Thomas' formula:

$$\text{MPN per 100 ml} = \frac{\text{no. of positive tubes} \times 100}{(\text{milliliter sample in negative tubes}) \times (\text{milliliter sample in all tubes})}$$

The MPN per 100 ml of water sample is:

QUESTIONS

1 Does the standard plate count reveal all the microorganisms present in the water sample? Explain.

2 What factors might be responsible for a positive presumptive test in a sample which, on subsequent testing, does not yield a positive confirmed test?

3 In the analysis performed, why would it be unsatisfactory to streak the water specimen directly on to an EMB plate?

EXERCISE 73

Standard Method of Water Analysis: Membrane-Filter Technique

OBJECTIVE
To carry out a standard analysis of water using the membrane-filter technique.

OVERVIEW
The second standard method for measuring coliform density in water as recommended by the American Public Health Association is the membrane-filter (MF) technique.

The membrane-filter technique offers the following advantages over the multiple-tube fermentation technique (Exercise 72):

1 It permits a direct count of coliform colonies, or colonies of other kinds of bacteria, for example, those of the genus *Salmonella*.
2 Results are obtainable in a shorter time.
3 Larger volumes of water sample can be processed.
4 Results are readily reproducible and more accurate.

The widespread use of the technique has confirmed its value.

The principle of the method is as follows: A measured amount of the water sample is filtered through a membrane filter with a pore size of 0.45 μm or smaller. The filter traps the bacterial cells on its surface; it is then transferred from the filter assembly onto an absorbent pad (in a Petri plate) that has been saturated with the particular medium designed to grow and differentiate the particular bacteria of interest. For example, a modified Endo medium (M-Endo Broth MF) is used if total coliform bacteria are the microorganisms of interest. Other media, such as M-FC broth for detection of fecal coliforms and KF streptococcus agar for detection of fecal streptococci, are available commercially. After incubation, the colonies on the membrane filter are counted.

In determining total coliforms, the amount of water filtered should be sufficient to give 20 to 80 colonies and no more than a total of 200 bacterial colonies of any type. About 50 to 200 ml of unpolluted water is often enough for obtaining such bacterial counts. Polluted water may contain such a high number of coliforms that dilution of the sample before filtration may be necessary, for example, 1 ml of polluted water with 50 ml of sterile water. This is done to provide enough volume for uniform bacterial dispersion over the whole filter surface as well as to obtain an acceptable coliform count.

At this point it is well to mention that the U.S. Public Health Service has set standards for drinking water. Water is considered of good quality when fewer than 2.2 coliforms/100 ml are detected by the most-probable-number (MPN) technique (Exercise 72) or less than 1/100 ml with the membrane-filter (MF) technique.

REFERENCES
MICROBIOLOGY, Chap. 29, "Microbiology of Natural Waters, Drinking Water, and Wastewater."

Standard Methods for the Examination of Water and Wastewater, 17th ed., American Public Health Association, 1015 Eighteenth Street, N.W., Washington, D.C., 1989.

MATERIALS
1 ampule of M-Endo Broth MF
1 sterile absorbent pad
1 47-mm Petri dish (sterile)
1 pair of forceps in 95% ethanol
1 5-ml pipette
Bacteriological membrane-filter assembly with vacuum pump (or vacuum line or water aspirator)
Presterilized membrane filters (0.45 μm pore size)

FIGURE 73.1
The membrane-filter technique.

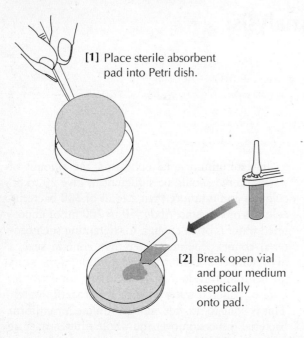

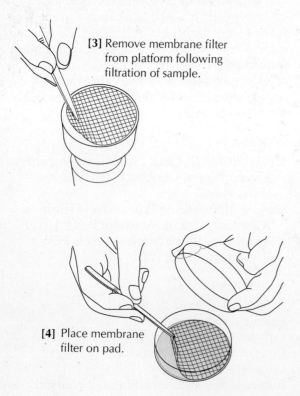

Water samples (100 ml each): tap water, well water, surface water, or swimming pool water (Assigned by instructor to different students.) (Equipment and materials, as well as literature, for the MF technique are obtainable from Millipore Corp., Bedford, MA 01730.)

PROCEDURE

The membrane-filter technique is outlined in FIGURE 73.1.

PREPARATION OF PETRI DISH

1 Place a sterile absorbent pad in a 47-mm Petri dish with a pair of forceps sterilized by immersion in 95% ethanol and then flamed to burn away the alcohol.

2 Break open an ampule of Endo medium and pour the entire contents (2 ml) onto the absorbent pad.

FILTRATION AND INCUBATION

1 Set up a bacteriological membrane-filter assembly as shown in FIGURE 73.2. (The membrane-

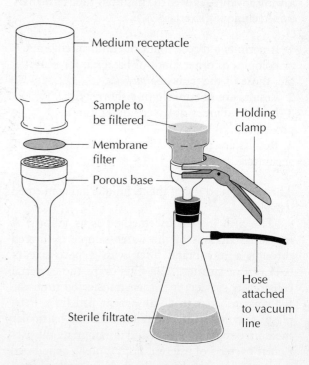

FIGURE 73.2
Bacteriological membrane-filter assembly.

filter unit covered with aluminum foil and its collection, or vacuum, flask should have been sterilized in the autoclave at 121°C for 15 min and then cooled to room temperature before use.)

2 Aseptically place a sterile membrane filter on the filtration platform of the filter assembly with the use of a pair of forceps (sterilized as before).

3 Filter 100 ml of water sample with the aid of vacuum.

4 Transfer the membrane filter with sterile forceps to the nutrient pad prepared previously.

5 Incubate the Petri plate, inverted, at 35°C for 22 to 24 h.

IDENTIFICATION AND COUNTING

Determine the number of coliform colonies per 100 ml of sample. (Coliform density is always expressed in terms of a 100-ml sample. If dilution of the sample was performed, the number of colonies must still be calculated for a 100-ml sample. Similarly, if less than 100 ml was filtered, the coliform density must also be expressed in terms of 100 ml.) The typical coliform colony has a pink to dark red color with a metallic sheen. [See PLATE 32.] The sheen area may vary in size from a small pinhead to complete coverage of the colony surface.

NAME _____

Standard Method of Water Analysis: Membrane-Filter Technique

RESULTS The total coliform count of the water sample (_____) analyzed

was: _____ per 100 ml. Show calculations below.

  NOTE: When there are excessive colonies on the membrane filter, the report should be "TNTC" ("too numerous to count"). If there is growth without well-defined colonies, the report should be "confluent." In either case, a new sample should be requested and more appropriate volumes selected or dilutions made for filtration.

QUESTIONS 1 Can you think of any advantages of the MPN technique over the MF technique?

2 The exercise as performed gave you the standard total coliform count. What other counts can be obtained by the MF technique?

EXERCISE 74

Bioluminescence

OBJECTIVE
To become acquainted with light emission by bacteria.

OVERVIEW
Some bacteria have the ability to emit blue-green light as a result of their metabolism. These bacteria include certain marine species belonging to the genera *Vibrio* and *Photobacterium*, which require Na^+ for growth, and also the species *Xenorhabdus luminescens*, a nonmarine parasite of nematode worms. Light emission by these bacteria depends on a minor bypass of the normal bacterial respiratory electron-transport chain. The sequence of reactions occurs as follows:

1 Oxidation of a reduced substrate (such as glycerol), with transfer of the electrons to nicotinamide adenine dinucleotide (NAD) and then to flavin mononucleotide (FMN).

2 The reduced FMN ($FMNH_2$) participates in a reaction involving oxygen and a long-chain aldehyde (RCHO; probably tetradecanal). The reaction is catalyzed by an enzyme named, appropriately, *luciferase*:

$$FMNH_2 + O_2 + RCHO \longrightarrow FMN + H_2O + RCOOH + light$$

In the above reaction, the chemical energy liberated by oxidation of the $FMNH_2$ is ultimately expressed as light energy rather than being conserved as ATP. (In a sense, bioluminescence is the reverse of photosynthesis; in the latter, light energy is ultimately converted into chemical energy.)

REFERENCES
MICROBIOLOGY, Chap. 9, "The Major Groups of Procaryotic Microorganisms: Bacteria."
MICROBIOLOGY, Chap. 29, "Microbiology of Natural Waters, Drinking Water, and Wastewater."
Nealson, K. H., and J. W. Hastings, "Bacterial Luminescence: Its Control and Ecological Significance," *Microbiological Reviews* 43: 496–518, 1979.

MATERIALS
2 250-ml Erlenmeyer flasks (cotton-stoppered) each containing 100 ml of *Photobacterium* medium
Photobacterium phosphoreum ATCC 11040 cultured in *Photobacterium* medium at 15 to 20°C
Incubator at 15 to 20°C

PROCEDURE
1 Inoculate two flasks of *Photobacterium* medium from the stock culture and incubate at 15 to 20°C.
2 Examine daily for 3 days to determine:
   a When maximum luminescence occurs
   b When maximal growth occurs
Take the flasks into a totally dark room and wait until your eyes become adjusted to the darkness before determining luminescence. Agitate each flask to obtain maximum brightness.

Bioluminescence

RESULTS   Record your observations in the chart provided.

GROWTH AND LUMINESCENCE IN THE BACTERIUM *PHOTOBACTERIUM PHOSPHOREUM*							
Amount of growth				Degree of luminescence			
Flask	Day 1	Day 2	Day 3	Flask	Day 1	Day 2	Day 3
1				1			
2				2			

+++ = abundant; ++ = moderate; + = slight; – = negative.

QUESTIONS   **1** When is luminescence best seen—after 1, 2, or 3 days? _____
Is there a correlation between cell density and degree of luminescence?

2 In statically incubated flasks, luminescence occurs only at the surface of the medium; yet when the flasks are swirled, the entire culture exhibits luminescence. Explain.

3 Where does *P. phosphoreum* occur in nature, and why is Na^+ required for its growth?

INDUSTRIAL AND APPLIED MICROBIOLOGY

Microorganisms have several characteristics which make them attractive to exploit for some practical benefits in terms of industrial processes and products. These may be summed up as follows:

1 Microorganisms have a tremendous capability for both the dissimilation of a wide variety of chemical substances (substrates) as well as for synthesis of a wide array of chemicals. Some examples of valuable products of microbial synthesis include enzymes, vitamins, antibiotics, and polysaccharides. Examples of dissimilation products include acids (lactic acid, acetic acid); alcohols (ethanol, butanol); and gases (carbon dioxide, methane).

2 Microorganisms grow rapidly, and the conditions for growth and metabolism can be manipulated for the maximum production of the desired product. For example, in the commercial production of monosodium glutamate (a food condiment), the yield of this substance can be greatly enhanced by limiting the biotin content of the medium. Other manipulations of the physical or chemical characteristics of the medium make it possible to improve the efficiency of the process.

3 Recent advances in recombinant DNA technology or genetic engineering make it possible to develop new strains of microorganisms that have the ability to produce a new substance. For example, a strain of *Escherichia coli* has been "genetically engineered" to produce human insulin.

4 Microorganisms are used for the preparation of many foods, such as fermented milk products (cheese, yogurt, buttermilk); fermented cucumbers, olives, and cabbage; bread; and soya and other sauces. Microorganisms are the key to the production of alcoholic beverages such as beer, wine, and liquors.

5 Developments in a process called *immobilized enzyme* (or *cell technology*) have increased the attractiveness of using microorganisms or their enzymes to bring about desired chemical changes in a substrate. FIGURE XIV.1 shows the principle of this technique which is based on keeping the microorganisms or enzymes stationary and allowing the substrate to make contact and be acted upon as it flows down a column.

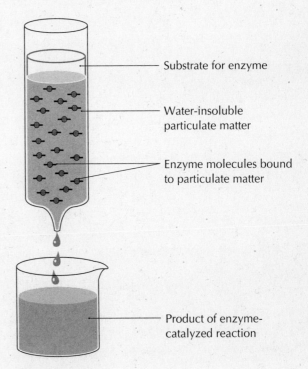

FIGURE XIV.1
Immobilized enzyme technology. The enzyme is bound to some inert material; it stays "fixed" as the substrate passes over it and is acted upon by the enzyme. This technique provides for an efficient use of enzymes or microbial cells.

Besides beneficial effects, microorganisms also have harmful effects. One dimension of the quality of food is its microbial content; this is particularly true of fresh, perishable foods like seafood, meats, and vegetables. For example, if shellfish are harvested from a polluted river, they may be contaminated with pathogenic microorganisms and be responsible for transmission of a disease. In another instance a food like milk that has an excessively high microbial population may be the result of faulty pasteurization, unsanitary conditions of production, or unsatisfactory holding conditions. Hence standard procedures for the determination of the numbers and kinds of microorganisms in foods have been developed.

EXERCISE 75

The Enumeration of Bacteria in Raw and Pasteurized Milk

OBJECTIVE
To carry out pasteurization of milk and enumerate bacteria in raw and in pasteurized milk.

OVERVIEW
Milk is well known for its exceptional nutritive value. It contains proteins, carbohydrates, fats, vitamins, and minerals. It has been called "nature's most nearly perfect food for humans." Milk is also an excellent medium for the growth of many bacteria. If undesirable bacterial growth is permitted, the milk will soon smell, look, and taste bad. When milk is produced within the udder, it is sterile. However, milk soon acquires an initial microbial flora at the time it is drawn because all mammals have microorganisms growing at the teat opening and within the teat canal. This is the reason that *first milk* has many bacteria and should be discarded. When milk is collected from the cow, all utensils employed in handling the milk serve as potential sources of additional microorganisms. The outside of the udders and teats may be tainted with manure, soil, and other materials, so that they are also a source of extraneous microbes. The air in the milking area may also contain many microorganisms that can contaminate milk. Therefore, cleanliness of utensils, the animal, and the environment is of paramount importance during the process of milking.

Also of importance is the state of health of the animals and of the humans performing the milking process. Otherwise, pathogens may be introduced into the milk. Diseases transmissible through milk include tuberculosis, brucellosis, typhoid fever, diphtheria, Q fever, and dysentery. However, the industry of milk and dairy products is so well regulated and surveyed by state (provincial) and federal agencies, and the production techniques are so sophisticated, that disease transmission by milk in North America is a rarity (although it still occurs). Information on the bacterial content of a milk sample may reflect the state of health of the cow and the conditions under which the milk was produced and stored. The *standard plate count* is one of the routine procedures widely used to enumerate the number of bacteria in milk. It is the most reliable indication of the milk's sanitary quality. A high count means that there is a greater likelihood of disease transmission and spoilage of the milk. It is for this reason that the American Public Health Association recognized the standard plate count as the official method in its *Milk Ordinance and Code*.

Milk sold commercially is *pasteurized*; this is the heat treatment that destroys all *pathogenic* microorganisms in milk. The minimal time-temperature conditions required for the pasteurization of market milk are:

62.8°C for 30 min (LTH, or low-temperature-holding pasteurization)
71.7°C for 15 s (HTST, or high-temperature–short-time pasteurization)
137.8°C for 2 s (ultrapasteurization)

These temperature and time combinations are used because the most heat-resistant pathogen likely to occur in milk, *Coxiella burnetii* (the etiologic agent of Q fever), does not survive these treatments.

In this exercise you will pasteurize milk and do a standard plate count on raw and pasteurized milk samples.

REFERENCES
MICROBIOLOGY, Chap. 30, "Microbiology of Food."
Richardson, G., ed., *Standard Methods for the Examination of Dairy Products*, 15th ed., American Public Health Association, Washington, D.C., 1985.

MATERIALS

250 ml of standard-methods agar
6 tubes of litmus milk
9 sterile Petri dishes
6 sterile 1.1-ml pipettes (or 12 1-ml pipettes)
1 sterile 10-ml pipette
4 99-ml dilution blanks
2 9-ml dilution blanks
1 sterile test tube
Water-bath and thermometer
Colony counter, for example, Quebec colony counter
Approximately 20-ml raw-milk sample in screw-cap tube
Small container of pasteurized milk purchased from local store

PROCEDURE

PASTEURIZATION OF MILK SAMPLE

1 Adjust a water-bath to a temperature of 62.8°C.
2 Shake the sample of raw milk vigorously to obtain uniform distribution of microorganisms. Use a sterile 10-ml pipette to transfer 10 ml of the raw-milk sample into a sterile tube, and place this in the water-bath maintained at 62.8°C. Allow this sample to remain in the water-bath for 30 min. During this time, make sure that the temperature is held at 62.8°C, and occasionally shake the tube of milk. At the end of 30 min, remove the tube of milk and cool it immediately by holding it under cold running tap water.

THE STANDARD PLATE COUNT

1 Prepare and plate 1:10, 1:100, and 1:1000 dilutions of milk sample that you pasteurized (see Exercise 19). Follow the protocol in FIGURE 75.1 for making dilutions and plating the dilution samples into the plates.
2 Add approximately 25 ml of melted, cooled (45 to 50°C) agar medium to the samples in each plate. Immediately upon addition of the medium, rotate each plate gently to effect uniform distribution of the samples.
3 Repeat this plating procedure with the pasteurized milk purchased locally.
4 Prepare and plate 1:100, 1:1000, and 1:10,000 dilutions of the raw-milk sample as shown in FIGURE 75.2. Add medium to each of these plates as described in step 2.
5 Incubate all plates at 35°C for 48 h.

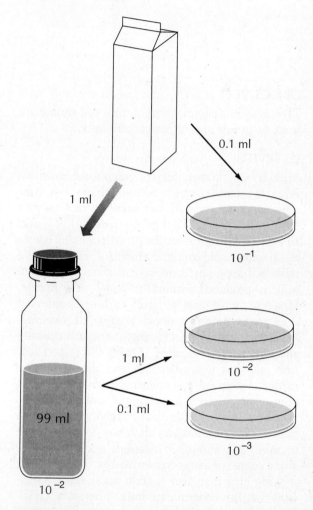

FIGURE 75.1
Procedure for diluting and plating pasteurized milk.

FIGURE 75.2
Procedure for diluting and plating raw milk.

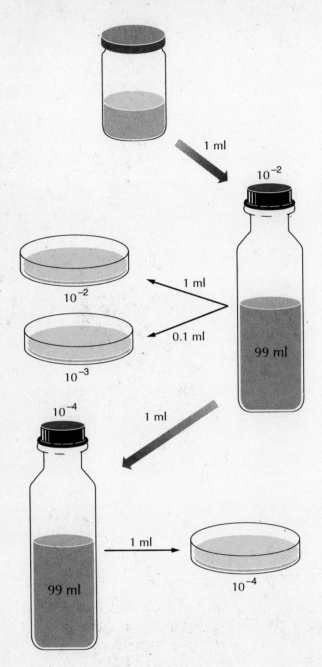

NAME _____

The Enumeration of Bacteria in Raw and Pasteurized Milk

RESULTS

1 Examine the plates and choose those containing between 30 and 300 colonies for counting. Use a colony-counting device, if available, to facilitate the counting. Calculate the standard plate count (as colony-forming units, CFU) per milliliter of milk by multiplying the number of colonies counted by the dilution that yielded the desirable number of counts. Enter the data and results in the chart provided.

STANDARD PLATE COUNTS OF MILK SAMPLES ANALYZED		
Sample	Colony counts from various dilutions	Colony-forming units per milliliter
Raw milk		
Pasteurized raw milk		
Commercial milk (pasteurized)		

2 Select one plate from the raw-milk series and one from the pasteurized raw-milk series that have well-isolated colonies. Make and examine a Gram-stain preparation from each colony. Transfer a portion of each colony into a tube of litmus milk. Incubate these tubes from 2 to 4 days at 35°C and record their reactions to the inoculated bacteria. Record these results in the chart provided on the following page.

	CHARACTERISTICS OF SOME BACTERIA IN RAW-MILK AND PASTEURIZED RAW-MILK SAMPLES		
Sample	Colonial morphology	Cell morphology in Gram reaction	Reaction in litmus milk
Raw milk			
Pasteurized raw milk			

QUESTIONS

1 Calculate the percentage reduction of bacteria accomplished by pasteurization of the milk sample.

2 Is pasteurized milk sterile? Explain.

3 Explain why it is more appropriate to report the plate count as the number of colony-forming units (CFU) rather than as the number of bacteria per milliliter.

EXERCISE 76

Microbiological Examination of Foods

OBJECTIVE
To determine the bacterial content of a food sample.

OVERVIEW
Results of microbiological analyses of foods provide information on the quality of raw material, the cause of spoilage if it has occurred, the present condition of the food, and whether it is the source of pathogenic microorganisms. A single method of analysis, that is, one medium and one condition of incubation, is not suitable for all samples. The method of choice will depend on the purpose of the investigation. For example, a microbiologist looking for food pathogens that are transmissible to humans might use *selective media* to allow for growth of one type of pathogenic microbe while inhibiting the growth of other microorganisms. He or she may also use *differential media* to make the pathogenic bacterial colonies look different from those of other types of bacteria on a solid agar medium.

However, the standard plate count is used to determine the total number of viable bacteria (colony-forming units, CFU) in a food sample. Standard plate counts are routinely performed on food by food-processing companies and public health agencies. The presence of large numbers of bacteria in fresh foods is undesirable because it increases the likelihood that pathogens might be present and also increases the potential for spoilage of the food.

Food may be contaminated in a variety of ways and from many sources. These include:

1 *The manner of preparation and storage:* For example, if the water used for washing the food is not clean, then spoilage and even pathogenic microorganisms may be added; where the food needs much processing, such as the mincing of meat to make hamburger, there are numerous opportunities for contamination to occur unless strict aseptic measures are followed. Refrigeration and freezing are good procedures to prevent growth of unwanted microorganisms.

2 *Food utensils:* If they are reused without having been properly sanitized, they are a source of microorganisms.

3 *Food handlers:* Those with poor personal hygiene and unsanitary habits are likely to contaminate foods with enteric and other pathogenic microorganisms, for example, staphylococci from the skin, respiratory tract, and open wounds.

REFERENCES
MICROBIOLOGY, Chap. 30, "Microbiology of Food."

Vanderzant, C., and D. Splittsoesser, eds., *Compendium of Methods for the Microbiological Examination of Foods*, 3rd ed., American Public Health Association, Washington, D.C., 1992.

MATERIALS
150 ml of Standard-methods agar
1 tube of litmus milk
1 tube of lactose broth
1 tube of nutrient gelatin
Scalpel, forceps, and spatula
Sterile squares of weighing paper in a Petri dish
Balance
Dilution blank: 9 ml of sterile saline in 100-ml screw-cap bottle containing enough small glass beads to form about two layers over the bottom
2 99-ml dilution blanks
4 sterile 1.1-ml pipettes or 8 sterile 1-ml pipettes
9 sterile Petri dishes
Food samples, for example, fresh hamburger, ham, fresh fish, frozen fish, canned food

PROCEDURE

1 Select one of the food samples for analysis.

NOTE: Before any manipulations are started, be sure that all necessary equipment is at hand for sampling of the specimen. Aseptic precautions must be observed. The scalpel, forceps, and spatula should be dipped in ethanol, flamed, and cooled before they are used to obtain your sample.

2 Weigh a 1-g portion of the food specimen, and transfer it to the 9-ml dilution blank that contains the glass beads. Shake vigorously to disintegrate the sample. (Some microbiologists prefer to blend the solid foods in a sterile blender with a measured volume of sterile saline for analysis.)

3 Prepare and plate further dilutions of 1:1000, 1:10,000, and 1:100,000 as follows:

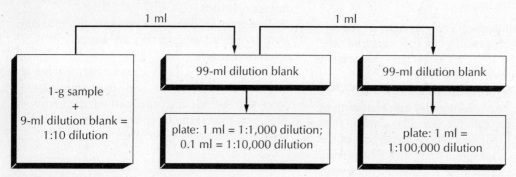

NOTE: Make triplicate platings of each dilution.

4 Incubate one set of plates in a refrigerator (approximately 7°C) for 7 days, one set at 35°C for 48 h, and one set at 55°C for 24 to 48 h.

5 Inoculate the tubes of litmus milk, lactose broth, and gelatin with a loopful of material from the 1:10 dilution. Incubate at 35°C for 2 to 4 days.

NAME _____

Microbiological Examination of Foods

RESULTS 1. Make colony counts from each of the plates that have isolated colonies. Calculate the standard plate count per gram of the food sample. Note changes in the litmus milk, gelatin, and lactose broth. Enter these results in the chart provided below.

MICROBIAL POPULATIONS AND CHARACTERISTICS OF ISOLATES FROM FOOD SAMPLE						
Incubation	Dilutions with countable colonies	CFU per plate	Standard plate count per gram	Litmus milk changes	Lactose broth changes	Gelatin changes
7°C for 7 days						
35°C for 48 h						
55°C for 24–48 h						

Food sample analyzed:

2 Prepare and examine Gram stains of representative colonies that developed at each temperature. Record descriptions of the microorganisms in the chart provided below.

DESCRIPTION OF MICROORGANISMS FROM REPRESENTATIVE ISOLATED COLONIES AT EACH TEMPERATURE		
Temperature of incubation of colony	Gram stain and morphology of microorganisms	Description of colony

Food sample analyzed:

NAME _____ 76 MICROBIOLOGICAL EXAMINATION OF FOODS *(Continued)*

QUESTIONS 1 From the results of the changes produced in the inoculated tubes of litmus milk, gelatin, and lactose broth, what can you conclude as to the biochemical types of microorganisms that are present in the food sample examined?

2 What incubation conditions would be appropriate for microbiological examination of a spoiled canned-food sample? Explain why.

3 In the microbiological examination of shellfish, the detection of coliforms is stressed. Why?

EXERCISE 77

Preparation of Fermented Foods: Sauerkraut and Yogurt

OBJECTIVE
To prepare various items of food by microbial fermentations.

OVERVIEW
Microbial fermentation is a time-honored method for the preparation and preservation of certain foods. Sauerkraut, brined pickles, pickled olives, soy sauces, yogurt, and buttermilk are but a few of the products of microbial fermentation. In addition to the feature of preservation, the chemical changes that occur during fermentation add flavor and aroma to the end product which has appeal to the consumer. Furthermore, and particularly in the case of buttermilk, acidophilus milk, and yogurt, the ingestion of large numbers of lactic-acid bacteria in these products is claimed to have some beneficial effect on one's health.

The bacteria responsible for the fermentation of a food, for example, cabbage or cucumbers, are normally present on these vegetables. The cultural conditions for the fermentation favor the growth of the desired types of microorganisms. For production of other fermented foods, such as yogurt, acidophilus milk, and some cheeses, it is customary to use *starter cultures* to initiate the fermentation. The starter culture contains large numbers of the specific microorganisms needed for the fermentation to result in a high-quality product. Starter cultures can be purchased from companies that specialize in making these products.

The bacteria responsible for the fermentation of shredded cabbage to produce sauerkraut are mainly lactobacilli and leuconostocs. These bacteria ferment the sugars in the cabbage juice and produce a variety of products that include lactic acid, acetic acid, ethanol, mannitol, and carbon dioxide. In addition to providing the flavor and aroma of the finished product, some of these substances contribute to the keeping quality of sauerkraut.

The fermentation of milk for the production of yogurt is brought about mainly by lactic-acid bacteria, particularly streptococci and lactobacilli. The production of lactic acid from the sugar (lactose) in the milk by these bacteria results in the coagulation of milk proteins—a custardlike curd is formed.

REFERENCE
MICROBIOLOGY, Chap. 30, "Microbiology of Food."

EXERCISE 77.A

Preparation of Fermented Foods: Sauerkraut and Yogurt
77.A Preparation of Sauerkraut

MATERIALS
One head of cabbage (approximately 2 to 3 lb)
NaCl
Plastic container, 1- to 2-gal size
Shredder
Large knife
Balance
pH meter

PROCEDURE

1 Remove the outer leaves of cabbage. Cut the head of cabbage into quarters, and remove the core portion from each quarter. Determine the total weight of the four quarters.

2 Select a clean circular plastic container that is approximately 15 to 20 cm in diameter and 30 to 40 cm in height. Shred the cabbage into the container in a manner that layers of shredded cabbage (approximately 100 g per layer) are interspersed with layers of salt (3 g) as shown in FIGURE 77.1. The total amount of salt used should be approximately 3 percent of the weight of the cabbage, and it should be equally divided among the layers of shredded cabbage.

3 Press down firmly on the column of shredded cabbage so that a layer of juice rises to the top. The shredded cabbage must be tightly packed to eliminate as much air as possible. Remove a 10-ml sample of the juice for test purposes as called for under the results section. Place a cover (something like a saucer or plate) on top of the shredded cabbage, and weigh it down to keep the cabbage tightly packed. Incubate at room temperature (21 to 25°C) for 3 weeks.

FIGURE 77.1
Container with layers of shredded cabbage and salt for the preparation of sauerkraut.

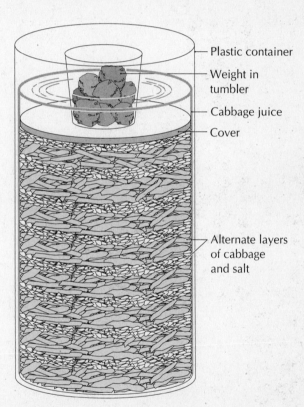

NAME _____

77.A

Preparation of Sauerkraut

RESULTS

1 At intervals of 0, 2, 4 to 6 days, then 1, 2, and 3 weeks, remove a 5-ml sample for the following tests: pH, flavor, aroma, and appearance.
2 Prepare and examine a methylene-blue-stained film of each sample.

0 days   2 days   4 to 6 days

pH: _____ _____ _____

Flavor: _____ _____ _____

Aroma: _____ _____ _____

Appearance: _____ _____ _____

1 week   2 weeks   3 weeks

pH: _____ _____ _____

Flavor: _____ _____ _____

Aroma: _____ _____ _____

Appearance: _____ _____ _____

QUESTIONS 1 What fermentation products contribute flavor to sauerkraut?

2 What bacterial species is the key to successful fermentation of cabbage to sauerkraut?

3 Sauerkraut has better keeping quality than raw shredded cabbage. Explain why this is so.

EXERCISE 77.B

Preparation of Fermented Foods: Sauerkraut and Yogurt
77.B Preparation of Yogurt

MATERIALS

Fresh plain yogurt (or commercial starter culture)
Fresh skim milk (100 ml)
Powdered, nonfat, skim milk (3 g)
LBS (*Lactobacillus* selection) medium
Beaker, 250 ml
Plastic cup
Hot plate
Thermometer
pH meter

PROCEDURE

1 Add 100 ml of fresh skim milk to a 250-ml beaker. Check and record the pH of the milk with a pH meter. With the use of a hot plate, heat this to 85°C for 15 min. During this heating period, stir the milk gently, and keep close check on the temperature with a nonmercury thermometer placed in the beaker.

2 After the 15-min heating at 85°C, remove the beaker, and add 3 g powdered nonfat skim milk and stir to dissolve.

3 Cool the milk to 40 to 42°C, and add either yogurt starter culture as suggested by your instructor or 2 teaspoons of fresh plain yogurt per 100 ml of milk.

4 Pour the contents of the beaker into a plastic cup, cover with foil, and incubate at 40 to 42°C from 18 to 24 h.

5 At the end of the incubation period, check the pH of the product (yogurt) and observe it for appearance, particularly the consistency of the curd, flavor, and aroma.

NAME _____

77.B

Preparation of Yogurt

RESULTS

1 Record the pH, physical appearance (consistency), flavor, and aroma of the product.

pH of milk:

pH of product:

Appearance of product:

Flavor of product:

Aroma of product:

2 Prepare and examine a Gram stain made from a sample of the product as well as from the starter culture or the purchased fresh, plain yogurt.

Yogurt starter culture, or fresh, plain yogurt

Laboratory-prepared yogurt

501

3 Prepare an agar-streak plate and incubate it for 48 to 72 h, preferably in a carbon dioxide incubator or a candle jar. After incubation, prepare and examine Gram-stained smears from different colonies.

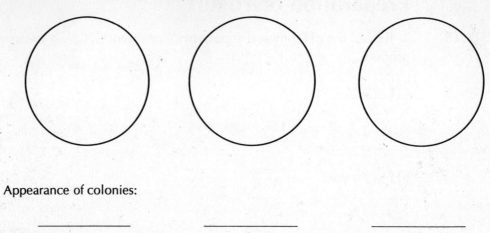

Appearance of colonies:

_____   _____   _____

QUESTIONS **1** Name the species of bacteria that are present in commercial yogurt starter culture.

2 Name the metabolic products of the starter culture bacteria that contribute to the flavor, aroma, and physical characteristics of yogurt.

EXERCISE 78

Fermentation of Grape Juice to Produce Wine

OBJECTIVE
To produce wine from grape juice.

OVERVIEW
Grape wine is the product of the fermentation of the sugars (glucose and fructose) in grape juice. The end products of this fermentation are ethanol and carbon dioxide according to the following reaction:

$$\underset{\substack{\text{Glucose/}\\\text{fructose}}}{C_6H_{12}O_6} \xrightarrow{\text{yeast cells}} 2\underset{\text{Ethanol}}{CH_3CH_2OH} + 2\underset{\substack{\text{Carbon}\\\text{dioxide}}}{CO_2}$$

Although we generally think of grapes as the source of fermentable sugars for wine making, other fruits such as cherries, apples, and berries may also be used.

There are two basic types of wine—red and white. Red wines are made from red grapes that are crushed with skins remaining in the fermentation mixture. White wines are prepared from either red or white grapes; however, the grape skins are removed prior to the fermentation. In natural grape fermentations the alcohol produced ranges between 11 and 14 percent by volume. Fluctuation in the amount of alcohol produced is due to the initial level of grape sugar as well as the conditions during the fermentation such as variations in temperature and time. The following general procedure for making red table wine at home is shown in FIGURE 78.1.

Step 1: The crush-stem process removes the grape stems, crushes the grapes, and recovers the skins and juice which is called *must*.

Step 2: The must is poured into plastic containers to approximately two-thirds full to avoid foaming over during fermentation. Wine yeast-starter culture is added to the must at this point. The

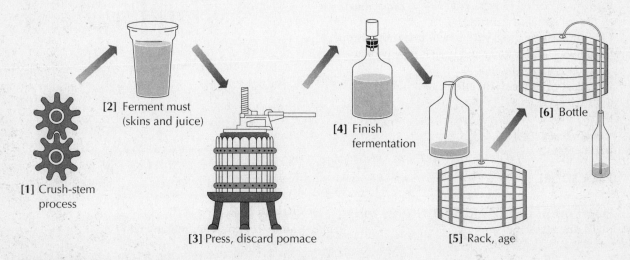

FIGURE 78.1
Essential steps in the making of red wine.

temperature for incubation ranges between 60 and 80°F (15.6 and 26.7°C). Initially, incubation is in the 60 to 70°F range, and in the latter stages the temperature is raised to 75 to 80°F.

Step 3: After several days' fermentation, the ferment juice is squeezed from the must either through several layers of cheesecloth or by use of a basket press.

Step 4: The fermented juice (press-wine) is poured into narrow-necked glass containers fitted with a fermentation trap. The trap allows carbon dioxide to escape while preventing air from entering. The containers are incubated at 65 to 75°F (18.3 to 23.9°C) for approximately 2 to 3 weeks to complete the fermentation. During this time yeast cells and particulate matter settle to the bottom.

Steps 5 and 6: In the final stages, the clear supernatant wine is siphoned off (a process called ***racking***), stored in barrels, and subsequently bottled.

REFERENCES

MICROBIOLOGY, Chap. 31, "Biotechnology: The Industrial Applications of Microbiology."

Cook, George M., and James T. Lapsley, *Making Table Wine at Home*, Division of Agriculture and Natural Resources, University of California, Oakland, Calif. (Publication 214340), 1988.

MATERIALS

Frozen grape juice concentrate sufficient to make 500 ml of reconstituted fresh juice
Dextrose broth
Culture of *Saccharomyces cerevisiae* var. *ellipsoideus* (25 ml of a 48-h culture)
One-liter bottle fitted with a two-hole rubber stopper with glass tubing and rubber tube connections as shown in FIGURE 78.2
1 graduate cylinder (100 ml)
4 10-ml pipettes
4 glass slides
4 dextrose agar plates
pH meter

PROCEDURE

1 Dissolve the contents of a 6-oz can of frozen grape juice in approximately 500 ml of tap water in a 1-liter beaker.

2 Inoculate the reconstituted grape juice with 25 ml of the *Saccharomyces cerevisiae* var. *ellipsoideus* culture. Gently agitate the grape juice to distribute the yeast cells uniformly. Remove a 10-ml sample for analysis as outlined in step 4 of this procedure.

3 Pour the inoculated grape juice into the 1-liter fermentor bottle which has been set up as shown in FIGURE 78.2. Incubate the fermentor bottle at 20 to 25°C.

4 At intervals of 2, 4, 7, and 14 days, remove a 10-ml sample and perform the tests listed below:

   a Determine the pH of each sample.

   b Make a "wet preparation" as described in Exercise 5, and estimate the number of yeast cells per microscopic field.

   c Prepare a streak plate on dextrose agar, incubate for 24 to 48 h and observe for colony characteristics.

   d Check the aroma and flavor of each sample.

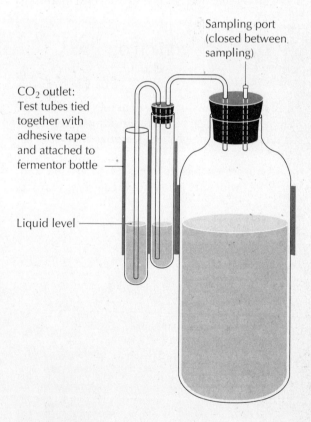

FIGURE 78.2
Fermentor bottle for making wine.

NAME _____

Fermentation of Grape Juice to Produce Wine

RESULTS Tabulate your results in the chart provided.

RECORD OF CHANGES DURING FERMENTATION OF GRAPE JUICE				
Type of change	Days of incubation			
	2	4	7	14
pH				
Yeast cell population estimate: Cells/field				
Characterization of colonies				
Aroma				
Flavor				

QUESTIONS 1 In the context of wine making, what is the meaning of the following terms?
Must:

Racking:

Starter culture:

2 What general observations can you make with respect to the changes that occurred in the grape juice from the beginning to the end of the fermentation?
pH:

Yeast population:

Appearance:

Aroma:

3 What is the difference in starting materials between a red wine and a white wine fermentation?

4 What is the usual concentration of alcohol that is produced in home-made table wine? What controls this?

EXERCISE 79

The Isolation of Antibiotic-Producing Microorganisms

OBJECTIVE
To search for and to isolate an antibiotic-producing microorganism.

OVERVIEW
One of the major industrial applications of microorganisms is their use in the manufacture of antibiotics. Hundreds of antibiotics have been discovered, and many of these are available for treatment of infections. However, the search goes on for other more effective antibiotics. The process of discovering a new antibiotic and developing it to the stage of being approved for therapeutic use is long and complex. Nevertheless, the first step in the process is to isolate a microorganism that produces an antibiotic substance. Past experience has shown that soil is a good source of potential antibiotic producers. In this exercise you will screen several cultures isolated from soil for their antibiotic activity.

REFERENCES
MICROBIOLOGY, Chap. 21, "Antibiotics and Other Chemotherapeutic Agents."
MICROBIOLOGY, Chap. 31, "Biotechnology: The Industrial Applications of Microbiology."

MATERIALS
FIRST LABORATORY PERIOD
Rich garden soil, approximately 10 g
Nutrient agar, approximately 200 ml
Dilution blank, 99 ml
6 Petri dishes
Sterile glass spreader
1-ml pipettes
Balance

SECOND LABORATORY PERIOD
4 Petri dishes
Nutrient agar, approximately 100 ml

THIRD LABORATORY PERIOD
4 Petri dishes inoculated in previous period
Diluted suspensions of test cultures *Staphylococcus aureus, Escherichia coli, Branhamella catarrhalis,* and *Saccharomyces cerevisiae*

PROCEDURE
FIRST LABORATORY PERIOD

1 At the beginning of the laboratory period, pour nutrient agar into each of six Petri dishes (approximately 25 ml per dish), and allow the medium to solidify. Label these 1, 1a, 2, 2a, 3, and 3a.

2 Weigh out approximately 1 g of soil and add it to a 99-ml dilution bottle. Shake this bottle vigorously. Make further dilutions from this suspension by transferring 1 ml to a second 99-ml dilution blank followed by 1 ml from the second dilution to a third 99-ml blank. This procedure is illustrated in FIGURE 79.1.

3 Using a 1.0-ml pipette, deliver 0.1 ml from dilution blank 1 on to the center surface of nutrient-agar plate 1. With the use of a sterile glass spreader, spread this inoculum uniformly over the entire surface (see Exercise 19); inoculate a second plate of nutrient agar (1a) *with this same glass spreader* (no additional inoculum). (**NOTE:** The purpose is to *dilute the inoculum* so that well-isolated colonies will develop on one or more plates. It is not the purpose of this exercise to make a quantitative determination of the microbial flora.)

4 Repeat step 3 with each of the other two dilutions, using plates 2 and 2a for the second dilution

FIGURE 79.1
Preparation of soil sample dilutions and inoculation of spread plates.

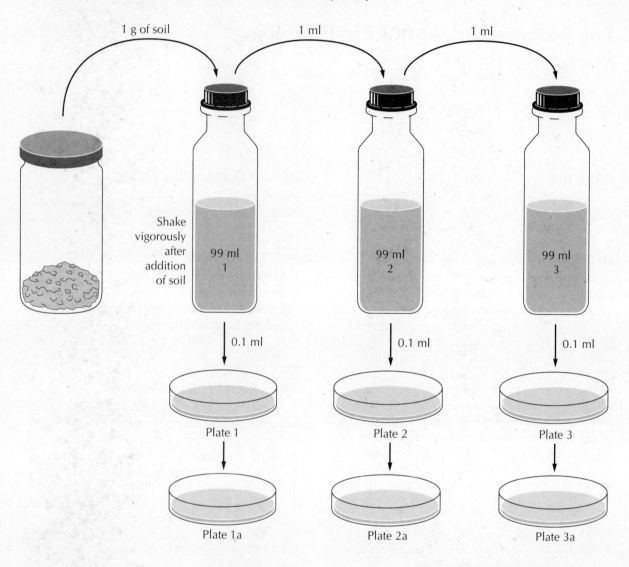

blank and plates 3 and 3a for the third dilution blank.
5 Incubate the inoculated plates at room temperature for 2 to 4 days.

SECOND LABORATORY PERIOD
1 At the beginning of the laboratory period, pour nutrient agar into four Petri dishes (approximately 25 ml per dish), and allow the medium to solidify.
2 Observe each of the plates that you inoculated with dilutions of soil during the previous laboratory period. Identify four well-isolated colonies, and with your marking pen, draw a circle on the bottom of the dish to indicate the location of each, and number the colonies 1 through 4.
3 Using your loop transfer needle, touch colony 1 and make a streak inoculation across the center of the plate. Label this plate 1. Repeat this process for colonies 2, 3, and 4. Incubate these plates for 2 to 3 days at room temperature.

THIRD LABORATORY PERIOD
1 Make a streak inoculation of each of the broth test cultures provided perpendicular to the streak of growth resulting from previous inoculation

79 THE ISOLATION OF ANTIBIOTIC-PRODUCING MICROORGANISMS

FIGURE 79.2
Cross-streaking with microbial culture suspensions (test cultures) on plates with prior growth of soil isolate.

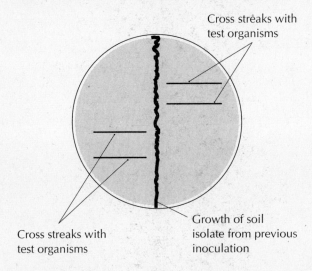

(during second laboratory period) with each colony as shown in FIGURE 79.2. Do not allow the transfer needle loop to touch the line of growth on the agar medium.

2 Incubate the four inoculated Petri dishes for 24 to 48 h, after which you should observe them to see whether growth of any of the test cultures are inhibited in the vicinity of streaked growth from the soil colonies (isolates).

3 Prepare and observe the Gram stains from the growth of each of the four soil colony types used in this exercise.

NAME _____

The Isolation of Antibiotic-Producing Microorganisms

RESULTS

1 Make a sketch of each of the four plates to illustrate the pattern of growth of each test culture.

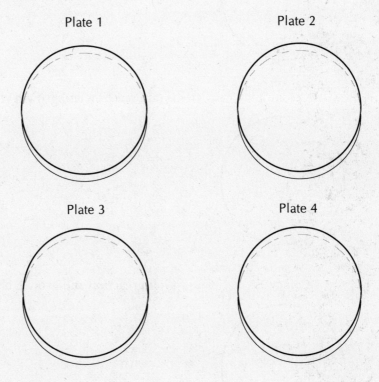

2 Was there any evidence of antibiotic activity on any of the plates?

3 Describe the Gram reaction, cell morphology (sketch several cells), and growth characteristics of the four soil colonies selected for testing antibiotic production.

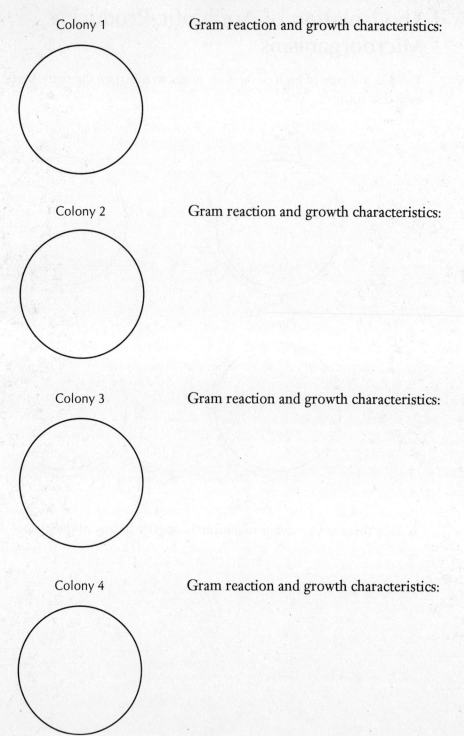

Colony 1 — Gram reaction and growth characteristics:

Colony 2 — Gram reaction and growth characteristics:

Colony 3 — Gram reaction and growth characteristics:

Colony 4 — Gram reaction and growth characteristics:

NAME _____ 79 THE ISOLATION OF ANTIBIOTIC-PRODUCING MICROORGANISMS *(Continued)*

QUESTIONS 1 What genera of microorganisms are frequently associated with production of antibiotics?

2 In this exercise, the assumption is made that if the growth of a test organism is inhibited in the vicinity of the streaked growth of the soil isolate, the soil organism has produced an antibiotic substance. What other reasons might account for this inhibition?

3 Assume that one of the microorganisms from one of the colonies you selected from soil in this exercise gave evidence of strong antibiotic activity. What would be the next steps in evaluating the potential of this culture for producing a new therapeutic agent?

EXERCISE 80

The Production and Assay of a Microbial Enzyme (Penicillinase)

OBJECTIVE
To produce penicillinase from bacteria and to assay the potency of the enzyme.

OVERVIEW
Penicillinase is an extracellular enzyme that acts upon the penicillin molecule by hydrolyzing the bond linking the nitrogen atom to the adjacent carbonyl carbon of the β-lactam ring as shown in FIGURE 80.1. Penicillinase is classified as a β-lactamase enzyme.

Over the last several decades, strains of many species of bacteria have developed resistance to penicillin by virtue of acquiring the ability to produce this enzyme.

A practical use for this enzyme is its application in testing the sterility of a pharmaceutical product that contains penicillin. Another application is in the clinical laboratory where cultural tests are performed to detect microorganisms present in the blood or secretions of a patient on penicillin therapy. In both instances, sterile penicillinase is added to the specimen to inactivate penicillin. If this is not done, the residual penicillin may prevent the growth of microorganisms in the specimens.

Some bacteria, particular species of *Streptomyces* and *Bacillus*, produce extremely large amounts of this extracellular enzyme. In this exercise, you will use a bacterial culture to produce penicillinase and determine its potency for inactivating penicillin.

REFERENCE
MICROBIOLOGY, Chap. 21, "Antibiotics and Other Chemotherapeutic Agents."

MATERIALS
200 ml of nutrient broth in a 1000-ml Erlenmeyer flask
1- and 10-ml sterile pipettes
Membrane-filter apparatus
Fluid thioglycollate broth, 15 ml per tube
1% phosphate buffer, pH 7.0
Freshly prepared penicillin solution, 50,000 units/ml
Nutrient-agar culture of *Streptomyces albus* or *Bacillus cereus*

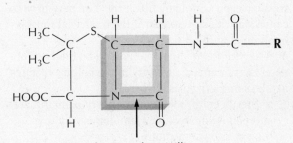

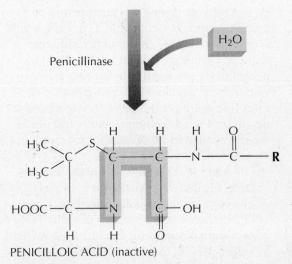

FIGURE 80.1

The mode of action of the enzyme penicillinase on penicillin.

Broth culture of *Staphylococcus aureus*
  (ATCC 6538P)
Rotary or reciprocating shaker
pH meter

PROCEDURE

1 Wash the growth off a 1-week-old nutrient-agar slant culture of *S. albus* (or a 48-h culture of *B. cereus*) with 3 to 5 ml of sterile nutrient broth. Use this to inoculate the 1-liter flask containing 200 ml of nutrient broth. Incubate the flask at 25°C on a shaking device.

2 Each day, over a 3-day period, remove a 10-ml sample from the flask culture; refrigerate the samples. On the fourth day, remove the flask from the shaker, and take another 10-ml sample. (Refrigerate the remainder of the culture in the flask.)

3 Prepare samples for assay of penicillinase activity. Determine the pH (use a pH meter) of each of the four samples; adjust pH to 7.2 if necessary; sterilize each sample by filtration. You may have to centrifuge the culture first if the microbial mass is likely to clog the filter.

4 Assay for penicillinase activity. The method for determining enzyme activity in the culture filtrates can be summarized as follows:

  a Sterile culture filtrate (source of penicillinase) and penicillin are added simultaneously to a tube of culture medium.

  b This mixture is incubated at 25°C for 2 h to allow enzyme action to occur.

  c The tube containing this mixture is then inoculated with *S. aureus* (ATCC 6538P), which is sensitive to about 0.1 unit of penicillin per milliliter; hence growth indicates no residual penicillin (i.e., the penicillin has been inactivated by penicillinase), and no growth indicates that penicillin still remains. Experimental details for testing the sterile culture filtrates for the ability to inactivate penicillin are shown in FIGURE 80.2.

  (1) Add 1.0, 0.5, and 0.1 ml of the sterile sample collected previously to individual tubes of fluid thioglycollate broth.

  (2) Prepare a sterile solution of sodium or potassium benzyl-penicillin in 1% phosphate buffer, pH 7.0, to contain 50,000 units/ml. (This solution must be prepared just before use.)

FIGURE 80.2

Procedure for estimation of penicillinase activity of culture filtrates.

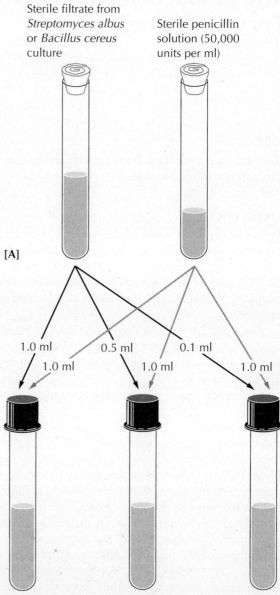

(3) Add 1 ml of this penicillin solution to each tube of thioglycollate broth that previously received culture filtrate.

(4) Shake the tubes vigorously to mix thoroughly. Allow to incubate at 25°C for 2 h. At the end of the 2-h period, inoculate each tube with 1 ml of a 1:100 dilution of a 24-h culture of *S. aureus*; incubate tubes at 37°C for 24 to 48 h; then observe for presence or absence of growth.

NOTE: The following controls should be included in this experiment: a tube of medium inoculated with each filtrate to ascertain sterility and a tube of medium inoculated with *S. aureus* (culture control).

5 If time is available, additional experiments, as suggested below, may be performed using the culture fluid remaining in the 1-liter flask; see step 2 above.

a Remove the microbial growth by filtration; precipitate the penicillinase by adding ammonium sulfate to saturation. Recover the enzyme, which is in the precipitate, by redissolving it in pH 7.0 phosphate buffer. Assay this enzyme preparation as previously described.

b The enzyme so recovered can be studied for its activity at various pHs and temperatures. To perform such experiments, one might allow the enzyme to be in contact with graded amounts of penicillin under specific conditions wherein temperature and/or pH would be the variable factor. The residual penicillin, after a period of exposure to the enzyme, could be determined as described in step 4 above.

NAME _____

The Production and Assay of a Microbial Enzyme (Penicillinase)

RESULTS

1 Describe the appearance of the culture in the shake flasks on each of the 4 days of incubation.

Day 1:

Day 2:

Day 3:

Day 4:

2 Tabulate results of the penicillinase assay in the chart provided.

\	RESULTS OF PENICILLINASE ASSAY		
Sample	Volume filtrate assayed	Growth, + or −	Penicillin inactivated, yes or no
Day 1	1.0 ml =		
	0.5 ml =		
	0.1 ml =		
Day 2	1.0 ml =		
	0.5 ml =		
	0.1 ml =		
Day 3	1.0 ml =		
	0.5 ml =		
	0.1 ml =		
Day 4	1.0 ml =		
	0.5 ml =		
	0.1 ml =		
+ = growth (penicillin inactivated); − = no growth (penicillin).			

80 THE PRODUCTION AND ASSAY OF A MICROBIAL ENZYME (Continued)

QUESTIONS

1 Assume that one of the inoculated penicillin-penicillinase culture tubes, as used in this exercise, showed no evidence of growth at 48 h but on further incubation, namely, after 7 days, growth of *S. aureus* did occur. How could this be explained?

2a Name other species of bacteria that produce penicillinase.

2b Are any of these pathogens?

2c If so, what does this mean in terms of their susceptibility to penicillin chemotherapy?

3 Discuss the importance and necessity of employing various controls in an experimental procedure such as the ones used in this exercise.

4 Why was a specific strain of *S. aureus* (ATCC 6538P) used in this exercise?

APPENDIXES

A List of Cultures and Sources of Cultures
B Stains, Staining Solutions, and Reagents
C List of Media and Sources of Media
D Cross Reference of Laboratory Exercises in This Manual to Chapters in Selected Introductory Microbiology Texts
E Audiovisual Aids Sources
F Selected Bibliography

APPENDIX A LIST OF CULTURES AND SOURCES OF CULTURES

For Fungi (Other Than Animal Pathogens and Wood-Rotting)

U.K. National Collection of Fungus Cultures
Commonwealth Mycological Institute
Ferry Lane
Kew
Surrey TW9 3AF
Tel.: 01-940-4086

Some other sources of microbial cultures (and prepared slides) are:

Carolina Biological Supply Company
2700 York Road
Burlington, NC 27214
Tel.: 919-584-0381 or 1-800-334-5551
FAX: 919-584-3399

Carolina Biological Supply Company
Powell Laboratories Division
19355 McLoughlin Boulevard
Gladstone, OR 97027
Tel.: 503-656-1641 or 1-800-547-1733
FAX: 503-656-4208

Connecticut Valley Biological Supply Company
82 Valley Road
(Post Office Box 326)
Southampton, MA 01073
Tel.: 413-527-4030 or 1-800-628-7748
FAX: 413-527-8286

Ward's Natural Science Establishment, Inc.
Post Office Box 1712
Rochester, NY 14603
Tel.: 716-359-2502
FAX: 716-334-6164

Ward's Natural Science, Limited
1840 Mattawa Avenue
Mississauga, Ontario
Canada L4X 1K1
Tel.: 416-984-8900
FAX: 416-984-5952

APPENDIX B

Stains, Staining Solutions, and Reagents

STAINS

Acid-Fast Stain

1 Carbol fuchsin (Ziehl's)
SOLUTION A
  Basic fuchsin (90% dye content), 0.3 g
  Ethyl alcohol (95%), 10 ml
SOLUTION B
  Phenol, 5 g
  Distilled water, 95 ml
Mix solutions A and B and let stand several days before use.

2 Acid alcohol
  Ethyl alcohol (95%), 97 ml
  Hydrochloric acid, concentrated, 3 ml

3 Methylene blue
  Methylene blue, 0.3 g
  Distilled water, 100 ml

Cell-Wall Stain

1 Cetylpyridinium chloride solution
Use M/100 or 0.34% aqueous solution. (Molecular weight = 357.99.)

2 Congo red solution
Use a saturated aqueous solution.

3 Methylene blue solution
As in acid-fast stain.

Flagella Stain (Gray's Method)

1 Flagella mordant (Gray's)
SOLUTION A
  Saturated aqueous solution potassium alum, 5 ml
  20% aqueous tannic acid, 2 ml. (This solution must be preserved in the refrigerator with a few drops of chloroform if a large amount is to be kept on hand.)
  Saturated aqueous solution of mercuric chloride, 2 ml
SOLUTION B
  Saturated alcoholic solution of basic fuchsin, 0.4 ml
Mix solutions A and B on the day the mordant is to be used. Filter through paper before flooding on slide.

2 Ziehl's carbol fuchsin
As in acid-fast stain.

Gram Stain

1 Crystal violet-ammonium oxalate (modified Hucker's)
SOLUTION A
  Crystal violet (90% dye content), 2 g
  Ethyl alcohol (95%), 20 ml
SOLUTION B
  Ammonium oxalate, 0.8 g
  Distilled water, 80 ml
Mix solutions A and B. Store for 24 h before use. Filter through paper into staining bottle.

2 Gram's iodine
  Iodine, 1 g
  Potassium iodide, 2 g
  Distilled water, 300 ml

3 Decolorizer
  Ethyl alcohol (95%)

4 Safranin (counterstain)
  Safranin O (2.5% solution in 95% ethyl alcohol), 10 ml
  Distilled water, 100 ml

Hollande's Stain for Spirochetes

1 Mordant
  Tannic acid, 5 g
  Ethyl alcohol (95%), 50 ml

Glacial acetic acid, 5 ml
Distilled water, 50 ml

2 *Pyridine-silver solution*
Silver nitrate, 5 g
Pyridine, 2 ml
Distilled water, 100 ml
Allow crystallization to take place (several hours), and decant the clear solution, which keeps well in the dark.

Spore Stain

1 *Malachite green solution*
Malachite green, 5 g
Distilled water, 100 ml

2 *Safranin (counterstain)*
As in Gram stain.

STAINING SOLUTIONS

Carbol fuchsin (dilute)
Carbol fuchsin (Ziehl's), 1 ml
Distilled water, 19 ml

Methylene blue, alkaline (Loeffler's)
SOLUTION A
Methylene blue (90% dye content), 0.3 g
Ethyl alcohol (95%), 30 ml
SOLUTION B
Dilute KOH (0.01% by weight), 100 ml
Mix solutions A and B.

Nigrosin solution (Dorner's)
Nigrosin, water-soluble, 10 g
Distilled water, 100 ml
Immerse in boiling water bath for 30 min; then add, as a preservative, formalin, 0.5 ml.
Filter the solution twice through double filter paper. Store in small tubes.

Kinyoun carbol fuchsin
Basic fuchsin, 4 g
Ethanol (95%), 20 ml
Phenol crystals, 8 g
Distilled water, 100 ml

REAGENTS

Bromothymol blue indicator solution
Bromothymol blue, 0.4 g
Ethyl alcohol (95%), 500 ml
Distilled water, 500 ml
Dissolve the bromothymol blue in alcohol, and then dilute with water.

Diphenylamine reagent (for detection of nitrate)
Diphenylamine, 0.7 g
Sulfuric acid (concentrated), 60 ml
Water, 28.8 ml
Hydrochloric acid (concentrated), 11.3 ml
Dissolve the diphenylamine in sulfuric acid, and then add the water.
Cool the mixture, and add the hydrochloric acid slowly.
Allow to stand overnight.

Kovacs' reagent (for detection of indole)
p-Dimethylaminobenzaldehyde, 5 g
Amyl alcohol, 75 ml
Hydrochloric acid (concentrated), 25 ml

Methyl red solution (for detection of acid)
Methyl red, 0.1 g
Ethyl alcohol (95%), 300 ml
Distilled water, 200 ml
Dissolve the methyl red in the alcohol. Dilute with distilled water.

Nitrite test reagents
SOLUTION A
Sulfanilic acid, 0.8 g
Acetic acid (5 N; 1 part glacial acetic acid to 2.5 parts water), 1000 ml
SOLUTION B
Dimethyl-α-naphthylamine, 0.5 g
Acetic acid (5 N), 1000 ml

Trommsdorf's reagent (for detection of nitrite)
Add 20 g zinc chloride to 100 ml water, boil, and add this boiling solution to a mixture of 4 g starch in water.
Dilute with water, and add 2 g zinc iodide.
Dilute to 1000 ml with water, filter, and store in tightly stoppered amber bottles.

Voges-Proskauer reagent (for detection of acetoin)
Potassium hydroxide, 40 g
Creatine, 0.3 g
Distilled water, 100 ml
Dissolve the alkali in water and add creatine. The reagent should be prepared frequently and should be refrigerated when not in use. The reagent may be used for 2 to 3 weeks but deteriorates rapidly thereafter.

APPENDIX B STAINS, STAINING SOLUTIONS, AND REAGENTS

Supplies for the preceding staining solutions and reagents are obtainable from various supply houses. Some of these are listed below.

Aldrich Chemical Company
1001 West Saint Paul Avenue
Milwaukee, WI 53233
U.S.A.
Tel.: 1-800-558-9160 or 414-273-3850
FAX: 1-800-962-9591 or 414-273-4979

Fisher Scientific Company
Corporate Headquarters
711 Forbes Avenue
Pittsburgh, PA 15219
U.S.A.
Tel.: 412-562-8300

Baxter Healthcare Corp.
Scientific Products Division
8855 McGaw Road
Columbia, MD 21045
U.S.A.
Tel.: 1-800-234-8401 or 301-290-8418

Baxter Diagnostics Corporation
Canlab Division
2390 Argentia Road
Mississauga, Ontario
Canada
L5N 3P1
Tel.: 1-800-668-4666 or 416-821-9660

Fisher Scientific Company
Headquarters
112 Colonnade Road
Nepean, Ontario
Canada
K2E 7L6
Tel.: 613-226-8874
FAX: 613-226-8639

ICN Biomedicals, Inc.
3300 Hyland Avenue
Costa Mesa, CA 92626
U.S.A.
Tel.: 714-545-0100
FAX: 714-641-7216

ICN Biomedicals Canada Ltd.
925 McCaffrey Street
Saint Laurent, Quebec
Canada
H4T 9Z9
1-800-268-9943 or 1-800-268-9925

Sigma Chemical Company
Post Office Box 14508
Saint Louis, MO 63178
U.S.A.
Tel.: 1-800-325-3010 or 314-771-5750
FAX: 1-800-325-5052 or 314-771-5757

APPENDIX C

List of Media and Sources of Media

Formulas of the media used in the basic exercises of this manual are listed alphabetically in this appendix. Most of the media, with the exception of the chemically defined or synthetic ones, are available commercially in prepared or dehydrated form. Specific instructions for their rehydration and use are provided on the labels of the individual containers. If the media are to be prepared from the individual ingredients, the general procedures described in Exercise 14 may be followed. Prepared media in disposable plastic Petri dishes and tubes are also available from commercial sources.

Formulas are listed with ingredients in grams per liter of distilled water unless otherwise indicated. Media noted with an "*" are available commercially in powdered (dehydrated) form. Required pH's may be adjusted with 1 N HCl or 1 N NaOH prior to sterilization.

When agar is present in a medium, the preparation is heated gently with a magnetic spinner on a heated magnetic stirrer to dissolve the agar before autoclaving. Alternatively, the agar in the medium may be dissolved by placing the preparation briefly in a microwave oven. Sterilization of the media is usually carried out by autoclaving at 121°C for 15 to 20 min (depending on bulk volume) as mentioned in Exercise 12.

Ames test minimal medium
Glucose, 20
$MgSO_4 \cdot 7H_2O$, 0.3
Citric acid, 3
K_2HPO_4, 15
$Na(NH_4)HPO_4 \cdot 4H_2O$, 5.2
Agar, 15

Ames test top layer
NaCl, 5
Agar, 6.5
Dissolve agar by heating, and dispense 100-ml amounts into separate flasks. Autoclave for 15 min at 121°C, and cool to about 50°C. Add, after autoclaving, per 100 ml of top agar, 10 ml of a sterile stock mixture of:
L-Histidine·HCl (0.5 mM), 10.5 mg/100 ml
Biotin (0.5 mM), 12.2 mg/100 ml

Azotobacter medium
Saccharose, 10.0
K_2HPO_4, 0.5
$MgSO_4 \cdot 7H_2O$, 0.2
NaCl, 0.2
$MnSO_4 \cdot H_2O$, trace
$FeSO_4$, trace
$Na_2MoO_4 \cdot 2H_2O$, trace
$CaCO_3$, 5.0
Agar, 15

Blood-agar base (heart infusion agar)*, pH 7.3
Beef heart muscle, infusion from, 375
Tryptose or thiotone peptic digest of animal tissue, 10
Sodium chloride, 5
Agar, 15
Cool to 50°C after autoclaving and add 5% defibrinated sheep or rabbit blood. (Defibrinated blood may be prepared by putting fresh blood into a sterile flask with glass beads and shaking vigorously for about 5 min.)
NOTE: Prepoured disposable blood-agar plates and slants are also available commercially.

Brain-heart infusion broth*, pH 7.4
Calf brain, infusion from, 250
Beef heart, infusion from, 250
Peptone, 10
Sodium chloride, 5
Disodium phosphate, 2.5
Dextrose, 2.5

Brain-heart infusion salt agar
Brain-heart infusion, 10

Sodium chloride, 15
Starch, 5
Agar, 15

Brain-heart infusion salt broth
Brain-heart infusion, 10
Sodium chloride, 15

Calcium nitrate-salts agar (for algae)
Calcium nitrate, 1.0
Magnesium sulfate, 0.25
Monopotassium phosphate, 0.25
Potassium chloride, 0.25
Ferric chloride, 0.01

Dissolve chemicals in tap water. Dilute further (1 part of salt solution with 2 parts of tap water), and add agar to a concentration of 1.5%.

Casein solution
Sodium caseinate, 2
Glucose, 1
Dipotassium phosphate, 0.2
Magnesium sulfate, 0.2
Ferrous sulfate, 0.01

NOTE: Use tap water for preparation of this medium.

Chocolate agar
Pancreatic digest of casein, 7.5
Peptic digest of animal tissue or other peptone, 7.5
Cornstarch, 1.0
Dipotassium phosphate, 4.0
Monopotassium phosphate, 1.0
Sodium chloride, 5.0
Agar, 10.0

Prepare sterile base above. Add sterile 5 to 10% defibrinated blood, and heat at about 80°C with gentle agitation for 15 min or until color is chocolate brown.

Cooked meat medium*
Heart tissue granules, 98
Peptic digest of animal tissue, 20
Dextrose, 2
Sodium chloride, 5

NOTE: This medium is also available in prepared tubes. It is also available in the enriched form containing hemin and vitamin K (for enhancing growth of anaerobes).

Cornmeal agar
Cornmeal, 50
Agar, 15

Heat to boiling to dissolve. Tween 80 may be added at 1% concentration to promote chlamydospore formation by *Candida albicans*.

Cystine trypticase agar (CTA) medium*, pH 7.3
Cystine, 0.5
Trypticase, 20
Sodium chloride, 5
Sodium sulfite, 0.5
Agar, 2.5
Phenol red, 0.017

For fermentation reactions, carbohydrates may be incorporated at 0.5 to 1.0% as preferred.

Desoxycholate agar*, pH 7.2
Polypeptone, 10
Lactose, 10
Sodium chloride, 5
Dipotassium phosphate, 2
Ferric citrate, 1
Sodium citrate, 1
Sodium desoxycholate, 1
Agar, 16
Neutral red, 0.033

Add 46 g of the dehydrated medium per liter of distilled water. Allow the material to stand for 5 min, and mix thoroughly. After the suspension is homogeneous, heat gently with occasional agitation. Boil for 1 min or until solution is complete. The dissolved medium is ready to use when it has cooled to 42 to 44°C. Do not autoclave.

Desoxycholate solution
Sodium desoxycholate, 1
Distilled water, sterile, 9 ml

Use fresh solution for bile-solubility test.

Dextrose agar, pH 7.3
Beef extract, 3
Tryptose, 10
Dextrose, 10
Sodium chloride, 5
Agar, 15

Dextrose broth*, pH 7.2
Beef extract, 3
Dextrose, 5

APPENDIX C LIST OF MEDIA AND SOURCES OF MEDIA

Sodium chloride, 5
Tryptose, 10

Endo agar*, pH 7.4
Dipotassium phosphate, 3.5
Peptone, 10
Lactose, 10
Sodium sulfite, 2.5
Basic fuchsin, 0.5
Agar, 15

Eosin methylene blue agar*, pH 7.1
Peptone, 10
Lactose, 10
Dipotassium phosphate, 2
Eosin Y, 0.4
Methylene blue, 0.065
Agar, 15

Gelatin agar, pH 7.0
Peptone, 5
Beef extract, 3
Gelatin, 30
Agar, 15

Glucose-acetate medium, pH 4.8
Glucose, 1
$CH_3COONa \cdot 3H_2O$, 8.2
Yeast extract, 2.5
Agar, 15

Glucose broth, pH 7.3
Peptone, 10
Glucose, 5
Sodium chloride, 5

Glycerol minimal medium
Glycerol, 5
NH_4NO_3, 1
K_2HPO_4, 7
KH_2SO_4, 3
NH_4Cl, 5
Na_2SO_4, 2

After autoclaving and cooling the preceding solution, add 0.4 ml separately autoclaved 1.0 M $MgSO_4$.

Halobacterium medium
NaCl, 250
$MgSO_4 \cdot 7H_2O$, 10
KCl, 5

$CaCl_2 \cdot 6H_2O$, 0.2
Yeast extract, 10
Tryptone, 2.5
Agar, 20

The quantities given are to be used in preparation of 1 liter, final volume, of the medium. In preparation, make up two solutions, one involving the yeast extract and tryptone, and the other the salts. Adjust pH of the organic nutrient solution to 7.0. Sterilize separately. Mix and dispense aseptically.

Heart infusion broth*
Same as blood-agar base but without the agar.

Iron agar stabs
Tryptone agar, 500 ml
Ferric ammonium citrate, 0.25
Sodium thiosulfate, 0.25

Make sure all salts are dissolved. Distribute into tubes, and autoclave at 121°C for 15 min.

Lactose broth*, pH 7.0
Peptone, 5
Beef extract, 3
Lactose, 5

NOTE: Double-strength lactose broth contains twice the amounts of the preceding ingredients in 1 liter.

Litmus milk, pH 6.5
Skim milk, dehydrated, 100
Azolitmin, 0.5
Sodium sulfite, 0.5

Autoclave at 12 lb pressure for 15 min, and cool in water immediately after removal from the autoclave.

NOTE: Litmus milk is also available commercially in prepared tubes.

Loeffler's medium*
Mammalian serum, nonhuman, 750 ml
Infusion broth, 250 ml
Dextrose, 2.5

There are several procedures for the preparation of the medium. The following method seems to be the most straightforward.

a Dissolve the ingredients in distilled water warmed to 42 to 45°C.
b Dispense into tubes with screw caps.
c Slant the tubes in a steamer (or inspissator), and steam for 10 min to coagulate the medium.

d Sterilize for 15 min at 15 lb steam pressure (121°C) in an autoclave.

MacConkey agar*, pH 7.1
  Sodium taurocholate, 5
  Peptone, 20
  Sodium chloride, 5
  Lactose, 10
  1% aqueous neutral red solution, 5 to 7 ml
  Agar, 15

a Heat with frequent agitation, or put on magnetic stirrer.
b Boil for 1 min to completely dissolve the powder.
c Sterilize by autoclaving at 121°C for 15 min.

Malt-extract broth, pH 4.7
  Malt extract, 20

Mannitol salt agar, pH 7.4
  Beef extract, 1
  Peptone, 10
  Sodium chloride, 75
  D-Mannitol, 10
  Agar, 15
  Phenol red, 0.025

Milk agar, pH 7.2
  Beef extract, 3
  Peptone, 5
  Milk powder, 100
  Agar, 15

Milk-protein-hydrolysate agar, pH 7.0
  Milk-protein hydrolysate, 9
  Glucose 1
  Agar, 15

Minimal Media for Bacterial Genetics Exercises (Exercises 38 and 40)

A *Minimal agar*
  Water agar, 75 ml
  Minimal salts (×4), 25 ml
  Glucose (20%), 1 ml
Melt agar by placing into boiling water bath or microwave oven. Add warmed sterile salts and glucose. Medium is now ready to be dispensed into plates.

B *Minimal medium, liquid*
  Minimal salts (×4), 25 ml
  Glucose (20%), 1 ml
  Sterile distilled water to 100 ml
Mix the three components aseptically just before use.

C *Water agar*
  Agar powder, 20 g
  Distilled water to 1000 ml
Suspend and steam at 100°C until dissolved or place in microwave oven for dissolution. Adjust pH to 7.2 (to prevent hydrolysis). Dispense in 75-ml volumes into 100-ml bottles. Autoclave at 121°C for 20 min.

D *Minimal salts (×4 concentrate), pH 7.2*
  NH_4Cl, 20 g
  NH_4NO_3, 4
  Na_2SO_4, anhydrous, 8
  K_2HPO_4, anhydrous, 12
  KH_2PO_4, 4
  $MgSO_4 \cdot 7H_2O$, 0.4
  Distilled water to 1000 ml
Dissolve each salt in cold water in the order indicated, waiting until the previous salt is dissolved before adding next (a light precipitate will be formed). Filter 25-ml aliquots into storage bottles. Autoclave at 121°C for 15 min.

E *20% glucose (×100 concentrate)*
  D-Glucose, 200 g
  Distilled water to 1000 ml
Dissolve in warm water. Dispense into 100-ml bottles (for storage).
Autoclave at 5 lb (109°C) for 10 min.

Motility test agar, pH 7.0
  Peptone, 10
  Sodium chloride, 5
  Agar, 3.5

MR-VP broth, pH 6.9
  Peptone, 7
  Dipotassium phosphate, 5
  Dextrose, 5

Mueller-Hinton agar*, pH 7.4
  Beef, infusion from, 300
  Peptone, 17.5
  Starch, 1.5
  Agar, 17

Nitrate-salts medium
  Glucose, 10

Dipotassium phosphate, 0.5
Calcium chloride, 0.5
Magnesium sulfate, 0.2
Potassium nitrate, 1.0
   NOTE: Use tap water for preparation of this medium.

Nitrogen-free medium
Dipotassium phosphate, 1.0
Magnesium sulfate, 0.2
Sodium chloride, 0.01
Ferrous sulfate, 0.01
Manganese sulfate, 0.01
Glucose, 20
Calcium carbonate, 30

Nutrient agar*, pH 7.0
Peptone, 5
Beef extract, 3
Agar, 15

Nutrient broth*, pH 7.0
Peptone, 5
Beef extract, 3

Nutrient gelatin*, pH 7.0
Peptone, 5
Beef extract, 3
Gelatin, 120

Nutrient-tryptone soft agar
Nutrient broth*, 1 liter
Tryptone, 5
Sodium chloride, 5
Agar, 7.5

Peptone-iron agar*
Peptone, 15
Proteose peptone, 5
Ferric ammonium citrate, 0.5
Dipotassium phosphate, 1
Sodium thiosulfate, 0.08
Agar, 15

Peptone solution
Peptone, 2
Dextrose, 1
Dipotassium phosphate, 0.2
Magnesium sulfate, 0.2
Ferrous sulfate, 0.01
   NOTE: Use tap water for preparation of this medium.

Phenol-red dextrose broth*, pH 7.4
Glucose, 5
Sodium chloride, 5
Trypticase, 10
Phenol red, 0.018
Other carbohydrate-fermentation broths can be prepared using this same formula by replacing the glucose with the required carbohydrate, for example, sucrose, lactose, and maltose.

Phenylethyl alcohol (phenylethanol) agar*, pH 7.3
Trypticase, 15
Phytone, 5
Phenylethyl alcohol, 2.5
Sodium chloride, 5
Agar, 15

Photobacterium broth*, pH 7.0
Tryptone, 5
Yeast extract, 2.5
Ammonium chloride, 0.3
Magnesium sulfate, 0.3
Ferric chloride, 0.01
Calcium carbonate, 1
Monopotassium phosphate, 3
Sodium glycerol phosphate, 23.5
Sodium chloride, 30

Sabouraud's agar*, pH 5.6
Peptone, 10
Glucose, 40
Agar, 15

Simmon's citrate agar*, pH 6.9
Ammonium dihydrogen phosphate, 1
Dipotassium phosphate, 1
Sodium chloride, 5
Sodium citrate, 2
Magnesium sulfate, 0.2
Agar, 15
Bromothymol blue, 0.08

Skim milk
Skim-milk powder, 100

Sodium-acetate medium
Sodium acetate, 5
Agar, 15

Spirolate broth* (for spirochetes like Treponema pallidum, Reiter strain)
Pancreatic digest of casein, 15

Dextrose, 5
Yeast extract, 5
Sodium chloride, 2.5
Sodium thioglycollate, 0.5
L-Cysteine HCl, 1
Aseptically add sterile inactivated sheep, rabbit, or bovine serum, 10% by volume.

Standard methods agar, pH 7.0
Tryptone, 5
Yeast extract, 2.5
Dextrose, 1
Agar, 15

Starch-agar medium*, pH 7.0
Peptone, 5
Beef extract, 3
Soluble starch, 2
Agar, 15

Starch broth, pH 7.2
Soluble starch, 10
Beef-heart infusion broth, 1 liter

Thioglycollate medium, fluid*
Yeast extract, 5
Peptone, 15
Dextrose, 5
Sodium chloride, 2.5
L-cystine, 0.75
Thioglycollic acid, 0.3 ml
Agar, 0.75
Resazurin, certified, 0.001

Todd-Hewitt broth*, pH 7.8
Beef-heart infusion, 3.1
Peptone, 20
Dextrose, 2
Sodium chloride, 2
Disodium phosphate, 0.4
Sodium carbonate, 2.5

Tributyrin agar
Peptone, 5
Beef extract, 3
Tributyrin, 10
Agar, 15
NOTE: The tributyrin is added to the other ingredients after they have been dissolved and cooled to 90°C and then emulsified using a blender.

Trypticase agar, pH 7.3
Trypticase, 10
Sodium chloride, 5
Dipotassium phosphate, 2.5
Agar, 15

Trypticase broth, pH 7.3
Trypticase, 10
Sodium chloride, 5
Dipotassium phosphate, 2.5

Trypticase nitrate broth, pH 7.2
Trypticase, 20
Disodium phosphate, 2
Glucose, 1
Potassium nitrate, 1
Agar, 1

Trypticase soy agar*, pH 7.3
Trypticase, 15
Phytone, 5
Sodium chloride, 5
Agar, 15

Trypticase soy broth*, pH 7.3
Trypticase, 17
Phytone, 3
Sodium chloride, 5
Dipotassium phosphate, 2.5
Glucose, 2.5

Tryptone agar, pH 7.4
Tryptone, 20
Sodium chloride, 5
Agar, 15

Tryptone broth (tryptophan broth)
Tryptone, 10

Tryptone water sugar, pH 7.4
Tryptone, 10
Sugar, 5
Bromothymol blue (0.2%), 12 ml
Autoclave at 115°C (10 lb) for 12 min after dispensing into test tubes.

Yeast-extract broth, pH 7.0
Peptone, 5
Beef extract, 3

APPENDIX C LIST OF MEDIA AND SOURCES OF MEDIA

Yeast extract, 5
Sodium chloride, 5

Yeast-extract–peptone-glucose medium, pH 5 to 8
Yeast extract, 3
Peptone, 10
Glucose, 20
Agar, 20

Yeast-extract–tryptone (YT) broth
Tryptone, 8
Yeast extract, 5
NaCl, 5

YT agar
Same as YT broth but with 15 g of agar.

Useful manuals that may be consulted for media preparation and media information are the following:

Power, D. A., and P. J. McCuen: *Manual of BBL Products and Laboratory Procedures*, 6th ed., Becton Dickinson Microbiology Systems, Post Office Box 243, 250 Schilling Circle, Cockeysville, MD 21030, 1988.

DIFCO MANUAL: *Dehydrated Culture Media and Reagents for Microbiology*, 10th ed., Difco Laboratories, Detroit, MI 48232, 1984.

Sources of media and other laboratory teaching supplies are obtainable from the following companies. (See also lists in Appendix A and Appendix B.)

Becton Dickinson Microbiology Systems
Post Office Box 243
250 Schilling Circle
Cockeysville, MD 21030
U.S.A.

Difco Laboratories
Post Office Box 331058
Detroit, MI 48232
U.S.A.

Oxoid U.S.A., Inc.
9017 Red Branch Road
Columbia, MD 21045
U.S.A.

Scott Laboratories
Fiskville, RI 02823
U.S.A.

Carolina Biological Supply Company
2700 York Road
Burlington, NC 27214
U.S.A.

Ward's Natural Science Establishment
Post Office Box 1712
Rochester, NY 14603
U.S.A.

Ward's Natural Science, Limited
1840 Mattawa Avenue
Mississauga, Ontario
Canada
L4X 1K1

Connecticut Valley Biological Supply Company
Post Office Box 326
Southampton, MA 01073
U.S.A.

For good inoculating needles: calibrated, uncalibrated, looped, or straight:

Medical Wire & Equipment Co. USA
7.The Boardwalk
Sparta, NJ 07871
U.S.A.

APPENDIX D

Cross Reference of Laboratory Exercises in This Manual to Chapters in Selected Introductory Microbiology Texts

Exercise	Pelczar, Chan, and Krieg, McGraw-Hill, 1993	Alcamo, 3rd ed., Benjamin Cummings, 1991	Brock and Madigan, 6th ed., Prentice-Hall, 1991	Cano and Colome, West, 1988	Creager, Black, and Davidson, Prentice-Hall, 1990	Prescott, Harley, and Klein, W.C. Brown, 1990	Tortora, Funke, and Case, 4th ed., Benjamin Cummings, 1992
1 Microorganisms in the environment	2	3, 4	1, 17	1	1	1	1
2 Mixed microbial flora	2	3, 4	1, 17	1	1	1	1
3 Aseptic technique and the transfer of microorganisms	3	3	4	4	6	2	1
4 Principles & use of the bright-field microscope	3	3	3	2	3	2	3
5 Microscopic examination of microorganisms in wet mounts	3	3	3	2	3	2	3
6 Microscopic measurement of microorganisms	3	3	3	1, 2	3	2	3
7 Use of the dark-field microscope	3	3	3	2	3	2	3
8 Use of the phase-contrast microscope	3	3	3	2	3	2	3
9 Staining the whole cell	3	3	3	2, 7	3	2	3
10 Staining for cell structures	3, 4	3, 4	3	2, 7	3	2	3
11 Identification of a morphological unknown	3	3	3	7	3, 9	2, 20	3, 4
12 Preparation & autoclaving of nutrient broth & nutrient agar	5, 6	21	4	4	6, 13	5, 12	6, 7
13 Dry-heat sterilization	7	21	4	4	13	12	7
14 Preparation of a chemically defined medium	5	4	4	4	6, 13	5, 12	6
15 Evaluation of media to support growth of bacteria	5, 6	4	4	4	6	5, 6	6
16 Selective, differential, and enriched media	5, 6	4	4	4	6	5, 6	6

541

Exercise	Pelczar, Chan, and Krieg, McGraw-Hill, 1993	Alcamo, 3rd ed., Benjamin Cummings, 1991	Brock and Madigan, 6th ed., Prentice-Hall, 1991	Cano and Colome, West, 1988	Creager, Black, and Davidson, Prentice-Hall, 1990	Prescott, Harley, and Klein, W.C. Brown, 1990	Tortora, Funke, and Case, 4th ed., Benjamin Cummings, 1992
17 Streak-plate method for isolation of pure cultures	6	4	4	4	6	5	6
18 Pour-plate method for isolation of pure cultures	6	4	4	4	6	5	6
19 Enumeration of bacteria by plate-count technique	6	4	4, 9	4	6	5, 6	6
20 Turbidity measurement of broth cultures	6	4	4, 9	4	6	6	6
21 Anaerobic culture methods	6	4	9	4	6	5	6
22 Cultural characteristics	5	4	4	7, 8	5, 9	5, 6	6
23 Maintenance and preservation of pure cultures	7	4	4	4	6	5	6
24 Hydrolysis of polysaccharide, protein, & lipid	1, 3	5	4	3	5	7, 8	2, 5
25 Fermentation of carbohydrates	3, 11	5	4	3	5	7, 8	2, 5
26 Reactions in litmus milk	3	5	4	3	5	7, 8	5
27 Additional biochemical characteristics	3	5	4, 16	3	5	7, 8	5
28 Morphological, cultural, & biochemical characterization of an unknown culture	3	4, 5	4, 16	3, 7	6, 9	20	5, 6
29 Effect of temperature on growth	6, 7	4	9	4, 5	6	6	7
30 Resistance of bacteria to heat	7	21	9	4, 5	6	6	7
31 Bactericidal effect of ultraviolet radiations	7	21	9	5	13	12	7
32 Effect of osmotic pressure on microbial growth	6	21, 22	9	4	13	12	7
33 Comparative evaluation of antimicrobial chemical agents	8	23	9	5, 22	14	13	7
34 Antibiotics: agar-diffusion method	21	23	9	22	14	13	7
35 Effect of enzyme on bacteria: spheroplast formation by P. putida	4	4	9	3	4, 13	13	4
36 Bacterial variation due to environmental change	13	6	7	6	7	17	8
37 Bacterial variation due to genotypic change	13	6	7	6	7	17	8
38 Nutritional mutants	13	6	7	6	7	17	8

APPENDIX D CROSS REFERENCE OF LABORATORY EXERCISES

Exercise	Pelczar, Chan, and Krieg, McGraw-Hill, 1993	Alcamo, 3rd ed., Benjamin Cummings, 1991	Brock and Madigan, 6th ed., Prentice-Hall, 1991	Cano and Colome, West, 1988	Creager, Black, and Davidson, Prentice-Hall, 1990	Prescott, Harley, and Klein, W.C. Brown, 1990	Tortora, Funke, and Case, 4th ed., Benjamin Cummings, 1992
39 Isolation of streptomycin-resistant mutants of *E. coli*	13	6	7	6	7	17	8
40 Bacterial conjugation	13	6	7	6	7, 8	18	8
41 Regulation of enzyme synthesis: enzyme induction & catabolite repression	13	6	5	6	7	17, 18	8
42 Ames test: using bacteria to detect carcinogens	13	6	7	6	7	17	8
43 Bacterial lysis by bacteriophage: phage titer & plaque assay	15, 16	11	6	11	11	14, 15	13
44 Phage-typing	16	11	6	11	11	14, 15	13
45 Propagation of viruses by tissue culture	16	11	6	11	11	14	13
46 Isolation of DNA from bacterial cells	14, 31	6	5, 8	6, 24	7	19	9
47 Restriction enzyme analysis of DNA	14	6	5, 8	6, 24	8	19	9
48 Transformation of *E. coli* with plasmid DNA	14, 31	6	5, 8	6	8	18	9
49 Plasmid-mediated transformation of ampicillin-resistance & *lac* genes	14, 31	6	7	6	8	18	9
50 Characteristics of protozoa	4, 10	15	21	10	12	27	12
51 Morphology & cultivation of algae & cyanobacteria	4, 10	3	21	1	12	26	12
52 Morphology & cultural characteristics of molds	4, 10	14	21	9	12	25	12
53 Sexual reproduction of molds	4, 10	14	21	9	12	25	12
54 Dimorphism of *M. rouxii*	4, 10	14	21	9	12	25	12
55 Morphology of yeasts	4, 10	14	21	9	12	25	12
56 Normal flora of the human body	17, 18	17	11	12	15	28	14
57 Precautions & methodology of specimen collection & processing	18	17	13	12	15	33	Appendix D
58 Immunology & serology: immunoprecipitation	19, 20	19	12	13	18	30	17, 18
59 Immunoagglutination: bacterial agglutination tests	20	19	12	13	18	30	17, 18

Exercise	Pelczar, Chan, and Krieg, McGraw-Hill, 1993	Alcamo, 3rd ed., Benjamin Cummings, 1991	Brock and Madigan, 6th ed., Prentice-Hall, 1991	Cano and Colome, West, 1988	Creager, Black, and Davidson, Prentice-Hall, 1990	Prescott, Harley, and Klein, W.C. Brown, 1990	Tortora, Funke, and Case, 4th ed., Benjamin Cummings, 1992
60 Streptolysin O neutralization	19	19	12	13	18	30	18
61 Complement-fixation	18, 20	19	12	13	18	30	18
62 Complement-mediated lysis in agar	18, 20	19	12	13	18	30	18
63 Bacterial infection of a plant: demonstration of Koch's postulates	Prologue, 18	1, 17	1, 11	12	15	33	14
64 Airborne infections	24	17	15	15, 17	16, 21	36	24
65 Foodborne & water-borne diseases	25	17	15	15, 20	16, 22	36	25
66 Contact diseases	23, 27	17	15	15, 19	16, 25	36	26
67 Bacteriological analysis of urine	22, 27	17	13	15, 19	16, 25	33	26
68 Commercially marketed bacterial identification kits	3	17	13	7	16, 17	33	10
69 Microbiology of soil	28	26	17	23	26	41	27
70 Biological succession in a mixed microbial flora	31	25	17	23	26	40	27
71 Enrichment culture technique: isolation of phenol-utilizing microorganisms	31	26	17	23	26	40	27
72 Standard method of water analysis: multiple-tube fermentation	29	26	17	23	26	41	27
73 Standard method of water analysis: membrane-filter technique	29	26	17	23	26	41	27
74 Bioluminescence	9, 29	26	17	23	5	40	27
75 Enumeration of bacteria in raw & pasteurized milk	30	25	10, 14	23, 24	27	43	28
76 Microbiological examination of foods	30	24	10, 14	23, 24	27	43	28
77 Preparation of fermented foods	30	24	10	23, 24	27	43	28
78 Fermentation of grape juice to produce wine	31	24	10	23, 24	27	43	28
79 Isolation of antibiotic-producing microorganisms	21, 31	27	10	22	27	44	28
80 Production & assay of a microbial enzyme	21	27	10	3, 22	27	44	15

APPENDIX E

Audiovisual Aids Sources

Following is a list of audiovisual sources that may be used to obtain material to supplement laboratory exercises. Catalogues are available on request from each source.

Audio-Visual Services
Pennsylvania State University
Special Services Building
1127 Fox Hill Road
University Park, PA 16803-1824
Tel.: 814-865-6314 or 800-826-0132

British Universities Film and Video Council
55 Greek Street, 1st Floor
London, England W1V 5LR
Tel.: 071-734-3687

Canadian Learning Company, Inc.
2229 Kingston Road
Suite 203
Scarborough (Toronto), Ontario
Canada
M1N 1T8
Tel.: 416-265-3333

Carle Medical Communications
110 West Main Street
Urbana, IL 61801-2700
Tel.: 217-384-4838

CBC Enterprises
Media for Minds
Educational Resources from CBC Enterprises
Canadian Broadcasting Corporation
Box 500, Station A
Toronto, Ontario
Canada M5W 1E6
Tel.: 416-975-3500

Centers for Disease Control
Division of Media and Training Services
Still Picture Archives
Atlanta, GA 30333
Tel.: 404-689-2142

Connecticut Valley Biological Supply Co.
Post Office Box 326
82 Valley Road
Southampton, MA 01073
Tel.: 800-628-7748

Carolina Biological Supply Company
2700 York Road
Burlington, NC 27214
Tel.: 919-584-0381 or 800-334-5551

Educational Images, Ltd.
Post Office Box 3456
West Side
Elmira, NY 14905
Tel.: 607-732-1090

Films for the Humanities, Inc.
Post Office Box 2053
Princeton, NJ 08543
Tel.: 609-452-1128 or 800-257-5126

Industrial Biotechnology Association
1625 K Street, N.W.
Washington, DC 20036
Tel.: 202-857-0244

National Geographic Society
Educational Services
Washington, DC 20036
Tel.: 800-368-2728

National Science Resources Center
Room 1201
900 Jefferson Drive, S.W.
Washington, DC 20560
Tel.: 202-357-2555

Optical Data Corporation
30 Technology Drive
Post Office Box 4919
Warren, NJ 07060
Tel.: 800-524-2481

Sloane Audio Visuals for Analysis and Training
Post Office Box 3670
Fullerton, CA 92634
Tel.: 714-870-7880 or 800-472-8268

Taped Technologies
Post Office Box 384
Logan, UT 94321
Tel.: 801-753-6911

University of California
Extension Media Center
2176 Shattuck Avenue
Berkeley, CA 94704
Tel.: 415-642-0460

Video Britannica
10 South Michigan Avenue
Chicago, IL 60604
Tel.: 800-554-9862

Videodiscovery
1515 Dexter Avenue North
Seattle, WA 98109
Tel.: 800-548-3472

Ward's Natural Science Establishment, Inc.
Post Office Box 1712
Rochester, NY 14603
Tel.: 716-359-2502

Ward's Natural Science, Limited
1840 Mattawa Avenue
Mississauga, Ontario
Canada L4X 1K1
Tel.: 416-984-8900

Wm. C. Brown Publishers
(Software/Video Catalogue)
2460 Kerper Boulevard
Dubuque, IA 52001
Tel.: 319-588-1451

APPENDIX F

Selected Bibliography

The literature listed here can serve as resource material for added information for both student and instructor.

Balows, A., W. J. Hausler, Jr., K. L. Hermann II, D. Isenberg, and H. J. Shadomy, eds.: *Manual of Clinical Microbiology*, 5th ed., American Society for Microbiology, Washington, D.C., 1991. *Comprehensive coverage of clinical laboratory diagnostic techniques for all microorganisms. Section III covers procedures for control of infections, sterilization, disinfection, and antisepsis.*

Balows, A., H. G. Truper, M. Dworkin, W. Harder, and K. -H. Schleifer: *The Prokaryotes: A Handbook on the Biology of Bacteria: Ecophysiology, Isolation, Identification, Applications*, 2d ed., Springer-Verlag, New York, 1992. *Although not primarily concerned with classification, the four volumes of this monumental reference work provide a wealth of descriptive information and illustrations concerning the various genera of bacteria.*

Bergey's Manual of Systematic Bacteriology: N. R. Krieg and J. G. Holt, eds., vol. 1; P. H. A. Sneath et al., eds., vol. 2; J. T. Staley et al., eds., vol. 3; S. T. Williams and J. G. Holt, eds., vol. 4., Williams & Wilkins, Baltimore, 1984–1989. *The four volumes of this international reference work provide the most widely accepted classification of bacteria together with detailed descriptions of all established genera and species.*

Block, S. S., ed.: *Disinfection, Sterilization, and Preservation*, 4th ed., Lea & Febiger, Philadelphia, 1991. *A comprehensive coverage of the subject of killing and inhibiting microorganisms.*

Boatman, E. S., M. W. Burns, R. J. Walter, and J. S. Foster: "Today's Microscopy: Recent Developments in Light and Acoustic Microscopy for Biologists," *Bioscience* 37:384–394, 1987. *As the title implies, this article reviews recent developments in microscopic technology.*

Bold, H. C., and M. J. Wynne: *Introduction to the Algae: Structure and Reproduction*, 2d ed., Prentice-Hall, Englewood Cliffs, N.J., 1985. *A comprehensive and well-referenced textbook written for the teaching of general phycology. Well-illustrated with both photomicrographs and electron micrographs.*

Clesceri, L. S., A. E. Greenberg, and R. R. Trussell: *Standard Methods for the Examination of Water and Wastewater*, 17th ed., American Public Health Association, American Water Works Association and the Water Pollution Control Federation, Washington, D.C., 1989. *An extensive compendium of physical, chemical, and biological "standard methods" for the examination of waste and wastewater. The laboratory procedures are presented in considerable detail.*

Conte, J. E., Jr., and S. T. Barriere: *Manual of Antibiotics and Infectious Diseases*, 7th ed., Lea and Febiger, Baltimore, 1992. *Current information on antibiotics; therapy and susceptibilities; prophylactic antibiotics; and immunologic agents and antiparasitic drugs.*

Crueger, W., and A. Crueger: *Biotechnology: A Textbook of Industrial Microbiology*, 2d ed., Sinauer Associates, Inc., Sutherland, Maine, 1990. *This text provides the key principles and major processes of industrial microbiology. It is written for advanced undergraduate and graduate students.*

Demain, A. L., and N. A. Solomon, eds.: *Manual of Industrial Microbiology and Biotechnology*, American Society for Microbiology, Washington, D.C., 1989. *This volume brings together the biological and engineering methodology needed to develop industrial processes from the isolation of the culture to the isolation of the product.*

Grange, J. M., A. Fox, and N. L. Morgan: *Immunological Techniques in Microbiology,* Blackwell Scientific Publications, Oxford OX2 0EL, 1987. *Describes a number of the currently available immunological techniques for the detection and assay of a wide range of biological materials and to demonstrate how they have been successfully utilized by workers in many branches of microbiology.*

Harlow, E., and D. Lane: *Antibodies, A Laboratory Manual,* Cold Spring Harbor Laboratory, Box 100, Cold Spring Harbor, NY 11724, 1988. *Methods of modern immunochemistry are assembled together in this volume. There are protocols for raising, purifying, and labeling monoclonal and polyclonal antibodies, as well as description of ways of using antibodies to study antigens. Of interest to the student are the four introductory chapters that summarize the key features of the immune response, the structure of the antibody molecule, the activities of antibodies, and the mechanism of the antibody response.*

Hersgberger, C. L., S. W. Queener, and G. Hegeman: *Genetics and Molecular Biology of Industrial Microorganisms,* American Society for Microbiology, Washington, D.C., 1989. *This volume brings together the work of leading researchers in the field of recombinant DNA technology.*

Isenberg, H. D., editor-in-chief: *Clinical Microbiology Procedures Handbook,* American Society for Microbiology, Washington, D.C., 1992. *A comprehensive (two-volume) compilation of microbiological laboratory procedures which includes techniques for the cultivation of bacteria, fungi, viruses and parasites. These volumes also provide a good overview of the variety of techniques used in microbiology.*

Margulis, L., J. O. Corliss, M. Melkonian, and D. J. Chapman, eds.: *Handbook of Protoctista,* Jones and Bartlett Publishers, Boston, Mass., 1990. *A major comprehensive volume on eucaryotic microorganisms (the single-celled protists and their multicellular descendants). This volume on Protoctista is somewhat comparable to Bergey's manual for bacteria.*

Miklos, D. A., and G. A. Freyer: *DNA Science, A First Course in Recombinant DNA Technology,* Carolina Biological Supply Company, 2700 York Road, Burlington, NC 27215, 1990. *Finally there is a textbook on molecular biology that has been written for pedagogy! It is meant to be read from cover to cover and integrates successfully the theory, practice, and applications of recombinant DNA technology. Superbly illustrated with line art. This is a must book for the student going into the study of molecular biology.*

Perkins, J. J.: *Principles and Methods of Sterilization in Health Sciences,* 2d ed., Charles C Thomas, Springfield, Ill., 1983. *A good coverage of the fundamentals of sterilization and control of microorganisms.*

Rose, N.R., editor-in-chief: *Manual of Clinical Laboratory Immunology,* 4th ed., American Society for Microbiology, Washington, D.C., 1992. *Good guide to immunologic methods and their applications. Contains authoritative information on the best methods available for conducting specific immunologic tests, and laboratory techniques used in clinical diagnostic laboratories and in research in immunology. More than 240 contributing authors.*

Sambrook, J., E. F. Fritsch, and T. Maniatis: *Molecular Cloning, A Laboratory Manual,* 2d ed. (in three volumes), Cold Spring Harbor Laboratory Press, 10 Skyline Drive, Plainview, NY 11803. *A compendium of "consensus protocols" for use in the molecular biology laboratory. Gives detailed procedures with background information. A classic reference for molecular cloning.*

Vanderzant, C., and D. Splittstoesser: *Compendium of Methods for the Microbiological Examination of Foods,* American Public Health Association, Washington, D.C., 1992. *Detailed methods for the microbiological examination of foods compiled by an intersociety and interagency committee: very comprehensive. These methods represent what might be considered standard for the food industry.*

INDEX

Page numbers in **boldface** indicate illustrations.

Absorbance, 51, 144, 260
Achromatic-aplanatic condenser, 25
Acid, 175, 179, **198**, **426**, 479, 530
Acid alcohol, 75, 529
Acid-fast stain, 75, 94, 197, 352, 529
Acidic dyes, 59–60, 67, 91
Acinetobacter, 419
Acquired specific immunity, **357**
Adsorption, 287
Aerobe, 149, 197, 403–406, 447
Aeromonas, 404
Agar, 7, 99, 101, 203, **204**, 307, 387–390
  various forms of, **19**, 99–102, 155, 164
Agglutination, 358, 367–369, 371–372
Agrobacterium tumefaciens, 391
Air, 5, **8**, 27, 28
Alcaligenes, 419
  faecalis, 176, 177, 181, 183, **410**
Alcohol, 72, 75, 83, 137, 221, 451, 479
Algae, 5, 37, 307, 315–318, 435, 525, 534
Allosteric protein, 255
Alternaria, 320, 325
Ames test, 263–265, 267–268, 533
Amino acids, 99, 111, 171, 189, **198**, 243
Ammonia, 189, 195, 426, 439, 441–442
  test for, 439, 443, 445
Amoeba, 309, 310, 313, 314
Ampicillin, 225–228, 301, **302**, 305–306
Anaerobes, 149, 180, 193, 415, 443, 447
Anaerobic culture methods, 149–154, 534
Anaerobic jar, diagram of, 150, **151**
Antibiotic-impregnated disks, **204**, 226
Antibiotic inhibitory zones, 225–227
Antibiotic-producing microbes, 507–509
Antibiotic resistance, 225–228, 247, 301
Antibiotic susceptibility test, 225, 346
Antibiotics, 131, 169, 203, 204, 225, 479
Antibody, 357, 363–364, 373, 381, 387
Antibody titer, 346, 367
Antigen, 358, 363–364, 367, **378**, 381
Antigen-antibody complex, 377, **378**
Antigenic dose, 379–381, 384
Antimicrobics, 203–204, 221–224
Antiserum, **364**, 367
Antistreptolysin O (ASTO), 373–376
API 20E System, 423, **424**, 426–427
Aquaspirillum itersonii, 61, 65–67, 69
Arbovirus, 281
Arrangement of cells, 59, 61, 65, 94, 201
Arthrobacter globiformis, 125, 127, 128
Ascospore, 327, **339**
Aseptic technique, 1, 7, 17–19, 21–22, 61, 107, 123, 285, 488
Asexual reproduction, 337

Aspergillus niger, 37, 205, 207, 218, 220, 320, 324
Autoclave, 2, 3, 99, **103**, 203
Autoclaving process, 99, 101–102, 533
Autotroph, 115
Auxotroph, 246, 251, 263
Axenic culture, 123
Azotobacter, 447
Azotobacter medium, composition of, 533
Bacillus, 65, 79, **157**, 443, **515**
  cereus, 37, 72, 73, 91, 93, 171, 174, 515, 516
  coagulans, 79, 81
  stearothermophilus, 205, 207
  subtilis, 61, 65, 69, 79, 81, 107, 151, 153, 155, 158, 164, 167, 176, 193, 196, 209, 211, 221, 277, 301
Bacterial cell, **43**, 60, **217**
  isolation of DNA from, 287–291, 293
  stain of, 61–63, 65–68, 71–72, 75–76
Bacterial lysis by phage, 271–276
Bacterial variation, 235–242
Bacteriological filter, 203, 469, **470**
Bacteriostasis, 163
Bacterium(a), 1, 99, 461, 463, 525, 526
  biochemical activities of, 169
  capsulated, photomicrograph of, **83**
  characterization/identification of, 87, 94, 155–162, 169, 175, 179, 185, 189, 193, 277, 346, 359
    kits for, 423–427, 429–432
  for detection of carcinogens, 263–265
  disease-producing, 83, 391–394
  effect of enzymes on, 229–232
  enumeration of, 135–139, 141–142
  evaluation of media for, 115, 118
  light emission by, 475, 477–478
  microscopic examination of, 33, 37
  thermal death times of, 209
Bacteroides fragilis, 149, 352, 353, 407
*Bam*H I, 295–297
Basal medium ingredients solution, 456
Basic dyes, 59–63, 65–66
Bifidobacterium, 407
Bile solubility, 396
Binary fission, 337–**339**
Binocular microscope, 23, **24**, 32
Biochemical oxygen demand (BOD) test, 462
Biochemical testing, 423, **426**, **427**
  of unknown culture, 99, 197–**199**, 201
Biological succession, 451–454
Bioluminescence, 475, 477–478
Biomass, optical density and, 144
Bioremediation, 455, 456
Blood agar, 99, 119, 121, 347, 348, 373, **410**, **420**, 533

Blood-borne pathogens, 345
Bordetella, 352, 403, 404
Borrelia burgdorferi, 415, **416**
Brain-heart infusion media, 397, 533–534
Branhamella catarrhalis, 115, 117, 507
Brevibacterium linens, 218, 219
Bright-field microscope, 23, 25, 37, 59
  principles and use of, 29–33, 35–36
Brilliant green lactose-bile broth, **464**
Bromothymol blue indicator solution, 530
Broth culture, 19, 21, 269
Brucella, 403
Calcium nitrate-salts agar, 534
Campylobacter, 404
Cancellation, 51, **52**
Candida albicans, 337–**339**, 343, 534
CAP protein, 257
Capsule, 83–86, 94, 95, 181
Carbohydrates, 111, 452
  fermentation of, 175–178, **198**
Carbol fuchsin, 62, 75, 415, 530
  Kinyoun, 75, 76, 530
  Ziehl's, 61, 75, 87, 88, 529
Carbon dioxide, 179, **198**, 479, 503
Carbon dioxide incubator, 502
Carcinogens, detection of, 263–265, 267
Casein, 173, 179, 181, **198**, **199**, 243
Casein solution, 439, 441, 534
Catabolite repression, 255, 257
Catalase test, 348
Cell concentration, 143, 144, 260
Cell structures, 79–80, 83–84, 87–88, 119
Cell wall, 75, 83, 120, 218, 229
  staining of, 71, **91**, 93, 529
Cetylpyridinium chloride, 91, 529
Chemically defined medium, 99, 111, 113
Chlamydomonas, 315, 317, 318
Chloramphenicol, 225–228
Chlorella, 37, 315, 317, 318
Chocolate agar, composition of, 534
Chromobacterium violaceum, 164, 167
Chromophore, 59, 60, 67, 91
Chromosome, 233, 239, 251
Citrate test, 407, 408, 410, **424**, 462
Citrobacter, 407
Clear zone, 173, **198**, 203
Cloned gene, 285, 301
Clostridium, 79, 157, 179, 415, 447
  butyricum, 79, 81
  pasteurianum, 449
  perfringens, 149, 415, 416
  sporogenes, 151, 153
  tetani, 415, 416
  tetanomorphum, 79, 81
Coagulase, 410, 413
Coagulation, **198**, **199**
Cocci, 65, 98, 397–398, 415

549

Coenocytic hyphae, 319
Coliforms, 407, 461–464, 469, 471
Colonial morphology, 9–10, 155, 158, 201
Colony(ies), 7–10, 31, 131, 133, 239, 240
   isolation of, 125, 126, 131
Colony counter, 137, 139, 482, 485
Colony-forming units (CFU), 11, 131, 135, 138, 485, 487
Commensal, 327
Competent species, 301
Complement, 373, 377, 378, 381, 387
   titration of, 379, 380, 383
Complement-fixation test, 377–385
Complement hemolysis, 50% (CH50), 389
Complement-mediated lysis, 387–390
Completed test (water analysis), 462, 464
Complex culture medium, 99
Compound microscope, components of, 24
Concentration gradient, 247
Condenser, 24, 25, 27, 29–31, 47
Confirmed test (water analysis), 462–465
Congo red solution, 91, 309, 529
Conjugation, 239, 251–254, 301
Constitutive enzyme, 246, 255
Contaminants, 2, 3, 17, 19, 99, 123
Convection oven, 107
Cornmeal agar, 338, 339, 534
Corynebacterium, 347, 395–401
   diphtheriae, 395
   xerosis, 396
Cotton (stopper) plug, 7, 17, 100
Counterstain, 71, 72, 529
Coxiella burnetii, 481
Cross streaks with test organisms, 509
Crystal violet, 61, 65, 71, 72, 84, 529
Cultural characterization, 155, 169, 319
   of unknown culture, 197–199, 201–202
Culture medium(a), 3, 99, 111, 533, 539
   evaluation of, 115, 118, 119
   preparation and sterilization of, 7, 99
   streaking of cells on, 8, 19, 120, 125–130, 171
   various forms of, 100
Cultures, 3, 13, 525–527
Curd, 179, 180, 183, 198, 199
Cyanobacteria, 315–318
Cyclic AMP (cAMP), 257, 258
Cysteine, 150, 185, 198
Cystine-trypticase media, 123, 164, 396
Cytopathic effects (CPEs), 281, 282
Dark-field microscope, 23, 25, 59, 416
   use of, 47–49, 51
Decolorizer, 71, 72, 75, 529
Denitrification, 193, 443, 445–446
Desorption, 287–288
Desoxycholate media, 119, 121, 534
Dextrose media, 504, 534–535
Diatom, 48, 54
Differential media, 119, 239, 487
Differential stains, 59, 71–78
Diffraction grating, 143
Digest, 99, 101, 111
Dilutions, 125, 132, 136, 203, 369, 483
   of phage, 274
Dimorphism of *Mucor rouxii*, 333–336
Diphenylamine reagent, 443, 530
Diseases, 1, 2, 83, 169, 277, 345–346
   airborne, 403–406
   contact, 415–418
   enteric, 119, 461
   foodborne or waterborne, 407–414

Disinfectant, 1–3, 203, 223
Dissimilation, 169, 479
DNA, 233, 256, 285, 307
   isolation of, from bacteria, 287–291
   restriction enzyme analysis of, 295
   types of, 295, 296, 272, 301–304, 391
DNA ligase, 285
DNA replication, phenylethanol and, 119
Durham tube, 175, 198, 408, 409, 443, 462
Dye, 50, 59–60, 67, 189
*Eco*R I, 295–297
Electromagnetic spectrum, 27
Electron microscope, 23, 87, 91, 269, 273
Endo agar, composition of, 407, 535
Enriched media, 119
Enrichment culture technique, 455–457
Enterics, 119, 409–410, 423, 461
Enterobacter aerogenes, 83, 85, 190, 191, 407, 408, 410, 412, 425, 462–465
Enterotube II Multitest System, 423–426
Environment, 7–10, 235–238, 433, 455
Enzyme(s), 171, 243, 373, 455, 479
   constitutive, 246, 255
   degradation of cell wall by, 229–232
   inducible, 246, 255–257
   production and assay of, 515–517, 519
   regulation of synthesis of, 255–262
Eosin methylene blue agar (EMB), 407, 423, 461, 462, 464, 535
Escherichia coli, 72, 91, 93, 111, 115, 117, 119, 131, 133, 138, 155, 158, 171, 174, 176, 181, 183, 186, 187, 190, 191, 209, 211, 225, 228, 239, 240, 247–250, 255–262, 271, 273, 274, 277, 285, 301–306, 403, 407, 410, 419, 425, 461, 463, 464, 507
Ethanol, 71, 72, 479, 503
Ethidium bromide, 295–297
Eubacteria, 87, 205, 208
Eucaryotes, 307–344
Eyepiece, 23–25, 30, 32
F (fertility) factor, 251
Facultative microorganisms, 149, 180
Fermentation, 149, 175, 179, 198, 493
   of carbohydrates, 175–178, 202, 342, 411, 423, 503–506
   multiple-tube, 461–465, 467–468
Fermentation tubes, 175, 176
Fermentor bottle, 504
Fixation of smear, 61–63, 68
Flagella, 87–90, 94, 236, 529
Flaming techniques, 7, 18, 19
Flat field objective, 24
Flavobacterium capsulatum, 83, 85
Fluorescent microscopy, 23
Focal length, 25
Food, 479, 480, 487–491
   fermented, 493, 495, 497–499, 501–502
Four-way streak-plate method, 125, 126
Francisella, 403
Frankia, 447
Fungi, 1, 220, 307, 319, 335, 377, 435, 526, 527
Fusobacterium, 407
β-Galactosidase, 179, 255, 257, 261, 305
Gases, 198, 479
   production of, 424, 426, 462, 464, 465
Gel electrophoresis, 295–297
Gelatin agar, 156, 162, 198, 451, 535
Genes, 243, 255, 257, 285, 287
Genetics, bacterial, 233, 285, 536–539

Genotype, 233, 235, 239–242
Giemsa-stained monolayer, 282
Glove box, 150–151
Glucose, 101, 102, 111, 198, 255, 258, 410, 424, 426, 427, 536
   fermentation of, 169, 197, 202, 503
   transport of, PEP-PTS and, 257
Glucose media, 102, 115, 535
Glycerol, 163, 171, 173, 475
Gradient-plate technique, 247–248
Gram-negative bacteria, 71–72, 93, 120, 197, 229, 403, 415, 461, 464, 465
   phenylethanol inhibition of, 119
Gram-positive bacteria, 71, 72, 93, 119, 120, 229, 395–401, 455
   selective medium for, 119
Gram stain, 59, 71–75, 94, 197, 201, 529
   age of culture used in, 71
   color reactions of, 72
Gram-variable bacteria, 71
Gram's iodine solution, 171, 173, 529
Growth, 205–208, 217–220
   various types of, 9–10, 156
Growth rate, 205
Haemophilus influenzae, 301
Half-aperture angle, 27–28
Halobacterium salinarium, 218–220, 535
Hanging-drop preparation, 88, 95
Heart infusion media, 251, 533, 535
Heat resistance, 209, 211–212
Hektoen enteric agar, 423
Hemolysin, 119, 373, 377, 381, 382
   titration of, 377, 379, 383
Hemolysis, 377, 378, 380, 381, 389
   α or β or γ, 119, 347, 348, 373, 395
Hepatitis virus, 461
Heterologous antigen and antibody, 377
Heterotroph, fastidious, 115, 119
Hfr cell, 251
High-power objective, 24, 25, 37, 88
Histidine, 263
Hollande's stain, 415, 529–530
Homologous antigen and antibody, 377
Host-cell-virus interaction, 269, 271
Human immunodeficiency virus, 269, 345
Humans, microbes and, 1, 2, 5, 347–350
Hydrogen sulfide, 185, 198, 202, 411, 424
Hydrolysis, 171, 173–174, 189, 198, 452
Hypertonic or hypotonic medium, 217
Hyphae, 319, 323
Illumination, 25, 29–30, 32, 59
Immobilized enzyme technology, 479
Immunoagglutination, 367–369, 371–372
Immunoglobulins, 358
Immunology, 357–359, 363–366
Immunoprecipitation, 357–359, 363–366
IMViC tests, 407, 409–410, 412, 462
Indicator, 169, 175, 377, 378, 382, 461
Indole, 198, 243, 407, 424–427, 462, 530
   production of, 185, 189, 197, 198, 202
Induced enzyme, 246, 255, 256
Infection, 1–3, 203, 346, 377, 395–401
Inoculating needle, 8–9, 18–19, 120, 171
Inoculum, 137, 457
Inorganic compounds, 99, 111, 115
Inulin fermentation, 396
Iodine, 72, 198, 221
Ion-exchange chromatography, 287
Iris diaphragm, 24–26, 29, 30, 32, 33
Iron agar stabs, 535
Iron sulfide formation, 198
Ixodes dammini, 415

INDEX

KF streptococcus agar, 469
Kinyoun's cold procedure, 76
Kirby-Bauer antibiotic sensitivity, 225
Klebsiella pneumoniae, 425
Koch's postulates, 391–394
Koehler illumination, 29, 30, 33
Kovacs' reagent, 189, **198**, 407, **410**, 530
lac gene, **256**, 305–306
β-lactam ring, **515**
Lactic acid, 179, 479
Lactobacillus, 111, 407, 493
Lactose, 179, **198**, 255, 258, **410**
    fermentation of, 119, 179, 180, 199, 202, **239**, **424**, **426**, 461
Lactose broth, 462–**464**, 487, 489, 535
Lancefield groups of streptococci, 395
LBS (Lactobacillus selection) medium, 499
Leguminous plants, 447
Leptospira, 415
Leucolitmus, 179, **199**
Leuconostocs, 493
Light, 26–**27**, 29, 30, 48, 51–54, **53**
    emission of, by bacteria, 475, 477–478
Light intensity, 26, 33, 51
Light microscope, 23, 27, 29, 87, 91
Lipases, 171
Lipid hydrolysis, 171, 173–174, **198**, 202
Liquefaction, 162, 169, **198**, 201
Liquid media, 99, 101, 150
Litmus, 179, **199**
Litmus milk, 180, 197, 482, 487, 489, 535
    reactions in, 179–181, 183, **199**, 202
Loeffler's medium, 535–536
Loop-dilution procedure, 155
Low-power objective, 24, **25**, 37, **43**
Lowenstein-Jensen media, 345
Luciferase, 475
Luria broth (LB) agar, 305, **306**
Lymphocytes, 357
Lyophilization, 123, 163–**165**
Lysis, 218, 377, **378**, 380
Lysogenic or lytic cycle, 271, **272**
M-Endo broth MF, 469
MacConkey agar, **239**, 240, 407, 423, 536
Magnification, 23–**26**, 31
Malachite green stain, 79, 80, 339, 530
Malt-extract broth, 218, 536
Mannitol salt agar, 347, 536
Meat extract, 99, 101, 111, 151, 534
Membrane-filter technique, 203, 469–471
Mesophile, 205
Metabolism, 169, 175, 243
Metachromatic granules, 396
Methyl red test, 407, 408, 410, 462, 530
Methylene blue, 59, 75, 91, 339, 451, 529
    as anaerobic indicator, 151, 152, 529
    Loeffler's, 60–62, 65, 530
Microaerophile, 149
Microbial flora, 13, 15, 451–454
Microbial populations, control of, 203
Microbiology laboratory, 1–4, 7
Micrococcus, 119
    luteus, 125, 127, 128, 155, 158–162
    roseus, 205, 207
Micromanipulator, 123
Microorganisms, 1, 2, 5, 7–10, 169, 461
    beneficial/harmful aspects of, 169, 479
    identification of, 307, 346, 359, 423
    microscopic examination of, 33, 37, 39–43, 45, 47–49, 51–55, 57
    resident or transient, 347
    use of, for genetic experiments, 233

Microscope, 23–33, 41
Milk, 179, 451, 481
    count of bacteria in, 481–483, 485–487
    succession of microbes in, 451–454
Milk agar, 171, 174, **198**, 451, 536
Mineral oil, 107, 163, **164**
Mineral-salts medium, 456
Minimal hemolytic dose (MHD), 380
Minimal medium, 536
Mixed culture, 13, 15, 119, 123, 125, 127
Molds, 5, 31, 37, 212, 220, 233, 269, 307, 319–329, 331, 435, 526, 527
Mordant, 72, 87, 529
Morphology, 59–60, 65, 67, 169, 250, 307, 315, **316**, 319, 320, **322**, 337–344
    of unknown culture, 95, 97, 197–199
    (*See also* Colonial morphology)
Most probable number, 462, **463**, 468, 469
Motility, 87–90, 185, 201, 309
Motility-agar medium, 87, 88, 197, 536
MR-VP broth, 536
Mucor hiemalis, 327, **328**, 331
Mucor rouxii, 333–336
Mueller-Hinton agar, 225, 536
Multitest systems, 423
Must, 503
Mutants, 163, 239, 251, 263, 264, 319
    Ames, **263**
    nutritional, 243–246
    spontaneous, 233, 247
    streptomycin-resistant, 247–250
Mutation, 213, 239, 243, 247, 263
Mutualism, 407
Mycelium, 319, 323
Mycobacterium, 75
    phlei, 155, 158–162
    smegmatis, 75–77, 403–406
    tuberculosis, 352, 403–406
Mycophages, 269
Negative stain, 60, **67**–70
Neisseria:
    gonorrhoeae, 301, 352, 403–406, 415
    meningitidis, 352, 403–406
    subflava, 72, 73, 164, 167, 403, 404
Nessler's reagent, 439, 441, 443
Neurospora, 233
Neutral dyes, 59, 60
Nigrosin, 67, 83, 415, 530
Nitrate, detection of, 443, 445, 530
Nitrate reduction, 193, 195, **198**, 202, 426, **427**, 443, 445–446
Nitrate-salts medium, 443, 536–537
Nitrite, 193, 443, 445, 530
Nitrogen, 111, 163, 193, 195, 205, 435, 439, 443, 456
Nitrogen fixation, 447, 449–450
Nitrogen-free medium, 537
Nitrogenous compounds, 439, 441–442
Normal flora of human, 347–350
Nostoc, 315, 317, 318
Nucleic acid, 271, **272**, 287
Numerical aperture (NA), 25–28, 47
Nutrient agar, 7, **8**, 99–102, 111, **156**, 235, **236**, 269
    preparation of, 101–102, 537
    various forms of, 123, 159, 161, 197
Nutrient broth, 99–102, 111, 155, 197
    characterization of growth in, **156**, 160
    preparation of, 101–102, 537
Nutrient gelatin, 155, 197–**198**, 487, 489
Objective, 23–**25**, 27, 29–32, 54
Ocular micrometer, **41–43**

Oil-immersion objective, 24–**25**, 27, 32
    numerical aperture of, 28, 47
    procedure for use of, 31, 37
Operator, 255, **256**
Operon, 243
Opsonizing bacteria, 357
Optical characteristics of colony, 9
Optical density (OD), 144, 145
Organic compounds, 99, 115, 169, 462
Osmotic pressure, 203, 217–220
Ouchterlony immunodiffusion, 363–366
Oxidase test, 404
Oxidation-fermentation test medium, 175
Oxidation-reduction (O/R) indicator, 179
Oxidation-reduction potential, 149, 150, 175, 179, 185
Oxygen, 149, 150, 175, 193
Paraboloid condenser, 47
Paramecium caudatum, 37, 309, **310**, 314
Parfocal objectives, 23, 25, 33, 37
Pasteurella, 404
Pasteurization of milk, 480–483, 485–487
Pathogens, 119, 345, 455, 461, 481, 487
    opportunistic, 347
*p*BLU plasmid, 305–306
*p*BR322 plasmid, map of, 301, **302**
Penicillin, 225–228, **515**
Penicillin resistance, 515
Penicillinase, **515**–517, 520–521
Penicillium notatum, 320, 323
Peptone, 99, 101, 111, 179, 185, 189
Peptone-iron agar, 185, 197, **198**, 537
Peptone solution, 439, 441, 537
Peptostreptococcus, 407
Petragnani media, 345
Petri plates, use of, 7, **8**, 32, 126, 139
Petroleum-utilizing microbes, 455, 456
pH indicator, 179
pH of media, 99, 102, 111, 203, 287, 533
Phage, 269, 271–277, 279, 280
Phase-contrast microscope, 23–25, **52**, 59
    use of, 51–55, 57
Phenol, 221, 235–238, 456
    microbe utilization of, 455–**457**, 459
Phenol-coefficient technique, 203
Phenol-mineral-salts medium, 456, **457**
Phenol-red fermentation broths, 176, 177, 197, **198**, 537
Phenotype, 233, 235
Phenylethyl alcohol medium, 119–121, 537
Phosphate buffer saline, 138, 373, 387
Phosphoenolpyruvate phosphotransferase system (PEP-PTS), 257
Photobacterium phosphoreum, 475, 537
Photoreactivation, 213
Photosynthetic microbes, 315, **316**
Phycology, 315
Pigmentation, 9, 21, 157, 207, 235
Pipetting of solutions, 2, 3, 135–**137**
Pityrosporum, 347
Plants, infection of, 1, 391–394
Plaque (teeth), 55, 131
Plaque assay, 269, 271–276
Plasmid, 239, 251, 263, 285, 301–306, 391
Plasmodium vivax, 309, 310, 313
Plasmolysis, 217, 220
Plasmoptysis, 218, 220
Plate-count technique, 135–139, 141–142
Polysaccharides, 83, 169, 171, 173, 479
Pond water, stagnant, 310, 312
Pour-plate method, 131–134, 137, 419

Precipitation, 185, **198**, 358, 363, 377, 411
Precipitin reaction patterns, 363–**364**
Precipitin ring test, 359, 361
Presumptive test for coliforms, 462–**464**
Primary culture, 352
Primary producers, 315
Primary stain, 71, **72**, 75, 79
Procaryotes, 1, 233, 251–254, 315, 319
Promoter, 255–257
Prophage, 271, **272**
Propionibacterium, 347
Protease, 171
Proteins, 62, 189, 257, 260, 439
    hydrolysis of, 171, 173–174, 179–181, **198, 199**, 452
Proteus, 407, 419
    vulgaris, 83, 85, 88, 89, 176, 177, 186, 187, 190, 191, 235–237, 425
Protoplast, 220, 229
Prototroph, 246, 252
Protozoan(a), 5, 37, 307, 309, 435, 526
Pseudohypha, 337
Pseudomonas, 404, 419
    aeruginosa, 115, 117, 155, 158–162, 181, 183, 193, 196
    fluorescens, 88, 89, 171, 174, 205, 207, 443, 445
    putida, 229–232
Psychrophile, 205
Pure culture, 13, 17, 123, 131, 451
    isolation of, 1, 123, 125–130, 131–134
    maintenance/preservation of, 123, 163
Pyridine-silver solution, 530
Quellung phenomenon, 83
Racking process in wine production, 504
Radiant-field diaphragm, 26, 29–**31**, 33
Recognition sequences, 295
Recombinant DNA technology, 285, 295
Recombination, 239, 251–254
Red blood cells, 218, 358, 373, **378**, 382
    hemolysis of, 119, 373
    sheep (SRBC), 377–379, 381, 387
Reduction, 185, 193, 195–196, **198, 199**
Refractive index, 27, 28, 51, 54
Reinforcement, 51, **52**
Rennet, 179, 181, **199**
Repressor proteins, 255, **256**
Resazurin, 152
Resolution, **25**–27, 47, 87
Restriction endonucleases, 285, 295–297
Revertant, 265
Rhizobium, 447
Rhizopus stolonifer, 320, 324
Ribonuclease, 287, 288
RNA polymerase, 255–257
Rods, 65, 197, 347, 403–406
Root nodules, 447
Rosindole dye, 189
Route of infection, 395
Sabouraud's agar, 205, 319–321, 329, 338, 342, 435, 537
Saccharomyces cerevisiae, 37, 205, 207, 218, 220, 337–**339**, 341, 343, 507
    var. *ellipsoideus*, 504
Safety in lab, 1–4, 48, 135–136, 214
Safranin, 71, **72**, 79, 80, 529
Salmonella, 407, 455, 469
    typhimurium, 263–265, 408, **410**, 425
Saprophyte, 347
Sauerkraut, 493, 495, 497–498
Schizosaccharomyces octosporus, 337–**339**
Selective media, 119, 487

Selenite broth, 455
Semisolid medium, 99, 101
Septate hyphae, 319
Serology, 357–359, 363–366
Serovar, 367
Serratia marcescens, 18, 19, 125, 127, 128, 144, 186, 187, 193, 196, 214, 235, 237
Serum, 75, 367, 380, 381, 387
Sexual reproduction of molds, 327–329
Shape of microbes, 5, 59, 61, 65, 66, 94
Shigella, 407, **408**, 455
    dysenteriae, **410**, 425
    flexneri, **410**
Simmon's citrate agar, 537
Size of microbes, 5, 59, 61
Slide culture, preparation of, **320**
Slime layer, 83, 181
Smear, 59, 60, 75
    preparation of, **62, 63, 67**, 75
Sodium-acetate medium, 338, **339**, 537
Sodium chloride, 111, 219
Soil, **13**, 131, 439, **508**
    microbes in, 5, 433, 435, 437–438, 507
Solid medium, 99, 101, 150
Solute, 217–220
Solvent, 169, 217
Specimen collection/processing, 351–353
Spectrophotometer, **143**–**145**, 258, 261
Spheroplast, formation of, 229–232
Spirochetes, 47, 58, 415–**416**, 529–530
Spirolate broth, 537–538
Sporangium, 79, 80
Spore, 79, 80, 94, 98, 109, 209
    staining of, 79–82, 95, 197, 201, 530
Spore formers, 79, 98, 203, 212, 449
Spread-plate method, 137, 435
Stab inoculation, 88, 115
Stage, **24**, 25, 41–**43**
Staining methods, 29, 59–63, 65–68, 529
Standard-methods agar, 482, 487, 538
Standard plate count, 451, 462–463, 467, 481, 482, 485, 487, 489
Staphylococcus, 407, 415
    aureus, 61, 65–67, 69, 72, 73, 88, 115, 117, 131, 133, 151, 153, 176, 203, 207, 209, 211, 218, 225, 227, 277, 347, 348, 407–**410**, 507, 516
    epidermidis, 347, 407, 408, **410**, 412
Starch hydrolysis, 173, 197, **198**, 202
Starch media, 171, 173, 174, **198**, 538
Starter culture, 493, 503
Stereomicroscopy, 31–32
Sterilization, 1, 3, 7, 99, 101, 203, 533
    various types of, 101, 107, 109, 213
Sticky ends, 285, 295
Streak inoculation, 8, **19, 120**, 125, **171**
    calibrated loop-direct, 419
Streptococcus, 119, 347, 359, 373, 395, 396, 415, 469, 493
    equi, 119, 120
    faecalis, 396, 397, 419
        liquefaciens, 181, 183, 218, 219
    faecium, 119, 120
    lactis, 115, 117, 118, 181, 183
    pneumoniae, 301, 395–397
    pyogenes, 395–**397**
    salivarius, 396, 397
Streptolysin O (STO), 373–376
Streptomyces, 515
    albus, 155, 158–162, 515, 516
Streptomycin, 225–228

Streptomycin-gradient plate, **248**
Streptomycin-resistant mutants, 247–250
Subculture, 17–19, 21, 163, 164, 319
Sucrose, fermentation of, 197, **198**, 202
Symbiosis, 13, 447
Synthetic culture medium, 99, 111
Temperate phage, 271
Temperature, 79, **205**
    effect of, on growth, 203, 205–208
Test organism, 72, 203, 221
Tetracycline, 225–227, 301, **302**
Thermal-death-point determination, 209
Thermophile, 205, 208
Thioglycollate medium, 152, 515, **516**, 538
Thymine dimer, 213
Tissue culture, 269, 281–284
Todd-Hewitt broth, 397, 538
Transduction, 239, 301
Transfer of microbes, 8, 17–19, 21–22
Transformation, 239, 301–306
Transmittance, 144, 145
Transport methods, 149, 351
Treponema, 415
    denticola, 48, **155**
    pallidum, 352, 415, 537–538
    socranskii, **155**
Tributyrin agar, 171, 173, 174, 197, **198**, 451, 538
Trommsdorf's reagent, 443, 530
Trypticase media, 164, 197, **198**, 271, 288, 538
Tryptone media, 189, 190, 197, **198**, 538
Tryptophan, 189, **198**, 243–246
Tuberculous sputum, 75–77, 352, 481
Turbidity, 21, 107, 117, 118, 143
    of broth cultures, 143–145, 147–148
Turgor, 218
Ubiquity of microorganisms, 5
Ultraviolet (UV) light, 2, 213–216
Universal precautions, 345
Unknown culture, characterization of, 95, 97, 197–199, 201–202
Urease, **410**
Urine bacteriological analysis, 419–422
Variability, 233
Vegetative cells, 79, 80, 209
Viability, 51, 135, 149, 164, 435, 487
Viable plate-count, 135–139, 141–144
Vibrio, 475
Virulent organism, 83, 271
Viruses, 1, 2, 163, 269, 271–277, 279–280, 357, 373, 377, 461, 526
    propagation of, 281–284
    two replication cycles of, **272**, 281
Vitamins, 99, 111, 407, 455, 479
Voges-Proskauer test, 407, **424**, 462, 530
Water agar, 536
Water analysis, 461–465, 467–471, 473
Wavelength, 51, **52**, 213
Wet mounts, 37, 39, 87, 88, 307, 309, 341
Wild-type strain, 239, 243, 247
Wine production, 169, 337, 503–506
Xenorhabdus luminescens, 475
Yeast, 5, 212, 220, 311–312, 333, 337–344, 503, 526
Yeast extract, 99, 101, 119, 302, 333
Yeast-extract media, 102, 538–539
Yogurt, 479, 493, 499, 501–502
Z buffer, 258
Ziehl-Neelsen method, 75–76
Zone of inhibition, 119, 225–227, 389
Zygospore, 327–329, 331–332